AI Applications to Communications and Information Technologies

AI Applications to Communications and Information Technologies

The Role of Ultra Deep Neural Networks

First Edition

Daniel Minoli
DVI Communications, New York, NY, USA

Benedict Occhiogrosso
DVI Communications, New York, NY, USA

IEEE PRESS

WILEY

Published by John Wiley & Sons, Inc., Hoboken, New Jersey.
Published simultaneously in Canada.

For general information on our other products and services or for technical support, please contact our Customer Care Department within the United States at (800) 762-2974, outside the United States at (317) 572-3993 or fax (317) 572-4002.

Wiley also publishes its books in a variety of electronic formats. Some content that appears in print may not be available in electronic formats. For more information about Wiley products, visit our web site at www.wiley.com.

Library of Congress Cataloging-in-Publication Data
Names: Minoli, Daniel, 1952- author. | Occhiogrosso, Benedict, author.
Title: AI applications to communications and information technologies: the role of ultra deep neural networks / Daniel Minoli, Benedict Occhiogrosso.
Description: First edition. | Hoboken, New Jersey: Wiley, [2024] | Includes index.
Identifiers: LCCN 2023025361 (print) | LCCN 2023025362 (ebook) | ISBN 9781394189991 (hardback) | ISBN 9781394190027 (adobe pdf) | ISBN 9781394190010 (epub)
Subjects: LCSH: Artificial intelligence. | Neural networks (Computer science)
Classification: LCC Q335 .M544 2024 (print) | LCC Q335 (ebook) | DDC 006.3/2–dc23/eng/20231011
LC record available at https://lccn.loc.gov/2023025361
LC ebook record available at https://lccn.loc.gov/2023025362

Cover Design: Wiley
Cover Image: © Yuichiro Chino/Getty Images

Set in 9.5/12.5pt STIXTwoText by Straive, Pondicherry, India

Contents

About the Authors *xi*
Preface *xiii*

1   Overview *1*
1.1   Introduction and Basic Concepts *1*
1.1.1   Machine Learning *5*
1.1.2   Deep Learning *6*
1.1.3   Activation Functions *13*
1.1.4   Multi-layer Perceptrons *17*
1.1.5   Recurrent Neural Networks *21*
1.1.6   Convolutional Neural Networks *21*
1.1.7   Comparison *26*
1.2   Learning Methods *26*
1.3   Areas of Applicability *39*
1.4   Scope of this Text *41*
A.   Basic Glossary of Key AI Terms and Concepts *44*
        References *57*

2   Current and Evolving Applications to Natural Language Processing *65*
2.1   Scope *65*
2.2   Introduction *66*
2.3   Overview of Natural Language Processing and Speech Processing *72*
2.3.1   Feed-forward NNs *74*
2.3.2   Recurrent Neural Networks *74*
2.3.3   Long Short-Term Memory *75*
2.3.4   Attention *77*
2.3.5   Transformer *78*
2.4   Natural Language Processing/Natural Language Understanding
        Basics *81*
2.4.1   Pre-training *82*

2.4.2   Natural Language Processing/Natural Language Generation
         Architectures *85*
2.4.3   Encoder-Decoder Methods *88*
2.4.4   Application of Transformer *89*
2.4.5   Other Approaches *90*
2.5   Natural Language Generation Basics *91*
2.6   Chatbots *95*
2.7   Generative AI *101*
A.   Basic Glossary of Key AI Terms and Concepts Related to Natural
         Language Processing *103*
         References *109*

3   Current and Evolving Applications to Speech Processing *117*
3.1   Scope *117*
3.2   Overview *119*
3.2.1   Traditional Approaches *119*
3.2.2   DNN-based Feature Extraction *123*
3.3   Noise Cancellation *126*
3.3.1   Approaches *128*
3.3.1.1 Delay-and-Sum Beamforming (DSB) *129*
3.3.1.2 Minimum Variance Distortionless Response (MVDR) Beamformer *130*
3.3.1.3 Non-adaptive Beamformer *131*
3.3.1.4 Multichannel Linear Prediction (MCLP) *132*
3.3.1.5 ML-based Approaches *132*
3.3.1.6 Neural Network Beamforming *135*
3.3.2   Specific Example of a System Supporting Noise Cancellation *138*
3.4   Training *141*
3.5   Applications to Voice Interfaces Used to Control Home Devices
         and Digital Assistant Applications *142*
3.6   Attention-based Models *146*
3.7   Sentiment Extraction *148*
3.8   End-to-End Learning *148*
3.9   Speech Synthesis *150*
3.10   Zero-shot TTS *152*
3.11   VALL-E: Unseen Speaker as an Acoustic Prompt *152*
A.   Basic Glossary of Key AI Terms and Concepts *156*
         References *166*

4   Current and Evolving Applications to Video and Imaging *173*
4.1   Overview and Background *173*
4.2   Convolution Process *176*
4.3   CNNs *181*

4.3.1   Nomenclature   *181*
4.3.2   Basic Formulation of the CNN Layers and Operation   *181*
4.3.2.1 Layers   *181*
4.3.2.2 Operations   *188*
4.3.3   Fully Convolutional Networks (FCN)   *190*
4.3.4   Convolutional Autoencoders   *190*
4.3.5   R-CNNs, Fast R-CNN, Faster R-CNN   *193*
4.4   Imaging Applications   *195*
4.4.1   Basic Image Management   *195*
4.4.2   Image Segmentation and Image Classification   *199*
4.4.3   Illustrative Examples of a Classification DNN/CNN   *202*
4.4.4   Well-Known Image Classification Networks   *204*
4.5   Specific Application Examples   *213*
4.5.1   Semantic Segmentation and Semantic Edge Detection   *213*
4.5.2   CNN Filtering Process for Video Coding   *215*
4.5.3   Virtual Clothing   *216*
4.5.4   Example of Unmanned Underwater Vehicles/Unmanned
         Aerial Vehicles   *218*
4.5.5   Object Detection Applications   *218*
4.5.6   Classifying Video Data   *222*
4.5.7   Example of Training   *224*
4.5.8   Example: Image Reconstruction is Used to Remove Artifacts   *225*
4.5.9   Example: Video Transcoding/Resolution-enhancement   *228*
4.5.10   Facial Expression Recognition   *228*
4.5.11   Transformer Architecture for Image Processing   *230*
4.5.12   Example: A GAN Approach/Synthetic Photo   *230*
4.5.13   Situational Awareness   *231*
4.6   Other Models: Diffusion and Consistency Models   *236*
A.   Basic Glossary of Key AI Terms and Concepts   *238*
B.   Examples of Convolutions   *246*
         References   *250*

5   **Current and Evolving Applications to IoT and Applications to Smart
         Buildings and Energy Management**   *257*
5.1   Introduction   *257*
5.1.1   IoT Applications   *257*
5.1.2   Smart Cities   *258*
5.2   Smart Building ML Applications   *275*
5.2.1   Basic Building Elements   *275*
5.2.2   Particle Swarm Optimization   *276*
5.2.3   Specific ML Example – Qin Model   *279*
5.2.3.1 EnergyPlus™   *281*

5.2.3.2   Modeling and Simulation   *282*
5.2.3.3   Energy Audit Stage   *286*
5.2.3.4   Optimization Stage   *287*
5.2.3.5   Model Construction   *289*
5.2.3.6   EnergyPlus Models   *289*
5.2.3.7   Real-Time Control Parameters   *290*
5.2.3.8   Neural Networks in the Qin Model (DNN, RNN, CNN)   *290*
5.2.3.9   Finding Inefficiency Measures   *294*
5.2.3.10   Particle Swarm Optimizer   *294*
5.2.3.11   Integration of Particle Swarm Optimization with Neural Networks   *296*
5.2.3.12   Deep Reinforcement Learning   *298*
5.2.3.13   Deployments   *298*
5.3   Example of a Commercial Product – BrainBox AI   *301*
5.3.1   Overview   *301*
5.3.2   LSTM Application – Technical Background   *302*
5.3.3   BrainBox AI Commercial Energy Optimization System   *305*
A.   Basic Glossary of Key IoT (Smart Building) Terms and Concepts   *314*
           References   *339*

6   Current and Evolving Applications to Network Cybersecurity   *347*
6.1   Overview   *347*
6.2   General Security Requirements   *349*
6.3   Corporate Resources/Intranet Security Requirements   *353*
6.3.1   Network and End System Security Testing   *358*
6.3.2   Application Security Testing   *360*
6.3.3   Compliance Testing   *362*
6.4   IoT Security (IoTSec)   *363*
6.5   Blockchains   *365*
6.6   Zero Trust Environments   *369*
6.7   Areas of ML Applicability   *370*
6.7.1   Example of Cyberintrusion Detector   *373*
6.7.2   Example of Hidden Markov Model (HMM) for Intrusion Detection   *374*
6.7.3   Anomaly Detection Example   *378*
6.7.4   Phishing Detection Emails Using Feature Extraction   *383*
6.7.5   Example of Classifier Engine to Identify Phishing Websites   *386*
6.7.6   Example of System for Data Protection   *388*
6.7.7   Example of an Integrated Cybersecurity Threat Management   *390*
6.7.8   Example of a Vulnerability Lifecycle Management System   *392*
A.   Basic Glossary of Key Security Terms and Concepts   *396*
           References   *400*

7 **Current and Evolving Applications to Network Management** *407*

7.1 Overview *407*

7.2 Examples of Neural Network-Assisted Network Management *408*

7.2.1 Example of NN-Based Network Management System (Case of FM) *413*

7.2.2 Example of a Model for Predictions Related to the Operation of a Telecommunication Network (Case of FM) *416*

7.2.3 Prioritizing Network Monitoring Alerts (Case of FM and PM) *419*

7.2.4 System for Recognizing and Addressing Network Alarms (Case of FM) *424*

7.2.5 Load Control of an Enterprise Network (Case of PM) *428*

7.2.6 Data Reduction to Accelerate Machine Learning for Networking (Case of FM and PM) *431*

7.2.7 Compressing Network Data (Case of PM) *435*

7.2.8 ML Predictor for a Remote Network Management Platform (Case of FM, PM, CM, AM) *437*

7.2.9 Cable Television (CATV) Performance Management System (Case of PM) *441*

A. Short Glossary of Network Management Concepts *446*

References *447*

Super Glossary *449*

Index *467*

About the Authors

Daniel Minoli, Principal Consultant, DVI Communications, New York, graduate of New York University, has published 60 technical books, 350 papers, and made 90 conference presentations. He started working on AI/expert systems in the late 1980s, when he co-authored the book *Expert Systems Applications in Integrated Network Management* (Artech House, 1989). Over 7,500 academic researchers cite his work in their own peer-reviewed publications, according to Google Scholar, and over 250 US patents and 40 US patent applications cite his published work. He has many years of technical and managerial experience in planning, designing, deploying, and operating secure IP/IPv6, VoIP, telecom, wireless, satellite, and video networks for carriers and financial companies. He has published and lectured in the area of M2M/IoT, network security, satellite systems, wireless networks, and IP/IPv6/Metro Ethernet, and has taught as adjunct for many years at New York University, Stevens Institute of Technology, and Rutgers University. He also has served as a testifying expert witness for over 25 technology-related litigation cases pertaining to patent matters, equipment forensics, and breach of contract. He also has served as a testifying expert witness for technology-related litigation pertaining to patent matters, equipment forensics, and breach of contract.

Benedict Occhiogrosso is a co-founder of DVI Communications, New York. He is a graduate of New York University Polytechnic School of Engineering. For over 25 years, he has served as the CEO of a multidisciplinary consulting firm and advised both technology producers and consumers on adoption and deployment of technologies. He is also a technology expert in various aspects of telecommunications, security, and information technology with concentration in speech recognition, video surveillance, and more recently IoT. He has supported high-tech litigation encompassing intellectual property (patent and trade secrets) as a testifying expert witness and advised inventors and acquirers on patent valuation. He has also served as both a technology and business advisor to several start-up and operating companies with respect to product planning, company organization, and capital formation.

Preface

Following decades of research, artificial intelligence (AI) technologies, particularly machine learning (ML) and deep learning (DL), are becoming ubiquitous in nearly all aspects of modern life. R&D and productization are proceeding at a fast pace at this juncture. For example, as a reference year, in 2021 IBM received over 2,300 patents dealing with AI (just about a quarter of all patents granted for the year); many other research organizations and firms are doing the same. In addition, and for example, ChatGPT, the popular generative AI chatbot from OpenAI, is estimated to have reached 100 million monthly active users just two months after launch (in late 2022), making it the fastest-growing consumer application in history.

With the widespread deployment of smart connected sensors in the IoT ecosystem in the smart city, smart building, smart institution, and smart home; data collection; and associated analysis, nearly all major industries have been impacted by and benefitted from AI. This is particularly true of systems operating as "narrow AI" performing objective functions utilizing data-trained models and algorithms in the context of deep learning and/or machine learning. AI is entering an advanced stage that encompasses perception, reasoning, and generalization.

AI applications include, but are certainly not limited to, autonomous driving and navigation, speech recognition, language processing, robotics, computer vision, pattern recognition, face recognition, predictive text systems, generative systems/chatbots, social networks and advertisement placing, behavior-predictive systems, data mining systems, financial systems, medical sciences, military systems, and telecommunications-related systems, to name just a few.

Technology heavyweights such as Google, Apple, Microsoft, and Amazon are currently investing billions of dollars to develop AI-based products and services. The US Department of Defense continues to sponsor extensive AI research and universities started offering AI-based programs their curricula.

This text focuses on AI and Neural Network applications to Information and Communications Technology (ICT), for example applications to voice recognition, video/situational awareness/face recognition, smart buildings and energy management, wireless systems, cybersecurity, and network management – it should be noted that each of these topics could, in fact, benefit by a dedicated text. We explore basic principles and applications in these disciplines, emphasizing recent developments and near-term commercial and deployment opportunities. This treatise, however, is not intended to be a basic tutorial on Neural Networks, or a review of the current applicable academic research, which as indicated is quite extensive; nonetheless, sufficient background is provided on the fundamental underlying topics to make the reading of this text reasonably self-contained. AI is an extremely well-documented science, and this text does not claim pure originality *per se* (or research innovation) but only a synthesis, an organization, a presentation, and a focused treatment of the field as applied to information technology and communications-related applications – other basic references can and should be used by the interested reader. The hundreds of researchers cited in the references deserve the credit of the original, but widely scattered, conceptualizations.

It is the authors' intent that this text can be used by industry observers, planners, investors, vendors, and students. It can serve as an ancillary text to an undergraduate or graduate course on AI, networking, or software development.

May 15, 2023

Daniel Minoli
Benedict Occhiogrosso

1

Overview

1.1 Introduction and Basic Concepts

Artificial intelligence (AI) is a subfield of computer science (CS) that focuses on the creation of computer-based systems, applications, and algorithms that mimic, to the degree possible, some cognitive processes intrinsic to human intelligence.[1] The field has had a long history and is now blossoming in an all-encompassing manner. AI technologies, particularly machine learning (ML) and deep learning (DL), are becoming ubiquitous in nearly all aspects of modern life. DL is a subfield of ML as discussed below. The goals of learning are (i) understanding a process or phenomenon and (ii) making prediction about outcomes, namely, inferring a function or relationship that maps the input to an output in such a manner that the learned relationship can be used to predict the future output from a future input. AI applications, and ML/DL-based systems in particular, are positioned to take over complex tasks generally performed by humans (decision-makers) or to provide added support to people. Siri, Alexa, augmented reality (AR), autonomous driving, and object recognition are just a few examples of AI applications. Driven in part off the massive data collection and associated analysis resulting from the widespread deployment of connected sensors in the Internet of Things (IoT) ecosystem in the smart city, the smart building, the smart institution, and the smart home, nearly, if not, all major industries have been impacted by and benefitted from AI.

---

1 Propitiously and/or with an anticipatory sense, the senior author's first technical book, of a number he has written, was a (co-edited) book on AI, *Expert Systems Applications in Integrated Network Management*, Artech House, 1989.

AI Applications to Communications and Information Technologies: The Role of Ultra Deep Neural Networks, First Edition. Daniel Minoli and Benedict Occhiogrosso.
© 2024 The Institute of Electrical and Electronics Engineers, Inc.
Published 2024 by John Wiley & Sons, Inc.

In practical terms, AI in a given application space (use case) is a set of very advanced algorithms pertaining to decision-making processes in that given application space, in conjunction with a large set of training data that is utilized to fine-tune the parameters of the algorithms, allowing them 'to learn'. In a number of applications (e.g. but limited to, object recognition, speech recognition, classification, and network management), classical non-AI algorithms for decision-making have been used in the past, however, the more recent utilization of AI-specific learning, training, and execution algorithms have significantly improved the quality, efficiency, timeliness, and trustworthiness of the results.

This text focuses on AI/ML/DL applications in the Information and Communications Technology (ICT) sector of the industry, which includes networking, telecommunications, and applications-supporting systems. As the name implies, ICT spans the fields of information technology (IT) and the fundamentally critical communication technologies that make all the modern business and personal applications possible. Some areas where AI is being used in ICT, telecommunications and support operations include but are not limited to the following: network operations monitoring, network management, predictive maintenance, network security and fraud mitigation, customer service, virtual assistants, chatbots, Intelligent Customer Relationship Management systems, and intelligent automation based on AI-supported Robotic Process Automation. Expert systems for network management have been used for decades [1, 2].

ML endeavors to give computers the ability to learn without being explicitly programmed. ML encompasses the study and construction of algorithms that may learn from existing data and make predictions about new data. These ML mechanisms operate by building a model from training data to make future data-driven predictions or decisions expressed as assessments or outputs. A large number of ML algorithms exist, for example, neural networks (NNs) (also called models), logistic regression, Naive Bayes, Random Forest (RF), matrix factorization, and support vector machines (SVMs), among others. The focus of this textbook is principally, but exclusively, on NNs.

One way to classify AI is to categorize it as *narrow AI* and *general AI* – general AI is also known as artificial general intelligence (AGI).

Narrow AI: seeks to identify and establish automated solutions to problems or functions that are typically undertaken by a human (say, an analyst or performer), while, hopefully, improving the performance of the manual tasks undertaken to solve the problems or support the functions, according to some metric – for example, efficiency and endurance. It aims at performing a task at a time while continuing to improve its operation. Narrow AI is implemented as software that automates an analytical activity typically performed by humans. At press time, most of the AI applications were examples of narrow AI.

General AI: Instead of focusing on a narrow or single task, the goal of AGI is to teach and to empower the machine to comprehend and assess a wide range of parameters, issues, and processes that characterize the underlying domain or eco-system. The goal is to enable the machine to make decisions based on dynamic learning instead of relying on prescriptive, earlier training. While utilizing a certain knowledge base and training, such a system has the ability to reach conclusions, decisions, and judgment based perhaps on another, possibly more appropriate paradigm; here, the independent learning is based on (or supplemented with) experience and pragmatics. Some researchers expect to achieve AGI through the advanced application of DL mechanisms. A number of tests to assess an AGI system's intelligence has emerged over the years. At press time there were nearly one hundred projects around the world focused on developing AGI.

Advanced AI fields include *cognitive computing systems* and *natural language processing* (NLP)/*natural language understanding* (NLU) under the rubric of generative AI. Cognitive computing systems are systems that endeavor to understand and emulate human behavior, while also providing intuitive and natural interface to the machine. NLP systems allow machines to understand written language or voice commands; they also support natural language generation that enables the machine to communicate in "spoken conversation." NLU is concerned with enabling computers to derive meaning from natural language inputs (such as spoken inputs). Related fields include *automatic speech recognition* (ASR), which is concerned with transforming audio data associated with speech into a token or other textual representation of that speech. ASR and NLU are heavily dependent on AI and are often used together as part of a language processing component of a system; see Figure 1.1 for an illustrative example. Additionally, *Text-to-Speech* (TTS) is a field concerned with transforming textual and/or other data into audio data that is synthesized to resemble quality human speech [3, 4]. AI is now also widely used for automated content generation systems. As a particular example, AI is now heavily used in medicine and healthcare; Table 1.1 lists just a few examples of usage and applicability.

This introductory chapter provides a brief overview of some key AI concepts but is not intended to provide an extensive tutorial on the AI field per se, for which there are many good sources. The chapter only lays out sufficient background to support the chapters that follow. However, a number of the basic concepts covered in this chapter are revisited and further expanded in the rest of the book, as appropriate. The Appendix at the end of this chapter provides a basic (non-exhaustive) glossary of key AI terms and concepts based on a variety of industry sources.

As noted, ML and DL are leading examples of AI; DL systems are considered to be a subset or subcategory of ML systems, as depicted graphically in Figure 1.2 and further discussed below. Systems such as IoT generate large amounts of data; the sheer volume of data requires AI, ML, and DL techniques for such volume of data to be mined properly and to allow one to reach accurate, useful, and timely data-driven insights. ML and DL techniques endeavor to examine and establish the internal relationships of a set of data; said data is typically collected from the

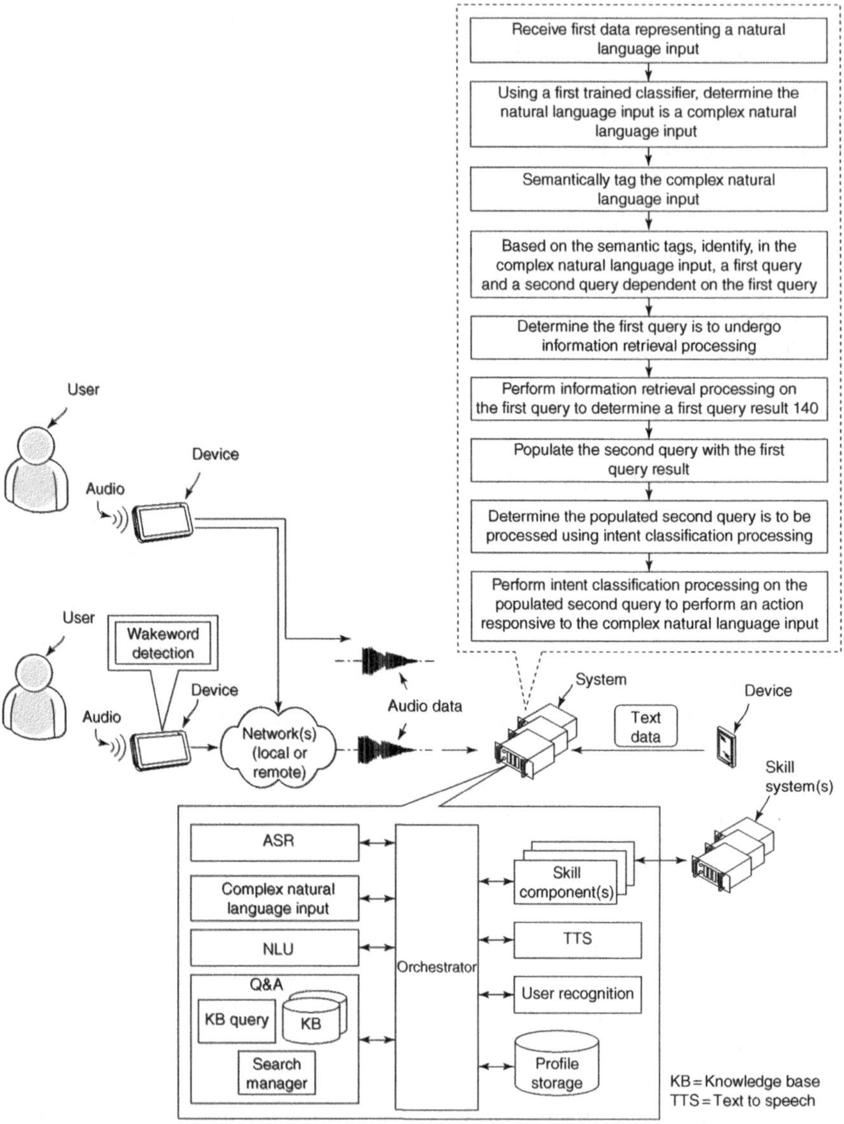

Receive first data representing a natural language input

Using a first trained classifier, determine the natural language input is a complex natural language input

Semantically tag the complex natural language input

Based on the semantic tags, identify, in the complex natural language input, a first query and a second query dependent on the first query

Determine the first query is to undergo information retrieval processing

Perform information retrieval processing on the first query to determine a first query result 140

Populate the second query with the first query result

Determine the populated second query is to be processed using intent classification processing

Perform intent classification processing on the populated second query to perform an action responsive to the complex natural language input

User

Device

Audio

User

Wakeword detection

Audio

Device

Network(s) (local or remote)

Audio data

System

Text data

Device

Skill system(s)

ASR

Complex natural language input

NLU

Q&A

KB query | KB

Search manager

Orchestrator

Skill component(s)

TTS

User recognition

Profile storage

KB = Knowledge base
TTS = Text to speech

Figure 1.1 AI in complex natural language processing. *Source:* Adapted from Ref. [3]).

variety of input devices aggregating visual, audio, image, and signal data from the field in question (the set of fields obviously being wide-ranging). ML and DL predict and classify data using various algorithms optimized to the dataset in question; in particular, DL-based systems are now routinely applied in recognition-, prediction-, and classification-related analyses.

Table 1.1 Artificial intelligence in medicine and healthcare (short list).

❖ Artificial intelligence in anesthesiology

❖ Artificial intelligence in drug design

❖ Artificial intelligence and machine learning in anesthesiology

❖ Artificial intelligence to deep learning: machine intelligence approach for drug discovery

❖ Artificial intelligence in medical imaging

❖ Artificial intelligence for diabetes care

❖ Artificial intelligence in nuclear medicine

❖ Artificial intelligence and ophthalmology

❖ Artificial intelligence in cardiovascular medicine

❖ Artificial intelligence and ophthalmic surgery

❖ Emerging role of deep learning-based artificial intelligence in tumor pathology

❖ Artificial intelligence for brain science

❖ Artificial intelligence and machine learning applications in musculoskeletal physiotherapy

❖ Machine learning combined with multispectral infrared imaging to guide cancer surgery

❖ Healthcare organizations seek to streamline their business processes and reduce insurance denials in real-time by leveraging AI and ML, data mining, analytics, and automated decision engines

❖ ChatGPT has potential to help cirrhosis and liver cancer patients

❖ Using machine learning in the electronic medical record to save lives of hospitalized children

---

Source: Adapted from https://pubmed.ncbi.nlm.nih.gov for the term artificial intelligence.

1.1.1 Machine Learning

ML entails advanced algorithms that process the data, form inferences (learns) from that data, and then apply what they have learned to make "informed" decisions. The goal of ML is acquiring skills or knowledge from experience, that is, synthesizing usable conclusions and decisions from historical data. ML techniques are increasingly utilized to analyze, cluster, associate, classify, and apply regression methods to the data in question (e.g. IoT data from smart city/building/ institution environments, medical data, video analytics data, and so on). Among other applications, ML techniques are applicable to computer vision, image processing and analysis, NLU, and speech recognition.

Two common groups of problem categories in ML are *classification-oriented problems* and *regression-oriented problems*. Classification problems, also known as categorization problems, seek solutions by classifying items into one of several category values (for example, is this object a pear or a peach). There are (very) many examples of classification tasks in ML settings; some include classifying

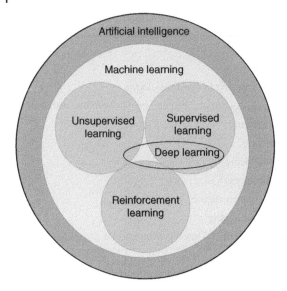

Figure 1.2 Positioning of various AI systems.

whether a neoplasm is benign or malignant, classifying whether an email is a spam or not, classifying whether a request for financial transaction is fraudulent or not, and so on. Regression learning algorithms seek solutions using the training data to uncover correlations among identified features that impact the outcome.

Traditional ML approaches generally entail the following steps: (i) defining domain-specific features of interest, (ii) undertaking computations related to features of interest using explicit algorithms (this being feature extraction), (iii) training a classifier to optimize performance based on ground-truth labels to determine the relationship between features and the decisioning or ultimate result being sought (training), and (iv) testing system performance on fresh data. The system performance can be impacted by any of these steps. For example, feature that did not correlate with ground-truth, methodology that did not accurately quantify the feature, classifier that imperfectly learned from the training data, or ground-truth labels that were too noisy or did not generalize [5, 6].

Refer to Tables 1.2 and 1.3 for a summary description of key concepts in this arena.

1.1.2 Deep Learning

DL (also known as deep neural learning or representation learning) is a special type of ML that imitates the learning approach humans utilize to gain knowledge using underlying NNs: an NN operates similar to a set of neurons in the human brain to perform a variety of computational tasks. DL is considered to be an elaboration of ML.

Table 1.2 Learning philosophies.

Term	Definition/explanation/concept
Deep learning (DL)	(aka as deep neural learning or representation learning): A subcategory of ML – and functions in a similar manner; the difference is that DL makes use of ANNs. DL algorithms include neural network algorithm, support vector machines (SVM) algorithm, Bayesian network (BN) algorithm, decision tree (DT) algorithm, hidden Markov model (HMM) algorithm, Random Forest (RF) algorithm, k-nearest neighbor (KNN, aka k-NN) algorithm, k-means algorithm, Gaussian mixture model (GMM) algorithm, among others.
Machine learning (ML)	An application of AI that allows systems to learn and evolve based on experience without the entire universe of functionality being explicitly programmed. An ML program can access the data in question and use it for its own learning. ML models (MLMs) are systems that accept an input and algorithmically generate an output predicted on the received input using some learning method. Systems that use an algorithm to learn to generate dynamic inferences based on a set of data.

DL is concerned with algorithms that analogize (some of) the structures and functions of the human brain utilizing NNs, also referred to as artificial neural networks (ANNs) or deep neural networks (DNNs). Effectively, a DNN is an NN with several/numerous hidden layers (discussed further below). At press time, some networks had as much as 1,000 layers and 200 million parameters; here, we refer to these networks as *ultra deep neural networks (UDNNs)* (some have called NNs of this type as Dense Convolutional Networks – DenseNet [7]); deeper networks are able to learn more complex functions. In what follows in this chapter, the terms NN and ANN are used interchangeably (in the rest of the book, the term NN is used exclusively).

As just stated, NNs are computational models that operate similarly to the functioning of a human nervous system. An NN is a collection of connected logical units called neurons; each connection is called a *synapse*. Each synapse carries a unidirectional signal with an activating strength that correlates with the strength of the connection. The receiving neuron can typically activate and propagate a signal to downstream neurons that are connected to it, normally based on whether the aggregated incoming signals – which can possibly be from several transmitting neurons – are of a sufficient (specified) strength. As implied, neurons are also often simply called nodes or units. NNs are at the core of the rich DL dénouement of recent years, providing momentum to the advancement of applications such as generative AI, speech recognition, face recognition, unmanned aerial vehicles, and self-driving cars, among many others. NNs may be constructed programmatically with software or utilizing specialized hardware.

There are several kinds of NNs, being distinguished and implemented based on the mathematical operations and the set of parameters required to determine the

Table 1.3 Types of NNs (partial list).

Term	Definition/explanation/concept
Artificial neural network (ANN)	A computational model that operates analogously to the functioning of a human nervous system; a model used in ML which is composed of artificial neurons (system nodes) that form a network by synaptic connections. Basically, a neural network (NN); synonymous (in this text) with NN.
Convolutional neural network (CNN)	(aka ConvNet) A DNN with a convolutional structure. A class of DNNs, see below, that deals with processing data that has a geometric grid/array-like topology, for example, an image. A given neuron in a CNN system only processes data in its receptive field. A CNN usually is comprised of three layers: a convolutional layer, a pooling layer, and a fully connected layer.
Deep neural network (DNN)	(aka multi-layer NN or n-layer neural network) Any NN with several hidden layers. A stacked NN that is composed of multiple layers; the layers are composed of nodes, which are logical locations where computation occurs. An ML model (MLM) has an input layer and output layer and one or more hidden layers; the hidden layers each apply a nonlinear transformation to the input it receives from the previous to generate an output for the next layer.
Feed-forward NNs	NNs where each neural node in a layer of the network is connected to every node in a preceding or subsequent layer (this being a "fully connected" NN), but the feed-forward NN does not include any connections that loop back from a node in a layer to a previous layer.
Multi-layer Perceptrons (MLP)	A feed-forward neural network where inputs are processed exclusively in the forward direction. Some equate with the term ANN: the term is used inconsistently, by some to loosely mean any feed-forward ANN and by others to refer to NNs composed of multiple layers of perceptrons. In our discussion, ANN is used as a general term and MPL as a specific type of ANN.
Neural network (NN)	(also simply called a "network" in the AI context) A computational model that operates similarly to the functioning of a human nervous system; a model used in DL which is composed of artificial neurons. NNs have origin in biology - here the network is made up of biological neurons. In the AI context the NN is composed of nodes that constitute artificial neurons that can store information. Thus, an NN is a system of software and/or hardware modeled after the operation of neurons in the human brain (an NN with some level of complexity, say with at least two layers, is an example of a DNN).
Recurrent neural network (RNN)	An NN where the algorithm using an internal memory "remembers" state information, making it well-suited for problems that involve sequential data. Typically, the hidden layer function in an RNN is an element-wise application of a sigmoid function. They are a family of feed-forward NNs that also include feedback connections between layers (they feed results back into the network). RNNs enable modeling of sequential data by sharing parameter data across different parts of the NN.
Ultra DNN (UDNN)	A DNN with a 1000 or more layers and with 100 million or more parameters.

solution, the output. Three types of NNs that were commonly used at press time for DL are: (i) multi-layer perceptrons (MLP) (which some equate with the term ANN[2] – these being feed-forward NNs), (ii) recurrent neural networks (RNNs), and (iii) convolutional neural networks (CNNs).

Thus, although *machine learning* and *deep learning* are occasionally used synonymously by some, there are key differences, as already implied. DL as a discipline is a subcategory of ML – and functions in a similar manner; however, DLs specifically make use of ANNs which autonomously allow machines to make "better" decisions without any help from humans. While ML models (MLMs) evolve and improve with more experience performing their function, they still need some guidance – for example, if an ML algorithm returns an inaccurate prediction, then a human intervention is required to make appropriate adjustments. With a DL model, the algorithm automatically assesses prediction accuracy relying on its own NN, continually analyzing data using logic similar to how a human would draw conclusions. DL systems use a layered structure of algorithms based on ANNs, using multi-layered processing [8]. DL models are much more complex than traditional decision-making algorithms, and they also ingest vastly larger amounts of data than their predecessors [9]. DL utilizes supervised, unsupervised, and reinforcement learning principles and algorithms (see Section 1.2).

As noted, DL is a *particularized form of ML: DL is ML implemented as a DNN with a (large) plurality of hidden layers.* See Figure 1.3 for a comparison of the two systems. The past few years have seen a shift from traditional feature extraction approaches to the routine utilization of ANNs (and CNNs in particular); this NN-based algorithmic approach constitutes the "deep learning" paradigm. Improvements in DNN model size and accuracy paired with the rapid increase in computing power have led to the adoption of DNN applications across multiple platforms, including resource constrained mobile and edge devices, also with the use of accelerators to deal with the typical memory bandwidth bottleneck issues. DNNs are now widely used in the domains of computer vision, image, video processing, cybersecurity, NLP, robotics, medical image/information processing, and speech recognition, because of DNNs' ability to achieve what observers call "humanlevel accuracy" [10]. Note that CNNs are a specific type of DNN, and RNNs are DNNs where the hidden activations from time step t are used as the

---

2 An MLP can be considered to be a fully connected feed-forward ANN. The term MLP is used inconsistently in the industry, by some to loosely mean any feed-forward ANN and by others to refer to NNs composed of multiple layers of perceptrons (with threshold activation). Even given this ambiguity in terminology, MLP does not refer to a single perceptron that has multiple layers, but to an NN that contains numerous perceptrons that are organized into layers – the term "multi-layer perceptron network" is also used by some. In our discussion, ANN is used as a general term and MPL as a specific type of ANN.

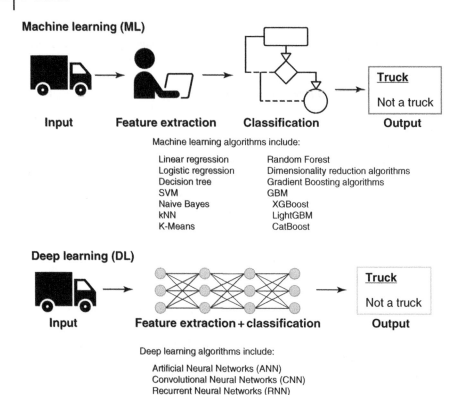

Figure 1.3 Comparing ML to DL. Note: the nodes in the hidden layers are also called "channels." Refer to the text for the definition of various acronyms. GBM: Gradient Boosted Machine.

context to compute the activations of time step $t + 1$; in particular, CNNs are the most commonly utilized DL ANNs in use at press time.

As stated, a DNN is an NN with several hidden layers; that is, a stacked NN that is composed of multiple layers; the layers are composed of nodes, which are logical locations where computation occurs. See Figure 1.4. A DNN uses a cascade of several layers of nonlinear processing units for feature extraction and transformation. Each successive layer utilizes the output from the previous layer as its input. Higher-level features are derived from lower-level features to form a hierarchical representation. The layers that follow downstream from the input layer may be convolution layers that produce feature maps that are filtering results of the inputs and are used by the next convolution layer [11]. In training of a DNN, a regression may be used that can include a minimization of a cost function; the cost function is a number representing how well the NN performed in mapping training examples to correct output.

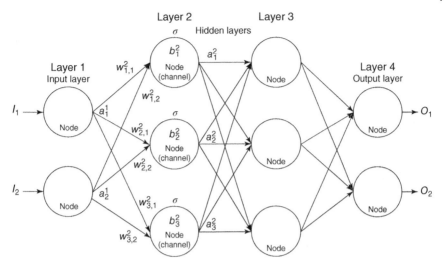

Figure 1.4 Example of DNN.

Contrary to traditional ML approaches, a DL classifier is able to learn without being explicitly told what to concentrate on (i.e. without user-defined features). While traditional programs construct linear networks, the hierarchical function of DL systems enables processing data in a nonlinear manner: DL utilizes a hierarchical level of ANNs that are assembled to mimic (in some basic fashion) the way the human brain operates; neuron nodes are interconnected with logical links in a "spider web" [12]. For example, in the context of image processing [13], a typical ML process begins with manually extracting selected features from images; the features[3] are then utilized to develop a model that can be used for categorizing the objects under scrutiny. Thus, in an ML-based process, one manually selects features and a classifier to sort images. In a DL-based process, on the other hand, the relevant features are extracted automatically (e.g. see Figure 1.5). As noted, ML uses algorithms to process data, learn from that data, and reach decisions based on what it has learned; DL organizes algorithms in layers to create an ANN that can learn and make intelligent decisions on its own.

---

3 DNNs typically include (i) a front-end network to perform feature recognition and (ii) a back-end network that has a mathematical model that performs operations (e.g. speech recognition, object classification) based on the feature representation given to it. Thus, DNNs enable ML without the need for human-crafted feature engineering: the DNNs can learn features based on statistical structure or correlation in the input data; thereafter, the learned features are provided to a mathematical model that maps detected features to an output.

The machine-learning algorithms utilize features for analyzing the data to generate assessments

Features

A feature is an individual measurable property of a phenomenon being observed. The concept of a feature is related to that of an explanatory variable used in statistical techniques such as linear regression. Choosing informative, discriminating, and independent features is important for effective operation of in pattern recognition, classification, and regression. Features may include one or more of words of the message, message concepts, communication history, past user behavior.

With the training data and the identified features the machine-learning tool is trained. The machine-learning algorithms utilize the training data to find correlations among the identified features that affect the outcome or assessment

In a supervised learning phase, all of the outputs are provided to the model and the model is directed to develop a general rule or algorithm that maps the input to the output. In an unsupervised learning phase, the desired output is not provided for the inputs so that the model may develop its own rules to discover relationships within the training dataset. In a semi-supervised learning phase, an incompletely labeled training set is provided, with some of the outputs known and some unknown for the training dataset.

New data is provided as an input to the trained machine-learning program and the machine-learning program generates the assessment as output.

The input is passed through a plurality of layers to arrive at an output. Each layer includes multiple neurons. The neurons receive input from neurons of a previous layer and apply weights to the values received from those neurons in order to generate a neuron output. The neuron outputs from the final layer are combined to generate the output of the neural network.

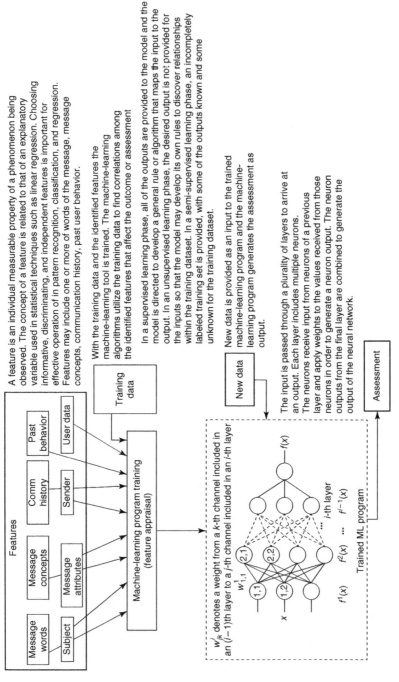

Figure 1.5 Feature usage approach in DL (example). *Source:* Adapted from Ref. [11]. Note: The input is a vector x that is processed through multiple layers, where weights $W_1, W_2, \ldots, W_n$ are applied to the input to each layer to derive $f_1(x), f_2(x), \ldots f_n(x)$, up to the point where the output $f(x)$ is computed.

1.1.3 Activation Functions

Each of the layers in an NN typically includes a plurality of channels, where each of the channels may include or represent a plurality of neurons (processing elements), configured to process data of the corresponding channel. An *activation is a function* creating a value that is an output of one channel and is also an input to channels included in the next layer, because of corresponding connection(s) with the next layer. Each of the channels calculates its own activation value, based on activations and weights received from the channels included in the previous layer. A *weight* is a parameter utilized to calculate an output activation in each channel, and thus is a parametric value assigned to a connection between channels. Thus, an output from a previous layer's channel may be provided as an input to a channel of a next or subsequent layer through a weighted connection between the previous layer's channel and the channel of the next layer. A neuron performs the activation function for a given node where the resulting value a_j^i denotes an activation of the j-th channel of the i-th layer is calculated as follows [14]:

$$a_j^i = \sigma\left(\sum_k \left(w_{jk}^i \times a_k^{i-1}\right) + b_j^i\right),$$

here σ denotes an activation function, w_{jk}^i denotes a weight from a k-th channel included in an $(i-1)$th layer to a j-th channel included in an i-th layer, b_j^i denotes a bias of the j-th channel included in the i-th layer. See Figure 1.6. For example, an activation of a first channel ($j = 1$) of Layer 2 (the second layer, $i = 2$) (with $k = 1, 2$) is denoted by $a_1^2 = \sigma\left(w_{1,1}^2 \times a_1^1 + w_{1,2}^2 \times a_2^1 + b_1^2\right)$. The weight of the *weighted connection is adjusted during the training* of the NN until the NN is trained for a desired objective: Connections and connection weights are altered to obtain an NN that is trained to support a given objective, for example, for an image recognition task.

Figure 1.7, based on [15], depicts an example of a multi-layer machine-trained feed-forward NN that has multiple layers of processing neuron nodes. Except for the first (input) and the last (output) layer, each node receives two or more outputs of nodes from earlier processing node layers and provides its output to one or more nodes in downstream layers. The output of the node (or nodes) in the last layer represents the output of the network. The output of the network can be a vector representing a point in an N-dimensional space (e.g. a 256-dimensional vector), a scalar in a range of values (e.g. 0–1), or a coded number representing one of a predefined set of categories (e.g. for a network that classifies each input into 1 of 16 possible outputs, the output could be a 4-bit value). In the example of Figure 1.7, the NN only has one output node. Other NNs can have several output nodes that provide more than one output value. Also, while the network in Figure 1.7 includes only a small number of nodes per layer, a typical NN may include a significant number of nodes per layer

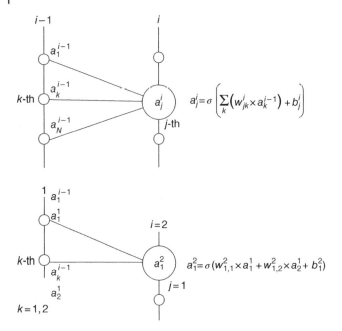

Figure 1.6 Activations and weights.

(with some layers including several 1,000 nodes) and significantly more distinct layers than shown (e.g. several dozen layers).

The developer of a DL system decides how many hidden layers there will be in the NN, how many nodes go into each hidden layer, and which activation functions to use at the nodes. See Figures 1.8 and 1.9. Also, there will be weights as multiplicative factors on the edges (synapses) and biases as additive values on the edges. These values are computed from a (training) dataset.

Three types of activation functions are: (i) binary step function, (ii) linear activation function, and (iii) nonlinear activation function (of which there are several in common use). Typical activation functions include Softmax, Softplus, sigmoid, and Rectified Linear Unit (ReLU), (Softplus can be perceived as a smooth version of ReLU). An NN operating *without* the activation function where a neuron only performs a linear transformation on the inputs utilizing the weights and biases is effectively a multi-step composed linear regression model; although this is a relatively simple learning (synthetizing) mechanism, a complex problem or task cannot be properly tackled with this method. Once an activation function is selected, one can start with a single layer and a node in each layer, and then determine how well the NN works; the NN can then be extended substantially. Weights and bias vectors for the various gates are adjusted during the training phase; after the training phase is complete, the weights and biases are utilized for normal operation.

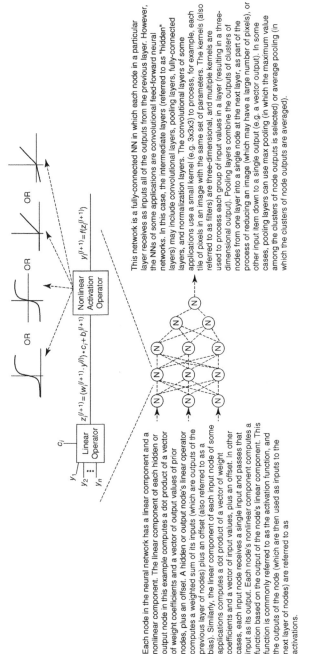

Each node in the neural network has a linear component and a nonlinear component. The linear component of each hidden or output node in this example computes a dot product of a vector of weight coefficients and a vector of output values of prior nodes, plus an offset. A hidden or output node's linear operator computes a weighted sum of its inputs (which are outputs of the previous layer of nodes) plus an offset (also referred to as a bias). Similarly, the linear component of each input node of some applications computes a dot product of a vector of weight coefficients and a vector of input values, plus an offset. In other cases, each input node receives a single input and passes that input as its output. Each node's nonlinear component computes a function based on the output of the node's linear component. This function is commonly referred to as the activation function, and the outputs of the node (which are then used as inputs to the next layer of nodes) are referred to as activations.

This network is a fully-connected NN in which each node in a particular layer receives as inputs all of the outputs from the previous layer. However, the NNs of some applications are convolutional feed-forward neural networks. In this case, the intermediate layers (referred to as "hidden" layers) may include convolutional layers, pooling layers, fully-connected layers, and normalization layers. The convolutional layers of some applications use a small kernel (e.g. 3x3x3) to process, for example, each tile of pixels in an image with the same set of parameters. The kernels (also referred to as filters) are three-dimensional, and multiple kernels are used to process each group of input values in a layer (resulting in a three-dimensional output). Pooling layers combine the outputs of clusters of nodes from one layer into a single node at the next layer, as part of the process of reducing an image (which may have a large number of pixels), or other input item down to a single output (e.g. a vector output). In some cases, pooling layers can use max pooling (in which the maximum value among the clusters of node outputs is selected) or average pooling (in which the clusters of node outputs are averaged).

Figure 1.7 Example of a multi-layer machine-trained feed-forward NN that has multiple layers of processing nodes. *Source:* Adapted from Ref. [15].

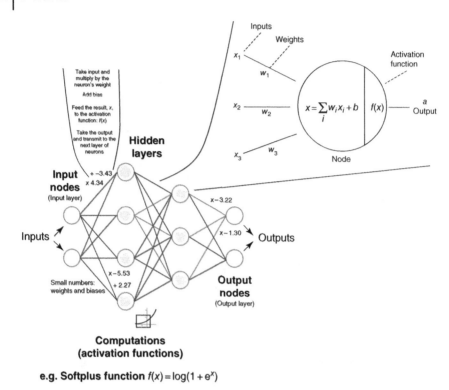

Take input and
multiply by the
neuron's weight

Add bias

Feed the result, *x*,
to the activation
function: *f(x)*

Take the output
and transmit to the
next layer of
neurons

**Hidden
layers**

**Input
nodes**

(Input layer)

Inputs

Small numbers:
weights and biases

**Computations
(activation functions)**

e.g. Softplus function $f(x) = \log(1 + e^x)$

or sigmoid function $f(x) = \dfrac{e^x}{e^x + 1}$

or ReLU function $f(x) = \max(0, x)$

Figure 1.8 Simplified neural network.

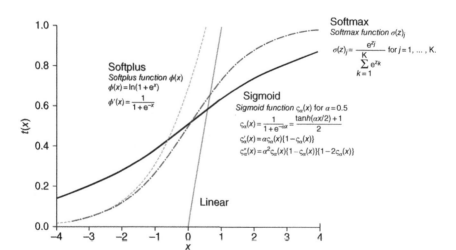

Figure 1.9 Typical activation functions.

1.1.4 Multi-layer Perceptrons

As indicated, DL can utilize MLPs, RNNs, and CNNs mechanisms (other mechanisms include but are certainly not limited to Long Short-term Memory [LSTM] and Generative Adversarial Network [GAN] techniques).

MLPs (which, as noted, some call ANNs) are typically used for problems that deal with tabular information, but they cannot optimally capture sequential data in the input points. An MLP utilizes three layers: an *input layer*, a *hidden layer*, and an *output layer*. The system entails a group of multiple perceptrons/neurons at each layer; however, inputs are processed only in the forward direction. Each layer seeks to learn appropriate weights. The input layer accepts the inputs (the input data), the hidden layer processes the inputs, and the output layer produces the result.

These DL systems have the ability to learn weights that map any input to the output. A key mechanism to achieve such universal approximation is the activation function mentioned above, that injects nonlinear properties to the NN, enabling it to learn any complex relationship between input and output. Inputs are processed with a weight, they are summed with a bias applied, and are further processed with the activation function – without an activation function the NN only learns a linear relationship from input to output and can never learn more complex relationships. The purpose of the activation function is to map (or transform) the aggregated weighted inputs from the input nodes into an output value to be transferred to the next hidden layer or as a final output, while adding some appropriate nonlinearity. However, there are limitations with MLP-based systems described as "vanishing and exploding gradient." The weights of an NN are updated by a backpropagation algorithm used for determining the gradients (see Figure 1.10), especially in the case of a DNN with a large number of hidden layers, the gradient may typically vanish as it propagates backward.

Taking a second view to the discussion of the previous paragraph, a *feed-forward network* is a basic NN that has an acyclic graph structure where the nodes are arranged in layers, specifically an input layer, at least one hidden layer, and an output layer. The NNs are fully connected[4] with logical edges to the nodes in downstream adjacent layers, but there are no logical edges between nodes within each layer. The hidden layer transforms input received by the input layer into a representation that is appropriate for generating output(s) in the output layer. Input (external) data received at the nodes of an input layer are "fed forward," that

---

4 For two adjacent layers (e.g. the i-th layer and the $(i+1)$-st layer) to be fully connected, each neuron in the i-th layer must be connected to every neuron in the $(i+1)$-st layer. A larger number of hidden layers in a DNN typically enables it to better model a complex situation, improving the ability of the model to fit a gamut of possible input data. Traditional NN layers are fully connected where every output neuron unit interacts with every input unit in the next layer over.

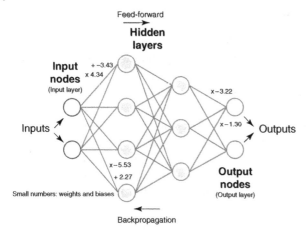

Feed-forward

Hidden layers

Input nodes
(Input layer)

Output nodes
(Output layer)

Inputs

Outputs

Backpropagation

Small numbers: weights and biases

$+ -3.43$
$\times 4.34$
$x - 3.22$
$x - 1.30$
$x - 5.53$
$+ 2.27$

Backpropagation: a training process where the weights are updated an algorithm that computes the gradient (direction and rate of increase of the data) of the loss function with respect to the weights in the NN for a paired input–output instance

Figure 1.10 Backpropagation.

is propagated downstream, to the nodes of the output layer via the nodes of the hidden layer. The arriving data at each neural node are summed and an activation function applies specific calculations to the combined data. The derived data is propagated to nodes of each successive layer in the network based on coefficients ("weights") associated with each of the logical edges connecting the layers. Figure 1.11 (loosely based on [16]) depicts what the input layer, hidden layers, and output layer look like, at the functional level, for one specific example (here for a NN system for gesture detection on a mobile device); training is also shown.

More formally, at a given layer i, the output from the computational unit node N at layer i will be $y(x) = h_{W,b}(x)$ with:

$$h_{W,b}(x) = f\left(W^T x\right) = f\left(\sum_{s=1}^{n} W_s x_s + b\right)$$

where $s = 1, 2, \ldots n$ represents the number of nodes from the previous layer that are feeding node j at layer i, x_s are the arriving signals, W_s are a weight of x_s, b is an offset (e.g. bias) of the neuron, and f is an activation function of the neuron; f is used to convert an input of the neuron to an output by introducing a chosen nonlinear feature to the NN (there is no parameter W at the input layer). The output of the activation function is used as an input to a neuron of a following convolutional layer in the NN: an output from one neuron is an input to another neuron. An input of each neuron may be associated with a local receiving area of a previous layer, to extract a feature of the *local receiving area* (also known as

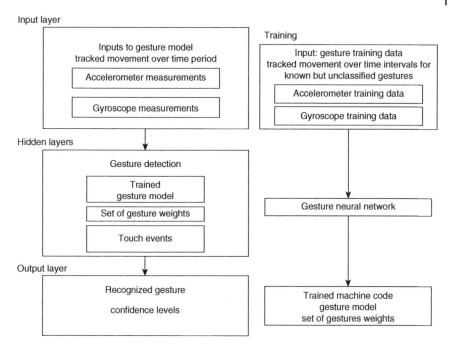

Figure 1.11 Example of input, hidden, and output layers in an NN used for gesture recognition.

local receptive field) [17]. The local receiving area may be an area consisting of several neurons.

The input data is typically a set of datapoints (rather than a single value), say x^v. The operation at each layer i can be expressed by the linear expression: $y^v(x) = f$ $(W^*x^v + b^v)$, where x^v is an input data vector, $y^v(x)$ is an output vector, b^v is an offset vector, W is a weight (or coefficient) matrix, and $f(\cdot)$ is an activation function. At each layer, the operation is performed on an input vector x^v, to obtain an output vector y^v. Given a number of layers and a number of nodes, the set of weights and biases can be substantial.[5] See Figure 1.12 for a rigorous mathematical definition of a DNN; however, a simpler formulation, as discussed above, can be kept in mind.

Training of the DNN is a process of establishing (learning) the appropriate weight matrix. Although DNNs have been used for many application domains,

---

5 As noted earlier generally, the weight from the k-th neuron at the $(i$ -1)-th layer to the j-th neuron at the i-th layer is denoted as w^i_{jk}, which clearly is captured as a complex matrix; for example w^3_{24} represents the weight from a fourth neuron ($k = 4$) at a second layer (i - 1 = 2) to a second neuron ($j = 2$) at a third layer ($i = 3$).

DNN acoustic models typically estimate the posterior distribution over a set of context-dependent tied states s of a hidden Markov model (HMM) given an acoustic observation o, $P(s|o) = \text{DNN}(o; \theta)$. The DNN is implemented as a nested function comprising L processing layers (non-linear transformation):

$$\text{DNN}(o; \theta) = f^L (f^{L-1}(...f^1(o; \theta^1)...; \theta^{L-1}) ; \theta^L)$$

The model is thus parametrised by a set of weights $\theta = \{\theta^l\}_{l=1}^{L}$ in which the lth layer consists of a weight matrix and bias vector, $\theta^l = \{W^l, b^l\}$, followed by a non-linear transformation ϕ, acting on arbitrary input x:

$$f^l (x; \theta^l) = \phi^l (W^{lT}x + b^l)$$

To form a probability distribution, the output layer employs a Softmax transformation $\phi_i^L(x) = \exp(x_i)/\Sigma_j \exp(x_j)$, whereas the hidden layer activation functions are typically chosen to be either sigmoid $\phi^l(x) = 1/(1 + \exp(-x))$ or rectified linear $\phi^l(x) = \max(0, x)$ units (ReLU).

Figure 1.12 A mathematical definition of a DNN. *Source:* Courtesy IEEE Ref. [18].

DL-based recognition systems typically require a large amount of memory and computing power, especially for models of a certain size in terms of layers, requiring large runtime memory and a large number of floating point operations per second (FLOPS) [19]. Low-end or smaller computing edge devices that are resource-limited may be incapable of supporting such resource-intensive NN, especially for real-time applications.

Backpropagation is a methodology for training a DNN. Backpropagation is used to adjust (i.e. update) the value of a parameter in the NN, such as a weight; the goal is to make the error (or a defined loss) in the output smaller. Typically, a loss is calculated with a loss function based on forward propagation of an input to an output of the NN. Backpropagation computes a *gradient of the loss function* with respect to the parameters of the NN; a gradient algorithm (for example, the gradient descent algorithm) is utilized to update the parameters endeavoring to reduce the loss function. Backpropagation is undertaken iteratively, seeking to have function converging to an acceptable value and/or minimized.

More specifically, in training, when an input is presented to the NN, it is propagated forward through the NN, layer by layer, until it reaches the output layer. The output of the NN is then compared to the desired output and then utilizing a pre-selected cost function an error value is calculated for each output layer node [20]. The error values are sent backward, propagated backward, starting from the output, until each node in the NN has an associated error value that approximately represents the node's contribution to the original output. The backpropagation methodology can utilize these error values to compute the gradient of the cost function with respect to the weights in the NN. The calculated gradient is presented to the selected optimization method to update the weights in an effort to minimize the cost function, and thus produce a usable final output.

1.1.5 Recurrent Neural Networks

RNNs capture the sequential information embedded in the input information, for example, dependency between words (or characters) in the text: each time step depends not only on the current value (e.g. word or character) but also on the previous value (e.g. word, character, or video frame). Thus, in an RNN, the previous state of the network influences the output of the current state of the network; RNNs typically employ mathematical models to predict the future based on a prior sequence of inputs – for example, as just noted, an RNN may be used to support statistical language modeling to predict an upcoming word given a previous sequence of words. See Figures 1.13 and 1.14. RNNs make the parameters available across different temporal steps, thus implying the need for fewer parameters to train (and thus reducing computational complexity). Nonetheless, deep RNNs (RNNs with a large number of time steps) also are negatively impacted from the vanishing gradient problem. Additionally, recurrent connections can be computationally intensive and difficult to parallelize.

1.1.6 Convolutional Neural Networks

Within the DL context, CNNs (a type of feed-forward networks) are a class of ANNs commonly applied to analyze visual information and imagery: they are ANNs that are designed to process pixel data that is typically found in image processing and image recognition. In CNNs most of the layers include computation nodes with a linear function followed by a nonlinear activation function (applied in turn to the result of the linear function); see Figure 1.15. The linear function is a mathematical dot product of input values (either the initial inputs based on the input data for the first layer, or outputs of the previous layer for subsequent layers) and predetermined (trained) weight values, in conjunction with bias (addition) and scale (multiplication) terms, which are also predetermined based on training.

Feed-forward neural network

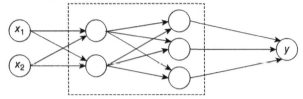

Recurrent neural network

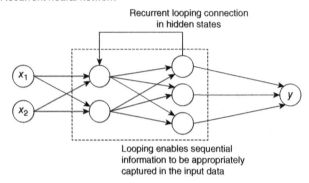

Figure 1.13 A looping constraint on the hidden layers of MLP makes it an RNN.

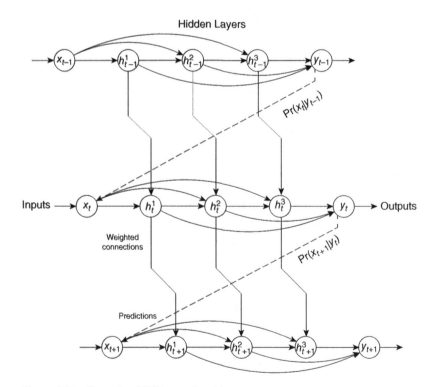

Figure 1.14 Example of RNN neural architecture.

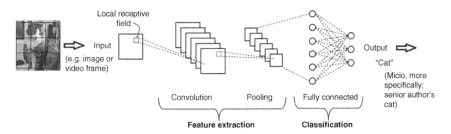

Figure 1.15 Basic architecture of a CNN.

A CNN is a class of NNs that focuses on processing data that has a lattice (cartesian)-like structure; for example, an image that comprises a block of pixels corresponding in a grid-like fashion (see Figure 1.16 for an example). In a CNN each individual neuron processes only data that corresponds to its receptive field.

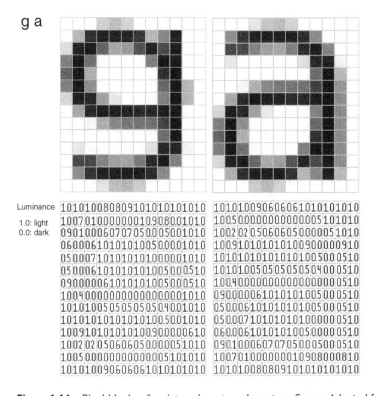

Figure 1.16 Pixel blocks of a picture, here two characters. *Source:* Adapted from Ref. [21].

Some applications of CNN include object detection, object recognition (especially in autonomous vehicle applications and facial recognition), semantic segmentation, and image captioning for images and videos. DL can also be used for classification (establishing patterns and assigning entities to classes), feature extraction, time series prediction, and regression functions such as localization and situational awareness applications. CNNs have characteristics that achieve invariance to the affine transformations of images that are presented to them; these characteristics enhance the ability to recognize image patterns that are shifted, tilted, or warped compared to a baseline.

In particular, DL is taking on an increasingly prominent role in computer vision analytics, which are relied upon, for example, by law enforcement agencies. Since 2014, DNNs, especially UDNNs, have entered the mainstream, particularly for computer vision, mobile vision, and big data use cases; CNNs are now the dominant ML approach for visual object recognition. Increased model size (but with ensuing computational complexity) typically translates into quality gains for most tasks compared to smaller models (assuming, however, that an adequate amount of labeled data is provided for training). Computational efficiency and low parameter count are still desirable goals for UDNNs/CNNs, particularly for edge/mobile computer vision (including robotics and drone applications).

CNNs were originally introduced over 25 years ago [22]. Recent advancements in computer hardware performance have given rise to truly deep CNNs, illustrative of UDNNs: while, for example, LeNet5 [23] consisted of 5 layers and VGG-19 had 19 layers [24], Residual Networks (ResNets) surpassed in recent years the 1,200-layer mark [25, 26]. However, as CNNs become increasingly deep and information about the input or gradient passes through many layers, the information can vanish and "wash out" by the time it reaches the end (or beginning) of the network [27]. A number of approaches have been advanced in the literature to address this and other problems, generally by creating short paths from early layers to downstream layers.[6]

CNNs are "feed-forward NNs" that utilize filters and pooling layers. More specifically, a CNN typically has three layers: (i) a convolutional layer, (ii) a pooling layer, and (iii) a fully connected layer. These layers are laid out in such a way so that they detect simpler patterns at first (e.g. lines, polygons, curves, boundaries, and so on) and more complex patterns further along (e.g. faces, objects, and so on).

---

6 Approaches include bypassing signal from one layer to the next via identity connections, using stochastic depth to shorten paths by randomly dropping layers during training, or repeatedly combining several parallel layer sequences with different number of convolutional blocks to obtain large depth while supporting short paths in the network

A basic description of the layers is as follows [21]:

- The convolution layer is the fundamental building block, undertaking the principal portion of the network's computational burden. In mathematical terms, this layer performs a dot product of two matrices, the first matrix being the set of learnable parameters (aka a kernel), and the second matrix being the restricted portion of the receptive field. If the image is composed of three groups of pixels (red, green, and blue – RGB), the kernel height and width may be spatially small, but the depth extends to all three streams. During the forward pass, the kernel slides across the width and height of the image producing the image representation of that region, namely, producing a 2-D representation of the image (aka activation map) that gives the response of the kernel at each spatial position of the image. Also note that nonlinearity layers are often injected after the convolutional layer (using a stage sometimes called a "detector stage") to introduce and/or enhance nonlinearity to the activation map.
- The pooling layer modifies the output of the network at various locations by deriving a summary statistic of the nearby outputs. This aims at reducing the spatial size of the representation, which, in turn, decreases the required amount of computation. The pooling layer uses a pooling function that substitutes the output of the convolutional layer with a summary statistic.
- The fully connected layer endeavors to map the representation between the input and the output. A classical CNN uses the fully connected layers after the convolutional layers in the network – a fully convolutional neural network (FCNN) can take input of arbitrary size.

In a CNN, units or neurons within a layer are directly tasked with learning visual features from a small region of the input data and the local receptive fields facilitate the process of recognizing desired visual patterns [28]. The local receptive field is a defined (segmented) area that is occupied by the portion (content) of input data that a neuron within a convolutional layer is exposed to during the process of the mathematical convolution. In the traditional fully connected feed-forward NNs, units/ neurons within a layer receives input from all units of the prior layer. In a CNN architecture, there are several layers which have within them a set of units or neurons; these units receive inputs from corresponding units from a similar subsection in the previous layer; however, generally *not all the neurons in a CNN are connected*. This is so not only because it would be impractical to connect all units from the previous layer to units within the current layer given that the computation resources needed to train such a network will be extensive and also require a large set of training data to utilize the full capacity of the network but, importantly, because *each neuron within a CNN is responsible for a well-defined region of the input data* (seeking, for example, to identify patterns such as lines). This defined region of space that a neuron or unit is exposed to in the input data is called the local receptive field.

CNN systems are being used across a gamut of different applications and domains, being prevalent in image and video processing applications; they perform equally well on sequential inputs.

1.1.7 Comparison

See Figure 1.17 for a simple comparison of the various NNs. Many of these concepts are revisited and further expanded upon in the appropriate chapter(s) to follow.

1.2 Learning Methods

A relatively large number of ML methods are currently in common use.[7,8] These are broadly discussed next. Figure 1.18 depicts pictorially the concept of learning/training.

Unsupervised learning methods (e.g. [30, 31]): analysis is undertaken by the "machine" on unlabeled data to find latent patterns in the data, after which the patterns are used to cluster populations into distinct groups. An unsupervised ML system has neither a training dataset nor a set of canonical outcomes: the system attacks the problem on a *tabula rasa* basis with only its logic to power it through the analysis and the decision-making. The usage of unlabeled data is preferred, given that it is generally easier and less expensive to acquire. Utilizing unlabeled data in the training process typically improves the quality of the final result while the processing and the absolute time and overall cost are reduced. There is a variety of unsupervised learning methods; three basic methods are (i) *clustering*, (ii) *density estimation*, and (iii) *data projection*. Clustering aims at identifying groups of comparable examples within the data; density estimation endeavors to determine the distribution of data within the input space; data projection seeks to project the data from a high-dimensional space to a lower dimensional space, preferable two or three dimensions. Many if not most unsupervised generative

DL NNs ->	MLPS	RNNs	CNNs
	Tabular/text data	Sequence data (time series, video, audio)	Image data

Figure 1.17 Comparison among various ANNs.

---

7 Some portions of the next few sections are based on reference published by these authors [29].
8 Google Scholar identifies over 3,100,000 papers on the topic of "machine learning" at the time of this writing.

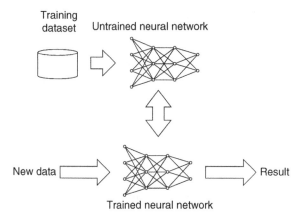

Figure 1.18 Generic concept of learning/training.

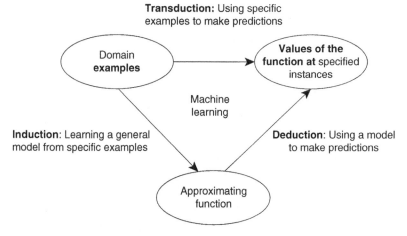

Figure 1.19 Comparison and interrelation of induction, deduction, and transduction. *Source:* References [47]/John Wiley & Sons.

models, for example deep belief networks (DBNs) or Boltzmann machines, need complex samplers – especially in voice processing – to train the generative model. Emerging AGI systems and concepts can make use of unsupervised ML principles.

Supervised learning methods (e.g. [32, 33]): the system in question, the "machine," undertakes the process of learning to utilize the best set of labeled examples and one of a number of data-driven learning algorithm; the goal is to produce an inference model where discrete data are categorized in an appropriate manner, and continuous data are related in the form of a regression function. Thus, supervised learning entails (i) *classification* and (ii) *regression*: the classification entails predicting a class label and regression that entails predicting a numerical value. Supervised ML learns by explicit example: it can utilize what the system has learned in the past and apply that knowledge to new data; it uses labeled examples to anticipate future events and patterns. In this system, the algorithm's range of possible outputs is already known and the data utilized to train the algorithm is pre-labeled with expected answers. Here, the training data comprises examples of the input vectors along with their corresponding target vectors [34–36]. A supervised learning algorithm learns by comparing its actual output with correct outputs to identify errors, such errors defined (identified) by the supervisor. The algorithm (the system) then modifies the model accordingly. In summary, supervised learning mechanisms operate to identify patterns where there is a dataset of "canonical answers" to learn from. Supervised ML is typically utilized in applications where historical data likely provides a stable prediction of future outcomes.

Semi-supervised (hybrid) learning methods (e.g. [37, 38]): analysis is performed by the "machine" using a small set of tagged data and larger (large) set of unlabeled data to refine the cluster classification. A semi-supervised system learns from the labeled (tagged) data, thereafter making a leap to make a judgment on the unlabeled data and find patterns, relationships, and structures [12, 39]. This offers higher accuracy with reduced human effort than traditional ML approaches to classification. Such a system is a hybrid of an unsupervised learning system and a supervised system. It has a "sweet spot" because in many instances the reference data needed for addressing the problem is available but could be incomplete or only partially accurate. Semi-supervised learning can utilize unsupervised learning methods such as *density estimation* and *clustering*: After patterns or groups are identified, supervised methods may be used to label the unlabeled examples or to apply labels to unlabeled data which is then utilized later on for prediction(s). A semi-supervised system is able to utilize the available reference data and then make use of other learning methods to mediate the gaps; namely, while supervised learning systems utilize labeled data and unsupervised systems are provided no labeled data, semi-supervised learning systems avail themselves of both types of data. Semi-supervised learning facilitates access and utilization to large amounts of unlabeled data without having to invest the time into the laborious (or at times impossible) task assigning labels to each datum; hence, when the expense or computational cost for labeling examples is an issue, semi-supervised learning methods can be used. A large number of semi-supervised learning techniques are discussed in the literature [40]. Such an algorithm starts with a standard labeled dataset, proceeds to keep only a portion of the labels (say, 10–20%) on that dataset, and thereafter treats the rest as unlabeled data [41]. Examples of applications span the fields of computer vision (with underlaying image data), NLP (with underlaying text data), and ASR (with underlaying audio data).

Self-supervised (hybrid) learning (e.g. [40, 42]): it refers to an unsupervised learning problem that is framed as a supervised learning problem to apply supervised learning algorithms to solve it. Such a system aims to learn representations from the data without explicit external supervision. It relies on pretext (surrogate) tasks that can be formulated using only unsupervised data. A pretext task is designed in a way that solving it requires learning of a useful image representation. Thus, supervised learning algorithms are utilized to solve an alternate or pretext task which is a model – or representation – that can in turn be utilized in the solution of the original (actual) modeling problem. A self-supervised learning framework utilizes unlabeled data to formulate a related pretext task (e.g. image rotation) for which an objective can be computed without supervision [36, 43]. Computer vision applications can make use of self-supervised learning techniques.

Reinforcement learning methods (also known as boost learning) (e.g. [44, 45]): the system in question, the "machine," assesses the environment; selects and executes certain activities (decisions) within its action domain; and depending on how it is "rewarded" or "punished," it then learns which strategy is best for future, comparable decisions. The system must literally *learn* to operate by making use of a feedback mechanism. That is, reinforcement learning entails algorithms that train using a feedback system of reward and punishment; the system learns by interacting with its environment in question and it is afforded rewards when performing correctly and is fed penalties for doing so incorrectly. Such a system learns by endeavoring to achieve the greatest reward and minimizing penalty (within a specific environment or context). A reinforcement learning entails three components: (i) the agent (the AI learner/decision-maker), (ii) the environment (everything the agent has interaction with), and (iii) the actions (what the agent can do) [12, 46]. The agent is motivated to reach the goal rapidly by finding the best way to do it, namely, endeavoring to maximize the reward, minimize the penalty and establishing the best way to do so. In many complex applications, reinforcement learning is the only practical way to train a system to perform at a measurably high level [35]. Reinforcement learning is often used in gaming applications, in robotics, and in navigation. Table 1.4, summarized in part from [36], provides some additional details of the various learning methods.

A relatively large number of ML algorithms are currently in common use to implement and support the learning methods described above. Figure 1.20 depicts key ML algorithms in common use for the various learning modes; some additional description for some of these decision algorithms follows below. Figure 1.21 provides another pictorial summary synthetized from reference [48]. Table 1.5 (partially inspired by reference [49]) provides a mapping of an expended set decision algorithms to decision method classes; these algorithms are either not further pursued in this text or they are only given a cursory treatment as/where needed

Table 1.4 Types of learning processes.

Learning processes	Supervised learning	A model is used to learn a mapping between input examples and the target variable. The algorithm learns by making predictions being provided with examples of input data; the model is supervised and, thus, corrected with an algorithm to better predict the expected target outputs in the training dataset. The model is tuned on training data comprised of inputs and outputs; the model is then utilized to make predictions on test sets where only the inputs are provided – the outputs from the model are compared to the withheld target variables and are used to estimate the skill of the model. Learning algorithms examples include, among many others, decision trees (DTs) and support vector machines (SVMs).
	Unsupervised learning	A model is used to describe or extract relationships in data. Unsupervised learning operates only on the input data without outputs or target variables. Unsupervised learning does not utilize a teacher correcting the model, as occurs in supervised learning. An example of a clustering algorithm utilized in unsupervised learning is k-Means; here, k is the number of clusters to discover in the data. An example of a density estimation algorithm utilized in unsupervised learning is Kernel Density Estimation; this involves using small groups of related data samples to estimate the distribution for new points in the problem space. An example of another unsupervised method is visualization that entails graphing or plotting data in different ways. Yet another example is a projection method that entails reducing the dimensionality of the data.
	Reinforcement learning	In this approach, there is no fixed training dataset; instead, there is one or more goals that a system is required to achieve, along with a feedback mechanism about performance toward the goal(s). The objective is to maximize a numerical reward score. Although it is similar to supervised learning by having some response from which to learn, the feedback may entail some amount of delay and/or be "noisy"; this may make it challenging for the model to correlate cause and effect. Examples include Q-learning and temporal-difference learning.
Hybrid learning processes	Semi-supervised learning	A supervised learning process where the training data contains a small set of labeled comparative examples and a large number of unlabeled comparative examples. The goal is to utilize in an effective manner all the available data. It is a class of algorithms that aims at learning from both labeled and unlabeled samples that are often assumed to be sampled from the same (or similar) distributions.

Self-supervised learning	A learning framework that relies on surrogate (pretext) tasks that can be formulated using only unsupervised data. An autoencoder is a type of a self-supervised learning algorithm, this being a projection method that reduces the dimensionality of input data. These algorithms are trained using a supervised learning method, but they aim at solving an unsupervised learning problem. An autoencoder has an encoder element and a decoder element; after being trained, the decoder is sidelined and the encoder is utilized to create compact representations of the input.
Multi-instance learning	A supervised learning environment where particular examples are unlabeled but where groups of samples (also called "bags") are labeled; that is to say, an entire collection of examples is labeled as containing or not containing an example of a class. Analysis entails making use of the knowledge that one or more of the instances in a "bag" are associated with a target label, and, by implication, predicting the label for new bags at a later instance given their composition of multiple unlabeled examples.
Statistical inference processes (aka, statistical learning) Inductive learning	Learning that entails utilizing evidence to determine the future outcome. Namely, it entails using specific cases to determine general outcomes. The process involves learning general rules from specific examples: that is, general rules (the model) are learned from provided historical examples (the data). A model is constructed about the problem or ecosystem (domain) at hand utilizing the training dataset, and that model is assumed to hold over new unseen (future) data. (See Figure 1.19.)
Deductive inference	Deductive inference refers to the top-down use of general rules to determine specific outcomes. (Inductive Learning involves going from the specific to the general); deductive inference entails going from the general to the specific. (See Figure 1.19.)
Transductive learning	Learning that aims at predicting specific examples given specific examples from a(nother) domain or ecosystem. No generalization is needed because specific examples are used directly. It is a statistical learning process; it entails estimating the value of a function at a specified instance, progressing from the particular to the particular. (See Figure 1.19.)

(Continued)

Table 1.4 (Continued)

Learning techniques	Multi-task learning	Learning that involves constructing a model on, or with, one dataset that addresses multiple related domain problems, in a supervised learning setting. It is typically used when there is a relatively large set of input data labeled for a particular task that can be shared with another task where there is much less labeled data (assuming that the data sharing is permitted or logically meaningful – for example in natural language processing).
	Active learning	Supervised (or possibly semi-supervised) learning method where the model can query a human during the learning process to improve performance by resolving potential ambiguities during the learning process. This approach facilitates the sampling of the domain in question in such a way that minimizes the number of data samples to be collected. Therefore, this approach is practical when there is a dearth of data available and when it is expensive to collect additional data or label it.
	Online learning	Often ML is performed in an offline mode, where one has a batch of data and the system seeks to optimize a metric. In some instances, one is able to receive real time streaming data; in this case one wants to perform *online learning* to update pertinent estimates as each new datum is received (as compared to waiting until such a time when al data is received, processed, and/or archived). Here learning is undertaken incrementally: the model updates itself frequently to maximize the freshness or relevance of its predictive capabilities.
	Transfer learning	A type of learning where a model is first trained on one task for a domain-related activity, and then some portion or all the model is utilized as the starting point for a related task, especially if the related task entails a large amount of data. A particular instance is image classification, where a model can be trained on a large set of generic images and where parameters of the model can then be used as a starting point a smaller but more specific dataset.
	Ensemble learning	An approach where several modes are trained on the same data and the predictions from these models are combined to attain improved performance.

Source: Adapted from Ref. [36].

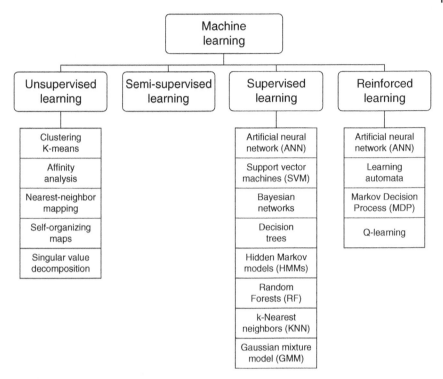

Figure 1.20 Commonly used machine learning methods and algorithms (figure is not exhaustive). Note: Only key acronyms are defined in this figure; see text and Glossary for fuller descriptions.

(since this text is not intended to be a comprehensive assessment of the AI topic, for which many other books are available).

An ML algorithm learns the mapping from an input realm to the output realm. In case of parametric models, the algorithm learns a function when provided with a few sets of weights:

$$\text{ML algorithm} : \text{Input} \rightarrow f\left(w_1, w_2, \ldots w_n\right) \rightarrow \text{Output}$$

In the case of simple classification problems, the ML algorithm learns processes that separate or distinguish two classes utilizing a *decision boundary*. The decision boundary enables the system to determine whether a given (future) data point belongs to a certain Class 1 or to a Class 2. Figure 1.22 depicts a logistic regression where the learning function at training is a ReLU; the function is trained to separate the two classes. At a future point in time, when a particular value is presented, it will be classified as a Class 1 or a Class 2 based on this boundary.

Reinforced Learning (RL) algorithms can be grouped into three different categories:

- *Value-based algorithms* – based on temporal difference learning to obtain value function.
- *Policy-based algorithms* – directly learn optimal policy or endeavor to obtain an approximate optimal policy based on the observation
- *Imitation algorithms* – endeavor to make decisions using demonstrations.

Hybrid algorithms combine value-based algorithms with policy-based algorithms, with the goal to represent the policy function by policy-based algorithms, where updates of policy functions depend on value-based algorithms.

Supervised Learning (SL) algorithms can be segmented into two major categories:

- *Classification algorithms* learn to predict a category as the output for a new observation, on the basis of labeled training data; examples: SVM, Adaptive Boosting (AdaBoost)
- *Regression algorithms* are utilized for regression problems whose output variable is a real, continuous value; examples: LR which endeavors to fit data with the best hyperplane going through the points of training data

Some algorithms are applicable to both classification problems and regression problems, such as k-nearest neighbors (k-NN), Random Forest (RF), and Boosted Regression Trees (BRT).

Unsupervised Learning (UL) algorithms can be segmented into two categories:

- *Clustering algorithms*, such as K-means clustering, discover the inherent groupings in the data.
- *Dimension reduction algorithms*

Deep Learning makes use of Artificial Neural Networks (ANNs)/Deep Neural Networks (DNNs). An ANN has a set of interconnected nodes designed to imitate the functioning of the human brain. Each node has a weighted connection to several other nodes in neighboring layers. Individual nodes take the input received from connected nodes and use the weights together with a simple function to compute output values. The main groups of NNs are:

- Fully-Connected Neural Networks (FNNs) – NNs with layers where neurons have full connections to all activations in the previous layer; every output unit interacts with every input unit; the output from the fully-connected layers is used to generate an output result from the network. Layers apply a linear combination and an activation function to the input operand and generate an individual partial sum.
- Convolutional Neural Networks (CNNs) – layers are often sparsely connected: the output of the convolution of a field is input to some nodes of the subsequent layer. These NNs have achieved dominant performance on visual tasks, such as exploiting fundamental spatial properties of images and videos;
- Recurrent Neural Networks (RNNs) – successful at characterizing the temporal correlations of data for time series tasks. The Long Short-term Memory (LSTM) methods, whose units are RNNs, are capable of learning order dependence in sequence prediction problems.

NNs have numerous extensions, such as Deep Belief Networks (DBN), Error-Feedback Recurrent Convolutional NNs (eRCNNs), Fully Convolutional Neural Networks (FCNs), and Spatio-Temporal Graph Convolutional NNs (STGCNs).

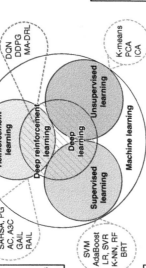

Figure 1.21 Summary pictorial for machine learning methods. *Source:* Synthetized from Ref. [48] and used with permission from Wiley.

Table 1.5 Decision algorithms mapped to decision method classes.

Decision method class	Key examples of specific decision algorithms or methods
Regression method	Ordinary Least Squares, Logistic Regression, Stepwise Regression, Multivariate Adaptive Regression Splines, Locally Estimated Scatterplot Smoothing
Instance-based methods	K-Nearest Neighbor (KNN), Learning Vector Quantization, Self-Organizing Map
Regularization methods	Ridge Regression, Least Absolute Shrinkage and Selection Operator, Elastic Net
Decision tree (DT) learning methods	Classification and Regression Tree, Iterative Dichotomiser 3, C4.5, Chi-Squared Automatic Interaction Detection, Decision Stump, Random Forest, Multivariate Adaptive Regression Splines, Gradient Boosting Machines
Bayesian methods	Naive Bayes, Averaged OneDependence Estimators, Bayesian Belief Network
Kernel methods	Support Vector Machines (SVM), Radial Basis Function, Linear Discriminate Analysis
Clustering methods	K-Means Clustering, Density-Based Spatial Clustering Of Applications with Noise, Expectation Maximization
Bidirectional encoder methods	Bidirectional Recurrent Deep Neural Network (BRDNN), Bidirectional Encoder Representation Form Transformers (BERT) for masked language model tasks and next sentence prediction tasks – and variations of BERT including ULMFiT, XLM UDify, MT-DNN, SpanBERT, RoBERTa, XLNet, ERNIE, KnowBERT, VideoBERT, ERNIE BERT-wwm, MobileBERT, TinyBERT, GPT, GPT-2, GPT-3, GPT-4, ELMo, content2Vec
Associated rule learning algorithms	Apriori Algorithm, Eclat Algorithm
ANN models	Perceptron Method, Backpropagation Method, Hopfield Network Method, Self-Organizing Map Method, Learning Vector Quantization Method
Deep learning methods	Restricted Boltzmann Machine (RBM), Deep Belief Network (DBN), Convolution Network Method, Stacked AutoEncoder Method, Variational Autoencoder (VAE), Generative Adversarial Network (GAN)
Dimensionality reduction methods	Principal Component Analysis, Partial Least Squares Regression, Sammon Mapping, Multidimensional Scaling, Projection Pursuit
Ensemble methods	Boosting, Bootstrapped Aggregation, AdaBoost, Stacked Generalization, Gradient Boosting Machine Method, Random Forest Method

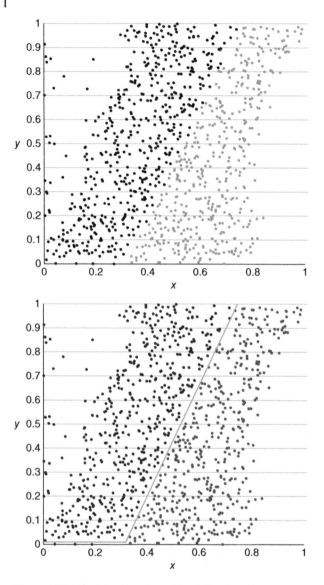

Figure 1.22 Decision boundary of logistic regression.

However, a more complex sample space where more complex relationships are at play may not be properly parseable by this basic learning method.

The AI/ML/DL systems under discussion use feature engineering, which comprises two steps: (i) *feature extraction* and (ii) *feature selection*. In the former, feature extraction, the developer extracts all pertinent features of the problem under study; in the latter, feature selection, the developer selects the relevant features that improve performance. Extracting features manually, for example, from an image, requires intrinsic knowledge of the subject matter and of the overall domain; it is typically an expensive and time-consuming process [50]. DL enables the automation of these processes using NNs to automate the process; as noted, these typically are MLP, CNN, and RNN. Some of the many algorithms in use are as follows (some described in [51]):

- For DL, NN algorithm (e.g. [52, 53]): a widely deployed methodology that enjoys good performance for estimation and approximation. NNs have origin in biology where the network is made of biological neurons; an ANN is composed of nodes that represent artificial neurons that can store information. Such an arrangement works very well for solving AI problems such as predictive modeling or adaptive control. The connections of the nodes are modeled as weights; inputs are modified by a weight and are summed, and an activation function controls the amplitude of the output; training occurs via a dataset.
- Support Vector Machines (SVM) Algorithm (e.g. [54–56]): a linear model for classification and regression assessment. The purpose of these algorithms is to identify and differentiate complex patterns in the data: they use learning algorithms to classify and analyze data regression. The basic concept of SVM is straightforward: the algorithm creates a line or surface (a hyperplane) that separates the data into classes, seeking a separation between the two classes that is as wide as possible. It injects the data as an input and then outputs a line that separates the data classes to the degree possible.
- Bayesian Network (BN) Algorithm (e.g. [57–59]) (also known as a unidirectional graph model or as a belief network): a probabilistic representation of a set of random variables and their conditional dependence. It is a probabilistic model that uses Bayesian inference for computation of probabilities. Such a network aims at modeling conditional dependence, which highlights causation, representing conditional dependence by edges in a directed graph.
- Decision Tree (DT) Algorithm (e.g. [60, 61]): a predictive mechanism used for solving regression and classification problems that make use of observations that represent graphs. Predictions are correlated with the traits studied by leaves and branches: it utilizes a DT (as a predictive model) to progress from observations about a phenomenon (represented in the branches) to conclusions about the item's target value (represented in the leaves). The aim is to create a training model that can be used to predict the class or value of the target variable by learning simple decision rules inferred from prior data (which is the training data). The DT can be designed with various ML algorithms.
- Hidden Markov Model (HMM) Algorithm (e.g. [62–64]): a statistical approach where the system is modeled as a Markov chain that has unseen (specifically, hidden) states or modes. The chain has a few "states" and the probability of being in a state j depends only on the previous state. Markov chains are often described by

a graph with transition probabilities. In an HMM, there is an underlying unobservable Markov chain in which each state randomly generates one out of k observations; the observations are visible. In practical settings, a series of observations are available from which the most probable corresponding hidden states can be found. In ML, the process entails parameter learning, namely one has some dataset, and one wants to find the parameters that best fit the HMM model.

- Random Forest (RF) Algorithm (e.g. [65, 66]): a structure consisting of a number of DTs for classification and regression that utilize a tree pattern to make a decision. The "forest" is an ensemble of DTs, usually trained with the "bagging" method where a combination of learning models is able to increase the overall quality of the result. Making use of, and merging, multiple DTs yields more accurate and stable predictions. RF algorithms are used for both classification and regression problems, which comprise the majority of ML systems; it is often used for clustering information.

- k-Nearest Neighbor (KNN) Algorithm (e.g. [67, 68]): an algorithm where a sample is categorized by the majority of its nearest k neighbors (k is typically a small integer).

- k-Means Algorithm (e.g. [69, 70]): an algorithm where samples are subdivided into categories whose members are similar: the k members are randomly distributed among n members selected as cluster centers; the remaining $n - k$ members are thereafter allocated to the nearest cluster. This algorithm is typically used for simple clustering of large datasets consisting of high dimensional numerical data; it is rather efficient when classifying similar data into the same cluster.

- Gaussian Mixture Model (GMM) Algorithm (e.g. [71]): an algorithm that probabilistically represents subpopulations (according to a normal distribution) within an overall population. An advantage of GMMs is that the subpopulation to which a data point belongs can be unknown. Thus, the model can learn the subpopulations automatically – a form of unsupervised learning; additionally, the methodology is usable in supervised learning. GMMs have been used in many applications including feature extraction from speech data and object tracking of multiple objects at each frame in a video sequence where the number of mixture components and their means predict object locations.

Some computer vision examples using NN algorithms are described in references [72–75] among others; using SVM algorithms, references [76–78]; using BN algorithms, reference [79] among others; using DT algorithms, reference [80]; using HMM algorithms, references [81–83]; using RF algorithms, references [84–86]; using k-nearest neighbor algorithms, references [87, 88]; using k-means algorithms, reference [89]; and using GMM algorithms, reference [90]. Figure 1.23 provides a generalized comparison of some (not all) of the available

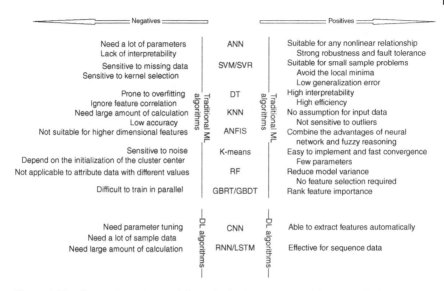

Figure 1.23 Comparison of some ML methods. *Source:* Adapted from Ref. [91].

algorithms, inspired by reference [91]. Many of these concepts are revisited and further expanded upon in the appropriate chapter(s) to follow.

1.3 Areas of Applicability

After several decades of evolutionary development, AI/DM/DL technology progressed to model/algorithm-based ML systems spanning almost every conceivable field. AI is now increasingly aiming at perception, reasoning, and generalization. The area of AI applicability is quite wide, for example,

- Transportation: autonomous cars
- Manufacturing: AI powered robots
- Healthcare: disease diagnostics and drug discovery
- Education: virtual tutors
- Journalism: e.g. systems to analyze complex financial reports
- Customer service: AI assistants
- Generative AI: AI systems that are able to generate (that is, create) NLP text, media, or image content, among other types of outputs

Many observers see AI as impacting the employment field by automating routine job tasks, even when these are quantitative in nature (e.g. analyzing MRI scans). It is relatively likely that by 2040 many workers in such routine-based functions will be displaced by AI.

More specific to this text, the deployment of 5G broadband services and the continued roll out of IoT services is fueling the rapid adoption of AI in the telecom industry, particularly for security and network management. The global AI in telecommunication market is forecast by some to reach US$15 billion by 2027, up from US$1.2 billion in 2020, at a CAGR at over 40% during 2021–2027, although all market forecasts are always subject to variability [92, 93].

Expert systems and ML algorithms have been widely used in telecommunication and networking. Table 1.6 depicts a few applications.

Table 1.6 Examples of general AI application in telecommunications.

Function	Details
Anomaly detection	AI models for detecting anomalous network conditions, intercepting deviations from the optimal functioning; examples: recognition of hardware or software malfunctionings, congestions, exceptional traffic, and security intrusions.
Interference management in cellular networks	Radio assets can be coordinated between the macro cells and micro cells. The parameters of the underlying algorithms can be "tuned" with AI. To optimize efficiency, algorithms are being used to dynamically determine what part of the spectrum should be used for which user and with which parameters.
Network optimization	AI algorithms used to detect and predict network anomalies and, in turn, to fix problems before customers are impacted. ● Routing algorithms to optimize the dynamic scheduling of optimal route in the network, based on advanced analytics to identify patterns ● Reinforcement learning to simulates the behavior of the users and optimizes operations ● Forecasting models: AI for the prediction of future traffic loads
Optimizing the parameters of a radio signal	ML has been utilized to optimize data flow to and from a Base Transceiver Station (BTS) in a mobile network. Aimed at determining the maximum amount of data that can be transmitted per quantity of spectrum per unit of time.
Power management	ML is used to achieve power savings in live mobile networks. Based on the number of users and their position, and based on meteorological data, antennas actively adjust their strength, direction, and radiation pattern to meet demand.

Table 1.6 (Continued)

Function	Details
Predictive maintenance	Techniques to predict when an equipment or a machine is likely to fail. This predictive maintenance task is a common challenge faced not only by manufacturing and operations-focused companies but also by telecommunications providers, given that telecommunications systems represent critical infrastructure.
Quality of transmission estimation – fiber links	ML methods to estimate how well the transmission over an optical connection will work. It determines the best path based on factors such as the cable length, other signals within the cable, and the equipment's age. The traffic is then routed based on this assessment.
Quality of transmission estimation – wireless links	ML has been applied to determine how much error correction or redundancy (e.g. retransmission) is used.
Traffic classification and flow clustering	Traffic profiling purposes, for example: • Identification of clusters of sessions based on their characteristics, and strategic management of sessions according to cluster features • Identification of traffic-flow-based clusters: identify and classify different network conditions based on the traffic flow characteristics • Detection of malicious sessions

Source: Summarized from Refs. [92, 94].

1.4 Scope of this Text

This text assesses AI/ML/DL applications in the telecommunications or telecommunications-related fields. In some cultures, the number 7 is identified with something being "finished" or "complete" or "perfect"; not to imply under any circumstances that this work has such characteristics, just that indeed it does have seven chapters; the topics covered are as follows:

- Chapter 2: Current and evolving applications to natural language processing and generative AI
- Chapter 3: Current and evolving applications to voice recognition
- Chapter 4: Current and evolving applications to image/video/situational awareness/ face recognition

- Chapter 5: Current and evolving applications to IoT and evolving applications to smart buildings and energy management
- Chapter 6: Current and evolving applications to network cybersecurity
- Chapter 7: Current and evolving applications to network management

Many of the applications discussed in this text (but not all) follow the architecture of Figure 1.24, where the client device obtains AI services from a cloud-based server, or it performs some basic function locally and then relies on a more powerful cloud-based AI; in some cases, the application operates as a standalone entity. Deploying DNNs on embedded devices is still a challenge given the extensive computation and storage requirements; the number of parameters and operations increases with the complexity of the model. The performance depends on the capacity of processors to support the DNN having multiple layers, multiple neurons per layer, and multiple filters (filter sizes and channels), while dealing with a large dataset. For example, the well-known CNN network ResNet50 requires up to 7.7 billion FLOPS and 25.6 million model parameters to classify a $224 \times 224 \times 3$ image; the more complex model VGG-16 with 138.3 million parameters model size requires up to 30.97 giga FLOPS (GFLOPs) [95].

This text does not focus on (physical) robots. We just make note that a robot is a machine that automatically processes or operates a given task by its own ability. A (physical) robot that recognizes an environment and performs a self-determination operation is identified as an intelligent robot. Robots may be classified into industrial robots, medical robots, home robots, military robots, and so on, according to

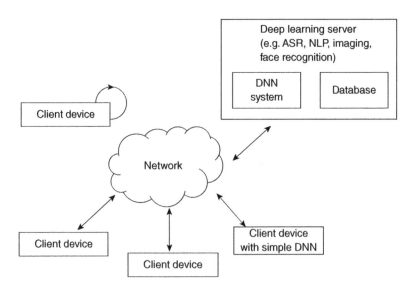

Figure 1.24 Very common network/cloud-based AI applications.

the use purpose or field. The robot may include a driving unit and may include an actuator or a motor and may perform various physical operations. In addition, a movable robot may include a driving unit with a wheel, a brake, and a propeller, and may travel on the ground through the driving unit or fly in the air [96].

Of late, technology entrepreneurs have been pursuing the development of AGI, which is expected to have impactful effects on society. These effects could, in theory, be positive, for example, they could release people from having to do mundane work, but the negative effects could be significant. The late Stephen Hawking and many AI academics have warned publicly that human-level AI could be problematic in the long run.[9] These issues are, fortunately, not of immediate concern within the context of this book.

---

9 For example, the Defense Advanced Research Projects Agency (DARPA) announced early in 2023 that it successfully tested an F-16 test aircraft with an AI software that flew the aircraft for 17 hours. Using the software, the plane was capable of taking off and landing by itself as well as performing against various simulated adversaries and with simulated weapons capabilities. The program Aerial Combat Evolution (ACE) was launched in 2019 to research and incorporate AI into its jet fighters.

A. Basic Glossary of Key AI Terms and Concepts

(Based on a variety of references, but specifically including [3, 4, 11, 19, 21, 28, 36, 48, 49, 96–107])

Term	Definition/Explanation/Concept
Activation Function	(aka Transfer Function) At a basic level, a mechanism for determining whether a neuron should be activated or not, that is, establishing if the neuron's input to the network is needed or not during the feed-forward propagation process; specifically, its role is to derive an output value from a set of input values fed to a node (or a layer) to facilitate the process of prediction using simpler mathematical operations. Stated differently, an activation function is a mathematical function that modulates the input to add nonlinearity and produces an output capturing nonlinearity; such output is then transferred forward into the next subsequent layer. Examples include but are not limited to Softmax, SoftPlus, and sigmoid functions. Regression-based decisioning approaches typically utilize the linear activation function, binary classification approaches typically make use of the sigmoid/logistic activation function, multiclass classification utilize Softmax functions, and multilabel classification utilize sigmoid functions.
Algorithm	Sequential list of rules to be utilized in solving a problem.
Artificial intelligence (AI)	Umbrella term capturing advanced computer-based intelligence, for example, simulating human cognitive thinking and decision-making. Three categories are sometimes recognized: narrow (or weak) AI, general (strong) AI (also known as artificial general intelligence – AGI), and conscious AI. AGI is the theoretical future achievement of AI where a computer-based system that is able to rival or exceed the ability of humans to perform a given intellectual task.
Artificial neural network (ANN)	A computational model that operates analogously to the functioning of a human nervous system; a model used in ML which is composed of artificial neurons (system nodes) that form a network by synaptic connections. Basically, a neural network (NN); synonymous (in this text) with NN. There are several kinds of ANNs which are defined based on the mathematical operations and the set of parameters needed to determine the output. Specifically, the ANNs are defined by (i) a connection structure (links, synapses) between neurons in different layers, (ii) a learning process for updating model parameters, and (iii) an activation function for generating an output value.

Term	Definition/Explanation/Concept
Autoencoder (AE)	A class of unsupervised NNs (AE networks) that are trained to reconstruct their input offered for representation learning tasks, by first mapping that input onto an intermediate representation, typically of lower dimension. AE networks are DL networks comprising two DNNs, one called an encoder and the other called a decoder. The function of an encoder is to generate, based on input data (say, an image), a compressed feature set, this being known as the latent space representation. The encoder is responsible for deriving a correct encoding from the input data; this encoding is a compressed type of information that is a smaller version/dimension of the input data (the encoding will be used to reproduce the data). The function of the decoder is to recreate the input data based on the features it is presented with. The decoder learns to reproduce the input data; the decoder uses the encoding to generate an output of the AE similar to the input of the same AE. Thus, an AE is a type of feed-forward NN utilized to learn efficient data coding in an unsupervised manner; AEs can be trained to recreate relevant input data without the associated noise that may present in the input.
Backward propagation (BP)	(also known as backpropagation or backprop or BP) A process where the weights of the network connections are iteratively adjusted to minimize the discrepancy (error) between the actual output value of the NN and the desired output value. A training process where the weights of an NN are updated by an algorithm that computes the gradient (direction and rate of increase of the data) of a specified loss function with respect to the weights in the NN for a paired input–output instance.
Bias	Bias as a *technique*: endowing a model with parameters whereby a model weights certain features or elements of a dataset more heavily than others.
	Bias as an *issue*: bias can occur when certain elements of a dataset are given more weight than others or when training data do not properly represent the use case at hand, causing problematic output.
Boltzmann Machine (BM); also, Deep Boltzmann Machine (DBM); also, Restricted Boltzmann Machine (RBM)	A Boltzmann Machine is a network of symmetrically connected (coupled) stochastic binary units; specifically, it contains a set of visible units v and a set of hidden units h. A Deep Boltzmann Machine (DBM) is an undirected probabilistic graphical model that contains one visible layer and several hidden layers. A Restricted Boltzmann Machine (RBM) is a thin two-layer neural network, an undirected graphical model with no hidden-to-hidden and no visible-to-visible connections; the first layer of the RBM is known as the visible or input layer; the second is known as the hidden layer. In RBM the interconnections between visible units and hidden units are established using symmetric weights. RBMs can be used for dimensionality reduction, classification, and regression; RBMs are the basic block of deep belief networks (DBNs). RBMs are efficient at learning; multiple hidden layers can be learned by treating the hidden activities of one RBM as the data for training a higher-level RBM.

(*Continued*)

(Continued)

Term	Definition/Explanation/Concept
Classification	One of two key types of supervised learning techniques (regression being the other). Models based on classification predict a class label, such as whether a certain image is a truck or not. This approach is useful for use cases (problems) that present themselves with large amounts of historical data, including labels, that establish if something belongs to one group or another. Clustering algorithms that group data in some empirical fashion are a type of classification technique – these are typically used in unsupervised learning.
Clustering	A method for placing members of a cohort (a population), based on specified observable features into subgroups, where members of a given group are more similar to each other than to members of other subgroups. Clustering is an unsupervised learning problem where one endeavors to find clusters of members (points) in the dataset that share some common characteristics.
Convolutional neural network (CNN)	(aka ConvNet) A DNN with a convolutional structure. A class of DNNs that deals with processing data that has a geometric grid/array-like topology, for example, an image. A given neuron in a CNN system only processes data in its receptive field. A CNN usually comprises three layers: a convolutional layer, a pooling layer, and a fully connected layer. The layering is arranged in such a way so that the layers detect simpler patterns first, such as boundary lines, curves, and more complex patterns later on, such as objects or faces. Only the last layer of a CNN is fully connected; by contrast in a general ANN, each neuron is connected to every other neuron in the network. See Chapter 4.
Deep belief network (DBN)	A generative NN that is composed of multiple layers of stochastic (random) variables. A stack of Restricted Boltzmann Machines (RBMs) having a two-layer network model, consisting of visible units and hidden units.
Deep learning (DL)	(aka as deep neural learning or representation learning) is a subcategory of ML and functions in a similar manner; the difference is that DL makes use of ANNs. Although "deep learning" and "machine learning" are occasionally used interchangeably, there are differences: DL models are much more complex than traditional ML methods, and they also process vastly larger amounts of data than the earlier systems. DL is considered to be an elaboration of ML. DL also utilizes supervised, unsupervised, and reinforcement learning principles and algorithms.
Deep learning (DL) algorithms	Procedures to achieve feature extraction and/or selection. Examples: Neural network algorithm, support vector machines (SVMs) algorithm, Bayesian network (BN) algorithm, decision tree (DT) algorithm, hidden Markov model (HMM) algorithm, Random Forest (RF) algorithm, K-nearest neighbor (KNN) algorithm, K-means algorithm, Gaussian mixture model (GMM) algorithm, among others. See text for descriptions.

Term	Definition/Explanation/Concept
Deep neural network (DNN)	(aka multi-layer NN or n-layer NN) Any NN with several hidden layers. A stacked NN that is composed of multiple layers; the layers are composed of nodes, which are logical locations where computation occurs. An ML model (MLM) that has an input layer, an output layer, and one or more hidden layers; the hidden layers each apply a nonlinear transformation to input it receives from the previous and to generate an output for the next layer. The output layer generates the final output. For example, CNNs are a specific type of DNN that are ideally suited for image recognition. For another example, RNNs are DNN in the "temporal" dimension rather than just being in a spatial dimension (in 2D or 3D); here, the hidden activations from time step t are used as the context to compute the activations of time step $t + 1$.
Dimension reduction algorithms	Algorithms that endeavor to identify the best representation of the data with fewer dimensions. Examples: Principal component analysis (PCA) and independent component analysis (ICA).
Distributed learning	A learning method that utilizes multiple distributed computing nodes to perform supervised or unsupervised training of an NN. Distributed learning can be performed utilizing model parallelism, data parallelism, or a combination thereof.
Feedback loop	(aka closed-loop learning) A process of comparing the output of an ML/DL system and ensuing end-user actions to retrain and improve models over time. During this process ML/DL-generated output (predictions or recommendations) are compared against the final decision and/or end-user action to provide feedback to the model, allowing it to learn from its suboptimal (or erroneous) output.
Feed-forward NNs	NNs where each neural node in a layer of the network is connected to every node in a preceding or subsequent layer (this being a "fully connected" NN); the feed-forward NN does not include any connections that loop back from a node in a layer to a previous layer. Feed-forward networks support an obvious direction for the information flow from the input layer(s) to the downstream classifier or regressor. These NNs can be represented by an acyclic graph. CNNs and DNNs are examples of "feed-forward" NNs. In a feed-forward NN, for any "cut" separating the inputs from the outputs, the designer is able to access the amount of information passing though the cut; considering any such cut, the designer seeks to avoid computational bottlenecks to maintain efficiency.
Feed-forward propagation	A system where the flow of information occurs in the forward direction. The input is used to calculate some value in the hidden layer, which is then utilized to determine the final output.

(Continued)

(Continued)

Term	Definition/Explanation/Concept
Generative Adversarial Network (GAN)	A class of AI algorithms used in unsupervised machine learning that are implemented by a system of two NNs contesting with each other in a zero-sum mathematical game environment. GANs include (at least) two modules: (i) a generative model (aka a generator) and (ii) a discriminative model (aka a discriminator). The two models compete with each other and learn from each other, such that a better output is generated. The generator and the discriminator may both be NNs (e.g. DNNs or CNNs).
	Generative AI: A subfield of AI. Generative AI systems are AI systems that are able to generate (that is, create) NLP text, media, or image content, among other types of outputs, where questions are asked (with NLP) and answers are generated and presented (with NLP or graphics or other outputs). Large language and text-to-image models are utilized. Generative AI is focused on creating new content, such as text, literature, graphics, or music, that uses ML techniques to generate new material that is similar to the training data. Massive training datasets are utilized. One example is the ChatGPT product.
Ground-truth	The actual nature of the problem, that is, the prevue of a machine learning model, instantiated by the relevant datasets associated with the use case under study. Supervised machine learning models are trained on labeled data that are considered "ground truths" for the model to identify patterns that predict those labels on a new set of data.
Hyperparameter	Operational parameters, such as node weights, have values that are derived through the training process; hyperparameters (model hyperparameters and algorithm hyperparameters), *on the other hand*, are parameters whose value is used to control the ML learning process. Various algorithms for model training utilize different sets of hyperparameters while simple algorithms (e.g. regression) do not require any. Upon specifying the hyperparameters, the training algorithm learns the operational parameters from the data. Examples of model hyperparameters are the size of an NN and the topology (neurons and connecting edges). Examples of algorithm hyperparameters are the learning rate and batch size of the sample set.
Hyperplane	Decision boundary used to classify the data points. Data points falling on either side of the hyperplane are then assigned to different classes. The dimension of the hyperplane depends upon the number of features; e.g. for two input features the hyperplane is a line, for three input features the hyperplane is a two-dimensional plane, and so on. Used with support vector machines (SVMs).

Term	Definition/Explanation/Concept
Imitation algorithms	(aka apprenticeship learning) Reinforced learning (RL) algorithms that endeavor to make decisions using demonstrations. Good performance is achieved when the reward function is difficult to specify and when it is challenging to directly optimize actions.
Incremental learning	A learning method that enables the trained NN to adapt to the new data without forgetting the knowledge acquired within the network during initial training.
Learning	The goals of (ML) learning are (i) understanding a process or phenomenon and (ii) making prediction about outcomes, namely, inferring a function or relationship that maps the input to an output in such a manner that the learned relationship can be used to predict the future output from a future input. A point in the training process is an input–output pair, with the input intended to map to a corresponding output. The learning process consists of inferring the relationship or function that maps between the input and the output, such that the learned relationship or function can be utilized to predict the new/future output from future input.
Linear regression (LR)/regression analysis	Mathematical methodology for establishing a relationship between input variables and an output variable, which in turn is used to predict the value of a response to a new set of inputs. In linear regression, a linear equation (a line) is fitted to previously collected data, specifically to pairs of an independent variable and a dependent variable, with the goal of minimizing the error between the observed values and the computed values using a given loss function, for example least squares. The linear equation has a value (a parameter) defining the slope and a "bias" the y-intercept.
Logistic regression	(aka logit regression) Logistic regression is a simple, commonly used ML classification algorithm (almost as common as linear regression – it can be used for binary or multivariate classification tasks). It is a statistical model that establishes the probability of an event; the log-odds for the event are taken to be a linear combination of one or more independent variables. A methodology used to estimate the parameters of a logistic model. Linear regression and logistic regression are similar; the distinguishing difference being what they are used for: linear regression algorithms are utilized to predict output values on new datasets while logistic regression is utilized for classification tasks.

(Continued)

(Continued)

Term	Definition/Explanation/Concept
	In logistic regression one needs the output of the algorithm to be class variable, i.e. 0 = Class 1; 1 = Class 2. Logistic regression algorithm uses a linear equation with independent variables x_1, x_2, and so on to predict a value. To map output of the linear equation into a range of [0,1], one can use the sigmoid function below (the value becomes asymptote to $y = 1$ for positive values of z and becomes asymptote to $y = 0$ for negative values of z): $$z = \theta_0 + \theta_1 \cdot x_1 + \theta_2 \cdot x_2 + \cdots$$ $$h = g(z) = \frac{1}{1 + e^{-z}}$$ $$0 < h < 1$$ One uses a logarithmic loss function to calculate the cost for misclassifying. $$\text{Cost}\big(h\,(x), y\big) = \begin{cases} -\log\big(h\,(x)\big) & \text{if } y = 1 \\ -\log\big(1 - h\,(x)\big) & \text{if } y = 0 \end{cases}$$ One can then take the (partial) derivatives of the cost function with respect to each parameter to obtain the gradients. The gradients are then used to update the values of the parameters.
Long short-term memory (LSTM)	NNs (specifically, RNNs) designed to overcome error back-flow problems by learning to bridge relatively large time intervals, even in the presence of noisy input data, without loss of short time lag capabilities. LSTM RNNs are able to learn long-term dependencies that are typically required for processing longer sequences of text. This is achieved using special internal units: the LSTM architecture allows for constant error flow using special, self-connected units; specifically, a multiplicative input gate unit (called *in-t*) to protect the memory contents stored in cell ct from perturbation by irrelevant inputs, and a multiplicative output gate unit (called *out-t*) to protect other units from perturbation by currently irrelevant memory contents stored in cell ct. The use of purpose-built memory cells to store information enables LSTM models to be better at finding and exploiting long range dependencies in the data.
Machine learning (ML)	An application of AI that allows systems to learn and evolve based on experience without the entire universe of functionality being explicitly programmed. An ML program can access the data in question and use it for its own learning.

Term	Definition/Explanation/Concept
Machine learning models (MLM)	Systems that accept an input and algorithmically generate an output predicted on the received input using some learning method. Systems that use an algorithm to learn to generate dynamic inferences based on a set of data. ML methods include: (i) supervised learning (e.g. employing logistic regression, back propagation NNs, Random Forests, decision trees), (ii) unsupervised learning (e.g. employing an apriori algorithm, K-means clustering), (iii) semi-supervised learning, (iv) reinforcement learning (e.g. employing a Q-learning algorithm, temporal difference learning); and (v) adversarial learning.
Machine learning (ML)-based processing	ML-based systems that take as input empirical data (such as, but not limited to network statistics and performance indicators) and recognize complex patterns in these data. One typical approach is the use of an underlying model M, whose hyperparameters are optimized for minimizing the cost function associated to M, given the input data. The learning process then operates by adjusting the hyperparameters such that the number of misclassified points is minimal. After this optimization phase (or learning phase), the model M can be used to classify new data points. Often, M is a statistical model, and the minimization of the cost function is equivalent to the maximization of the likelihood function, given the input data.
Machine learning programs	(aka machine learning algorithms or tools) computer programs that perform operations associated with machine learning tasks, such as machine translation or image recognition, among other tasks.
Model	A generalization of specific examples in the training dataset; generalized rules are learned from given historical cases (the data).
Multi-layer perceptrons (MLP)	A feed-forward neural network where inputs are processed exclusively in the forward direction. Some equate with the term ANN: the term is used inconsistently by some to loosely mean any feed-forward ANN, and by others to refer to NNs composed of multiple layers of perceptrons. In our discussion, ANN is used as a general term and MPL as a specific type of ANN.
Multi-task learning	A technique that allows the model to learn multiple related tasks simultaneously.
Neural network (NN)	(also simply called a "network" in the AI context). A computational model that operates similarly to the functioning of a human nervous system; a model used in ML which is composed of artificial neurons. NNs have origin in biology – here the network is made of biological neurons.

(Continued)

(Continued)

Term	Definition/Explanation/Concept
	In the AI context the NN is composed of nodes that constitute artificial neurons that can store information. Thus, an NN is a system of software and/or hardware modeled after the operation of neurons in the human brain (an NN with some level of complexity, say with at least two layers, is an example of a DNN). An NN has nodes and connections between nodes; the numbers associated with each connection represent parameter values that were estimated when the NN is fit to the data. NN operates with phases: training (or learning) and inference (or prediction); training deals more with development (but not always) and inference deals with production. In the development effort, the designer selects the type of NN and the number of layers; the training (phase) determines the weights.
Neural network layers	Input layer: accepts raw input from the domain. No computations are undertaken at this layer. Its nodes just forward the information (features) to the hidden layer.
	Hidden layer: its nodes (not exposed or part of the "black box") provide an abstraction to the NN by performing a wide range of computation on the data and/or feature descriptions entered by the input layer. The results are transferred to the output layer.
	Output layer: collects, collates, and displays the distilled information learned, processed, and generated by the hidden layer and delivers the final values to the stakeholder as a result.
Neuron	A functional element (node, unit) that includes memory and that has the ability to determine when to "retain" and when to "discard" values held in that memory, based on the weights of inputs provided to the given neuron. Each of the neurons is configured to accept a given number of inputs from other neurons to provide appropriate outputs for the content of the data being analyzed. The neurons may be interconnected and/or organized into graph-based structures in various connectivity configurations.
Parametric machine learning models	ML models (MLMs) that can generate the output based on (i) the input and (ii) values of the parameters.
Perceptron	A neuron in an NN.

Term	Definition/Explanation/Concept
Pooling	A process (a function) that replaces the output of a convolutional layer with a summary statistic of the nearby outputs. A number of pooling functions are available including but limited to average pooling, max pooling, and 12-norm pooling.
Pre-training	The (initial) training of an NN to establish parameters, weights, and/or variations thereof.
Receptive fields	(aka local receiving area, local receptive field) A portion of space containing units that provide input to a set of units within a corresponding CNN layer. The receptive field is defined by the filter size of a layer within a CNN. The receptive field is, thus, an indication of the extent of the scope of input data a neuron or unit within a layer can be exposed to.
Recurrent neural network (RNN)	An NN where the algorithm using an internal memory that "remembers" state information, making it well-suited for ML problems that involve sequential data. Typically, the hidden layer function in an RNN is an element-wise application of a sigmoid function. They are a family of feed-forward NNs that also include feedback connections between layers (they feed results back into the network). RNNs enable modeling of sequential data by sharing parameter data across different parts of the neural network. RNNs include cycles that represent the influence of a present value of a variable on its own value at a future time – some portion of the output data from the RNN is used as feedback for processing subsequent input in a sequence. This capability makes RNNs useful for language processing; they are used, for example, by Apple's Siri and Google's voice search.
Regression	Statistical processes used to estimate (derive) the relationships among variables; e.g. a process to determine values (parameters) that minimize a specified cost function when establishing a line/function/surface fitting the empirical data.
Reinforcement learning (RL)	Predictive systems that entail rewards and punishment. There are three broad categories: (i) value-based algorithms – based on temporal difference learning to obtain value function (e.g. Q-learning, SARSA [State, Action, Reward, State, Action], and DQN); (ii) policy-based algorithms – directly learn optimal policy or endeavor to obtain an approximate optimal policy based on the observation; and (iii) imitation algorithms that endeavor to make decisions using demonstrations.

(Continued)

(Continued)

Term	Definition/Explanation/Concept
Robotic Process Automation (RPA)	A process that uses software "robots" to replicate routine human-computer interaction to automate repetitive tasks. RPA is a way to introduce automation and gain business benefits at low cost and low risk; it bridges the gap between manual interaction and full automation. With basic RPA, a software robot literally does what a human would do; this includes routine tasks such as data retrieval and entry, button clicks, file uploads and downloads, or invoice processing; basic RPA is advantageous because it can improve the speed and accuracy of task completion while freeing humans to focus on higher-return work. Full automation, on the other hand, employs systems, processes, and even third-party services that are purpose-built for automation from the outset; here the potential benefit of full automation is higher, but so is the commitment. There is a middle ground: when integrated with other automation software to enhance its base capability, RPA can become a valuable component of an automation strategy that includes technologies such as process mining, AI, data capture, business rules, and workflow. For example, when RPA is integrated with AI, AI insights can be acted on by sending instructions directly to bots that complete tasks via other systems, such as an automation platform, this being achievable (i) with no lag time or human intervention and (ii) with improved efficiency as well as improved customer and employee experiences. When RPA is combined with AI, it is possible to identify the manual steps, quickly build an aligned RPA robot, and then add it to the pool of application processors. A workflow engine would gradually determine that the best way to handle certain types of invoices is to route them to the bot, sending recommendations directly to it. This automatic routing reduces response time, saves time for the invoice processor to focus on other work, and enables end-to-end automation of the process. RPA is useful where IT resources and budgets are limited, or for working with back-end applications that lack good APIs and would be difficult to automate without significant change to your systems [105].
Semi-supervised learning	A learning method where the training dataset includes a mix of labeled and unlabeled data of the same distribution.
Sigmoid layer	A layer that applies a sigmoid function to layers of the NN to modify a value of an output such that values that are in the middle of a range (e.g. values near 0.5 for a range of 0.0–1.0) are thus increased or decreased to be closer to the ends of the range. This mapping is used to more accurately undertake a yes-or-no decision (e.g. detection of a wakeword).

Term	Definition/Explanation/Concept
Softmax layer	A layer that receives outputs from previous layers of the NN and normalizes the outputs such that they acquire a probability distribution between two known values, such as 0 and 1 (the two values add up to a known sum, such as 1). The output of the Softmax layer is typically compared to a given threshold value, say 0.9, to determine a yes-or-no decision (e.g. if the audio data represents a wakeword or if a point in time represents a beginpoint or endpoint).
Statistical inference	Reaching a decision or outcome, specifically, applying a model and making a prediction. Examples of ML approaches are inductive learning/inference, deductive learning/inference, and transductive learning/inference.
Statistical learning	A methodology for ML that has foundations in the fields of statistics and functional analysis; it deals with the statistical inference problem of finding a predictive function based on a set of data; it refers to a set of tools drawing conclusions from complex datasets. These methods can be supervised or unsupervised. Supervised statistical learning involves building a statistical model for predicting, or estimating, an output based on one or more inputs; with unsupervised statistical learning, there are inputs but no supervising output, yet one can learn relationships and structure from such data. In the statistical learning context, supervised learning problems are problems of regression (when the output takes a continuous range of values) or classification (the output is an element from a discrete set of labels). An example of classification is facial recognition; here a photo of a person's face is the input (specifically a large multidimensional vector whose elements represent pixels in the picture), and the output label is that person's name. Statistical learning has applications in fields such as computer vision and speech recognition, among many others.
Supervised learning (SL)	A learning method (algorithm) where the NN processes the inputs and compares the resulting outputs against a set of expected or desired outputs. An algorithm (a model) that is used to establish a mapping between input examples and target variables. The algorithm learns by making predictions after being provided with examples of input training data, and is thus trained to make predictions on test sets when only the new inputs are provided.

(Continued)

Term	Definition/Explanation/Concept
	The training dataset includes input paired with the desired output for the input; alternatively, the training dataset includes input having known output and the output of the NN is manually graded. Discrepancies are then propagated back through the system and the training process can adjust the weights with the goal of converging toward a model that generates correct answers based on known input data. The training process continues iteratively, and the weights of the network are adjusted to refine the outputs until the NN reaches a statistically targeted accuracy. In summary, (i) the output generated by the NN in response to the input associated with an instance in a training dataset is compared to the appropriate labeled output for that instance, (ii) an error value representing the difference between the output and the labeled output is generated, (iii) the error value is backward propagated through the layers of the NN, and (iv) the weights associated with the various connections are adjusted to minimize the error.
Training	The process of utilizing a set of training data representing the problem being modeled and adjusting the NN weights until the model performs with an acceptable or minimal error for all instances of the training dataset. The desired target value is also known as a ground truth value. An NN can be trained with a training set of labeled data or unlabeled data using, for example, a backpropagation algorithm, and it can be trained by supervised learning algorithms or unsupervised learning algorithms (in the latter case using algorithm such as but not limited to Deep Boltzmann Machine [DBM] and a Generative Adversarial Network [GAN]). For a complex system (or a complex phenomenon under scrutiny), the training process often is computationally intensive. The accuracy of the results on the NN depends on the "quality" of the dataset used to train the algorithm. DNNs are generally computationally-intensive to train; however, the numerous hidden layers facilitate multistep pattern recognition that ultimately results in minimized output error.
Ultra DNN (UDNN)	A DNN with a thousand or more layers and with 100 million or more parameters.
Unsupervised learning (UL)	A learning method in which the NN attempts to train itself using unlabeled data (input data without any associated output data).

References

1 D. Minoli, E. Ericson, *et al.*, (Editors), *Expert Systems Applications in Integrated Network Management*, Artech House, 1989.

2 D. Minoli, "18-The Impact of Artificial Intelligence on Data Communications", in Bellcore Artificial Intelligence Symposium, Asbury Park, NJ, 1988.

3 H. Wang, T. Wang, *et al*, Complex Natural Language Processing, U.S. Patent 11,398,226, July 26, 2022. Uncopyrighted material.

4 S. Oh, S. Kim, *et al*, Artificial Intelligence Moving Agent, U.S. Patent 11,397,871, July 26, 2022. Uncopyrighted material.

5 R. H. Gensure, M. F. Chiang, J. P. Campbell, "Artificial Intelligence for Retinopathy of Prematurity", *Curr. Opin. Ophthalmol.* 2020; 31(5): 312–317, https://doi.org/10.1097/ICU.0000000000000680.

6 M. Mohri, A. Rostamizadeh, A. Talwalka, *Foundations of Machine Learning*, Adaptive Computation and Machine Learning series, 2nd Edition, The MIT Press, 2018. ISBN-10: 0262039400, ISBN-13: 978-0262039406.

7 G. Huang, Z. Liu, L. van der Maaten, "Densely Connected Convolutional Networks", in Proceedings IEEE/CVF Computer Vision and Pattern Recognition Conference (CVPR), 2017, arXiv:1608.06993, http://openaccess.thecvf.com. https://doi.org/10.48550/arXiv.1608.06993.

8 B. Grossfeld, "Deep Learning Vs Machine Learning: A Simple Way To Understand The Difference", Jan. 23, 2020, https://www.zendesk.com/blog/machine-learning-and-deep-learning. Accessed Dec. 1, 2020.

9 K. Nevala, *The Machine Learning Primer*, SAS Best Practices e-book, SAS Institute Inc., 100 SAS Campus Drive, Cary, NC 27513-2414, USA, 2017.

10 M.-T. Grymel, D. T. Bernard, *et al*, Write Combine Buffer (WCB) for Deep Neural Network (DNN) Accelerator, U.S. Patent 2023/0020929, Jan. 19, 2023. Uncopyrighted material.

11 B. Min, R. Zbib, Z. Huang, Cross-Lingual Information Retrieval and Information Extraction, U.S. Patent 11,531,824, Dec. 20, 2022. Uncopyrighted material.

12 D. Ankers, "Types of Artificial Intelligence: A Detailed Guide", Dec. 20, 2018, Corporate materials of Certes at https://certes.co.uk. Accessed Aug. 18, 2022.

13 D. Minoli, *Imaging in Corporate Environments – Technology and Communication*, McGraw-Hill, 1994.

14 S. Ha, G. Kim, D. Lee, Method and Apparatus with Neural Network Parameter Quantization, U.S. Patent 2023/0017432 Al, Jan. 19, 2023. Uncopyrighted material.

15 J. Ko, K. Duong, *et al*, Computation of Neural Network Node with Large Input Values, U.S. Patent 11,531,727, Dec. 20, 2022. Uncopyrighted material.

16 J. J. Robertson, Z. Mathe, Neural Network System for Gesture, Wear, Activity, or Carry Detection on a Wearable or Mobile Device, U.S. Patent 2023/0013680, Jan. 19, 2023. Uncopyrighted material.

17 Z. Yi, Q. Tand, *et al*, Methods and Systems for High Definition Image Manipulation with Neural Networks, U.S. Patent 20230019851, Jan. 19, 2023. Uncopyrighted material.

18 P. Swietojanski, S. Renals, "Differentiable Pooling for Unsupervised Acoustic Model Adaptation", *IEEE/ACM Trans. Audio Speech Lang. Process.* 2016; 24(10), https://doi.org/10.1109/TASLP.2016.2584700.

19 S. Yan, J. Li, Z. Liu, Slimming of Neural Networks in Machine Learning Environments, U.S. Patent 11,537,892, Dec. 27, 2022. Uncopyrighted material.

20 B. Min, R. Zbib, Z. Huang, Cross-Lingual Information Retrieval and Information Extraction, U.S. Patent 11,531,824, Dec. 20, 2022. Uncopyrighted material.

21 Mishra, M., "Convolutional Neural Networks, Explained", Aug. 26, 2020, https://towardsdatascience.com/convolutional-neural-networks-explained-9cc5188c4939#. Accessed Aug. 1, 2022.

22 Y. LeCun, B. Boser, *et al.*, "Backpropagation Applied to Handwritten Zip Code Recognition", *Neural Comput.* 1989; 1(4): 541–551.

23 Y. LeCun, L. Bottou, *et al.*, "Gradient-based Learning Applied to Document Recognition", *Proc. IEEE.* 1998; 86(11): 2278–2324.

24 O. Russakovsky, J. Deng, *et al.*, "ImageNet Large Scale Visual Recognition Challenge", *Int. J. Comput. Vis. (IJCV).* 2015; 115: 211–252, https://doi.org/10.1007/s11263-015-0816-y.

25 K. He, X. Zhang, *et al*, "Deep Residual Learning for Image Recognition", in IEEE/CVF Computer Vision and Pattern Recognition Conference (CVPR), 2016.

26 G. Huang, Y. Sun, *et al*, "Deep Networks with Stochastic Depth", in ECCV, 2016.

27 G. Huang, Z. Liu, *et al*, "Densely Connected Convolutional Networks", Jan. 28, 2018, arXiv:1608.06993, https://doi.org/10.48550/arXiv.1608.06993. Accessed Jan. 10, 2023.

28 Alake, R., "Understand Local Receptive Fields in Convolutional Neural Networks", June 12, 2020, https://towardsdatascience.com/understand-local-receptive-fields-in-convolutional-neural-networks-f26d700be16c. Accessed Mar. 15, 2023.

29 D. Minoli, B. Occhiogrosso, A. Koltun, "Situational Awareness for Law Enforcement and Public Safety Agencies Operating in Smart Cities – Part 1: Technology; Chapter in Springer's Book", in *IoT and WSN Based Smart Cities: A Machine Learning Perspective*, S. Rani, V. Sai, R. Maheswar, (Editors), Springer, 2022. ISBN: 978-3-030-84181-2.

30 T. Shi, S. Horvath, "Unsupervised Learning with Random Forest Predictors", *J. Comput. Graph. Stat.* 2006; 15(1): 118–138, https://doi.org/10.1198/106186006X94072.

31 M. Caron, P. Bojanowski, *et al*, "Deep Clustering for Unsupervised Learning of Visual Features", in Proceedings of the European Conference on Computer Vision (ECCV), pp. 132–149, 2018.

32 L. Collingwood, J. Wilkerson, "Tradeoffs in Accuracy and Efficiency in Supervised Learning Methods", *J. Inf. Technol. Polit.* 2012; 9(3): 298–318, https://doi.org/10.1080/19331681.2012.669191.

33 R. A. Caruana, A. Niculescu-Mizil, "An Empirical Comparison of Supervised Learning Algorithms", in ICML '06: Proceedings of the 23rd International Conference on Machine Learning, pp. 161–168, June 2006, https://doi.org/10.1145/1143844.1143865.

34 C. M. Bishop, *Pattern Recognition and Machine Learning*, Springer, 2006. ISBN-13: 978-0387310732, ISBN-10: 0387310738.

35 S. J. Russel, P. Norvig, *Artificial Intelligence: A Modern Approach*, 3rd Edition, Pearson, 2015. ASIN: 9332543518, ISBN-10: 9789332543515, ISBN-13: 978-9332543515.

36 Brownlee, J., "14 Different Types of Learning in Machine Learning", Start Machine Learning, Nov. 11, 2019, https://machinelearningmastery.com/types-of-learning-in-machine-learning. Accessed Jan. 13, 2023.

37 X. Zhai, A. Oliver, *et al*, "Self-Supervised Semi-Supervised Learning", in Proceedings of the IEEE/CVF International Conference on Computer Vision (ICCV), pp. 1476–1485, 2019.

38 N. N. Pise, P. Kulkarni, "A Survey of Semi-Supervised Learning Methods", in International Conference on Computational Intelligence and Security, Suzhou, pp. 30–34, 2008, https://doi.org/10.1109/CIS.2008.204.

39 I. Goodfellow, Y. Bengio, A. Courville, *Deep Learning*, Adaptive Computation and Machine Learning series, The MIT Press, 2016. Illustrated Edition. ISBN-10: 0262035618, ISBN-13: 978-0262035613.

40 X. Zhai, A. Oliver, A. Kolesnikov, "Self-Supervised Semi-Supervised Learning", in Proceedings of the IEEE/CVF International Conference on Computer Vision (ICCV), pp. 1476–1485, 2019.

41 A. Oliver, A. Odena, *et al.*, "Realistic Evaluation of Deep Semi-Supervised Learning Algorithms", *Adv. Neural Inf. Proces. Syst.* 2018: 3239–3250.

42 E. Jang, C. Devin, *et al*, "Grasp2Vec: Learning Object Representations from Self-Supervised Grasping", in Conference on Robot Learning, 2018.

43 S. Gidaris, P. Singh, N. Komodakis, "Unsupervised Representation Learning by Predicting Image Rotations", in International Conference on Learning Representations (ICLR), 2018.

44 V. Mnih, A. Puigdomènech, *et al.*, "Asynchronous Methods for Deep Reinforcement Learning", in *Proceedings of the 33rd International Conference on Machine Learning*, JMLR, New York, NY, USA, 2016. W&CP Volume.

45 C. Wirth, R. Akrour, *et al.*, "A Survey of Preference-Based Reinforcement Learning Methods", *J. Mach. Learn. Res.* 2017; 18(1): 1–46.

46 R. S. Sutton, A. G. Barto, *Reinforcement Learning: An Introduction*, 2nd Edition, Bradford Books, 2018. ISBN-13: 978–0262039246, ISBN-10: 0262039249.

47 V. N. Vapnik, *Statistical Learning Theory*, 1st Edition, September 30, 1998, Wiley-Interscience,. ISBN-13: 978-0471030034, ISBN-10: 0471030031.

48 T. Yuan, W. Da Rocha Neto, *et al*, "Machine Learning for next-generation intelligent transportation systems: a survey", *Trans. Emerg. Telecommun. Technol.*, 14 December 2021 https://doi.org/10.1002/ett.4427.

49 S. Woodward, S. Becker, S. Larson, Systems and Methods for Machine Learning Classification-Based Automated Remediations and Handling of Data Items, U.S. Patent 2023/0017384, Jan. 19, 2023. Uncopyrighted material.

50 A. Pai, "CNN vs. RNN vs. ANN – Analyzing 3 Types of Neural Networks in Deep Learning", Feb. 17, 2020, https://www.analyticsvidhya.com/blog/2020/02/cnn-vs-rnn-vs-mlp-analyzing-3-types-of-neural-networks-in-deep-learning. Accessed Jan. 30, 2023.

51 Djenouri, D., Laidi, R., *et al*, "Machine Learning for Smart Building Applications: Review and Taxonomy". *ACM Comput. Surv.* 52(2). Article No.: 24. March 2019. 10.1145/3311950.

52 F. Wu, K. Fu, *et al.*, "A spatial-temporal-semantic neural network algorithm for location prediction on moving objects", *Algorithms*. 2017; 10: 37.

53 A. Wang, W. Yuan, *et al.*, "A Novel Pattern Recognition Algorithm: Combining ART Network with SVM to Reconstruct a Multi-Class Classifier", *Comput. Math. Appl.* 2009; 57(11–12): 1908–1914, https://doi.org/10.1016/j.camwa.2008.10.052. ISSN: 0898-1221.

54 R. Pupale, "Support Vector Machines (SVM) — An Overview", Towards Data Sceince, June 16, 2018, https://towardsdatascience.com/https-medium-com-pupalerushikesh-svm-f4b42800e989. Accessed Dec. 1, 2020.

55 İ. Güven, F. Şimşir, "Demand Forecasting with Color Parameter in Retail Apparel Industry Using Artificial Neural Networks (ANN) and Support Vector Machines (SVM) Methods, 106678", *Comput. Ind. Eng.* 2020; 147, https://doi.org/10.1016/j.cie.2020.106678. ISSN: 0360-8352.

56 J. L. Awange, B. Paláncz, *et al.*, "Support Vector Machines (SVM)", in *Mathematical Geosciences*, Springer, Cham, 2018. https://doi.org/10.1007/978-3-319-67371-4_10.

57 D. Soni, "Introduction to Bayesian Networks", Towards Data Science, June 8, 2018, https://towardsdatascience.com/introduction-to-bayesian-networks-81031eeed94e. Accessed Dec. 1, 2020.

58 A. L. Madsen, F. Jensen, *et al.*, "A Parallel Algorithm for Bayesian Network Structure Learning from Large Data Sets", *Knowl.-Based Syst.* 2017; 117: 46–55, https://doi.org/10.1016/j.knosys.2016.07.031. ISSN: 0950-7051.

59 X. Sun, C. Chen, *et al.*, "Hybrid Optimization Algorithm for Bayesian Network Structure Learning", *Information*. 2019; 10: 294.

60 M. Somvanshi, P. Chavan, et al, "A Review of Machine Learning Techniques Using Decision Tree and Support Vector Machine", in International Conference on Computing Communication Control and automation (ICCUBEA), Pune, 2016, pp. 1–7, https://doi.org/10.1109/ICCUBEA.2016.7860040.

61 Y. Mu, X. Liu, *et al.*, "A parallel C4.5 Decision Tree Algorithm Based on MapReduce", *Concurr. Comput. Pract. Exp.* 2017, https://doi.org/10.1002/cpe.4015.

62 T. Amit, "Introduction to Hidden Markov Models", Towards Data Science, June 7, 2019, https://towardsdatascience.com/introduction-to-hidden-markov-models-cd2c93e6b781. Accessed Dec. 1, 2020.

63 S. Srinivasan, G. Gordon, B. Boots, "Learning Hidden Quantum Markov Models", in Proceedings of the Twenty-First International Conference on Artificial Intelligence and Statistics, Proceedings of Machine Learning Research (PMLR), Volume 84, pp. 1979–1987, 9–11 April 2018, Playa Blanca, Lanzarote, Canary Islands, 2018.

64 Y. Zhou, R. Arghandeh, C. J. Spanos, "Online Learning of Contextual Hidden Markov Models for Temporal-Spatial Data Analysis", in IEEE 55th Conference on Decision and Control (CDC), Volume 2016, pp. 6335–6341, Las Vegas, NV, 2016, https://doi.org/10.1109/CDC.2016.7799244.

65 M. Schonlau, R. Y. Zou, "The random Forest algorithm for statistical Learning", *Stata J. Promot. Commun. Stat. Stata.* 2020; 24; https://doi.org/10.1177/1536867X20909688.

66 J. Abellán, C. J. Mantas, *et al.*, "Increasing Diversity in Random Forest Learning Algorithm Via Imprecise Probabilities", *Expert Syst. Appl.* 2018; 97: 228–243, https://doi.org/10.1016/j.eswa.2017.12.029. ISSN: 0957-4174.

67 T. Cover, P. Hart, "Nearest neighbor pattern classification", *IEEE Trans. Inf. Theory*, vol. IT-13, no. 1, pp. 21–27 Jan. 1967.

68 Y. Wang, Z. Pan, Y. Pan, "A Training Data Set Cleaning Method by Classification Ability Ranking for the *K*-Nearest Neighbor Classifier", *IEEE Trans. Neural Netw. Learn. Syst.* 2020; 31(5): 1544–1556, https://doi.org/10.1109/TNNLS.2019.2920864.

69 S.-S. Yu, S.-W. Chu, *et al.*, "Two Improved K-Means Algorithms", *Appl. Soft Comput.* 2018; 68: 747–755, https://doi.org/10.1016/j.asoc.2017.08.032. ISSN: 1568-4946.

70 M. Ahmed, R. Seraj, S. M. S. Islam, "The k-Means Algorithm: A Comprehensive Survey and Performance Evaluation", *Electronics.* 2020; 9(8): 1295, https://doi.org/10.3390/electronics9081295.

71 J. McGonagle, G. Pilling, A. Dobre, "Gaussian Mixture Model", 2020, https://brilliant.org/wiki/gaussian-mixture-model. Accessed Dec. 1, 2020.

72 T. Gandhi, M. M. Trivedi, "Computer Vision and Machine Learning for Enhancing Pedestrian Safety", in *Computational Intelligence in Automotive Applications. Studies in Computational Intelligence*, Vol. 132, D. Prokhorov, (Editor), Springer, Berlin, Heidelberg, 2008. https://doi.org/10.1007/978-3-540-79257-4_4.

73 S. R. Khan, H. Rahmani, *et al.*, "A Guide to Convolutional Neural Networks for Computer Vision", *Synth. Lect. Comput. Vis.* 2018; 8(1): 1–207, https://doi.org/10.2200/S00822ED1V01Y201712COV015.

74 E. Lygouras, N. Santavas, *et al*, "Unsupervised Human Detection with an Embedded Vision System on a Fully Autonomous UAV for Search and Rescue Operations". *Sensors (Basel)* 2019; 19(16): 3542. Published 2019 Aug 14. 10.3390/s19163542.

75 M. S. Veillette, E. P. Hassey, *et al.*, "Creating synthetic radar imagery using convolutional neural networks", *J. Atmos. Ocean. Technol.* 2018; 35(12): 2323–2338, https://doi.org/10.1175/JTECH-D-18-0010.1.

76 Y. Ma, W. Chen, *et al.*, "EasySVM: a visual analysis approach for open-box support vector machines", *Comput. Vis. Media.* 2017; 3: 161–175, https://doi.org/10.1007/s41095-017-0077-5.

77 F. Alam, F. Ofli, M. Imran, "Processing social media images by combining human and machine computing during crises", *Int. J. Hum. Comput. Interact.* 2018, https://doi.org/10.1080/10447318.2018.1427831.

78 K. Lyu, M. Hu, *et al.*, (Editors), *AI-Based Services for Smart Cities and Urban Infrastructure*, Advances in Computational Intelligence and Robotics, IGI Global, 2020. ISBN: 1799850250, 9781799850250.

79 B. Mertens, L. Rothkrantz, P. Wiggers, "Dynamic Bayesian Networks for Situational Awareness in the Presence of Noisy Data". In ACM International Conference Proceeding Series, 2011, https://doi.org/10.1145/2023607.2023676.

80 S.C. Satapathy, K. S. Raju, *et al*, "Advances in decision sciences, image processing, security and computer vision", in International Conference on Emerging Trends in Engineering (ICETE), Volume 1, 2020. ISBN: 978–3–030-24321-0, 978-3-030-24322-7.

81 LeGland, F., Mevel, L., "Recursive Estimation in Hidden Markov Models", in The 36th IEEE Conference on Decision And Control, Volume 4, pp. 3468–3473, San Diego, USA, 1997.

82 V. E. Balas, L. C. Jain, B. Kovačević (Editors), Soft Computing Applications, Proceedings of the 6th International Workshop Soft Computing Applications (SOFA 2014), Volume 2, ISSN: 2194–5357, ISSN 2194–5365 (electronic). Part of Advances in Intelligent Systems and Computing. ISBN: 978–3–319-18415-9, ISBN: 978–3–319-18416-6 (eBook). https://doi.org/10.1007/978-3-319-18416-6.

83 M. Moussa, G. Beltrame, "On the Robustness of Consensus-Based Behaviors for Robot Swarms", *Swarm Intell.* 2020; 14: 205–231, https://doi.org/10.1007/s11721-020-00183-1.

84 N. Kausar, A. Majid, "Random Forest-Based Scheme Using Feature and Decision Levels Information for Multi-Focus Image Fusion", *Pattern. Anal. Appl.* 2015; 19(1), https://doi.org/10.1007/s10044-015-0448-4.

85 B. Arshad, R. I. Ogie, *et al.*, "Computer Vision and IoT-Based Sensors in Flood Monitoring And Mapping: A Systematic Review", *Sensors*. 2019; 19(22): 5012, https://doi.org/10.3390/s19225012.

86 Y. Wang, H. Xia, *et al.*, "Distributed Defect Recognition on Steel Surfaces Using an Improved Random Forest Algorithm with Optimal Multi-Feature-Set Fusion", *Multimed. Tools Appl.* 2018; 77(13): 16741–16770, https://doi.org/10.1007/s11042-017-5238-0.

87 N. García-Pedrajas, J. A. Romero, G. Cerruela, "A Proposal for Local k Values for k-Nearest Neighbor Rule", *IEEE Trans. Neural Netw. Learn. Syst.* 2015; 28(2): 1–6, https://doi.org/10.1109/TNNLS.2015.2506821.

88 A. S. Sohail, P. Bhattacharya, "Classification of Facial Expressions Using K-Nearest Neighbor Classifier", in Computer Vision/Computer Graphics Collaboration Techniques, Third International Conference, MIRAGE 2007, Rocquencourt, France, Proceedings, pp. 555–566, March 28–30, 2007. https://doi.org/10.1007/978-3-540-71457-6_51.

89 N. Naikal, "Towards Autonomous Situation Awareness", Technical Report No. UCB/EECS-2014–124, Electrical Engineering and Computer Sciences University of California at Berkeley, May 21, 2014, https://www2.eecs.berkeley.edu/Pubs/TechRpts/2014/EECS-2014-124.pdf. Accessed Dec. 1, 2020.

90 H. H. Santoch, P. Venkatesh, *et al.*, "Tracking Multiple Moving Objects Using Gaussian Mixture Model", *Int. J. Soft Comput. Eng. (IJSCE)*. 2013; 3(2): 113–119. ISSN: 2231-2307.

91 W. Xia, Y. Jiang, *et al.*, "Application of Machine Learning Algorithms in Municipal Solid Waste Management: A Mini Review", *Waste Manag. Res.* 2022; 40(6): 609–624, https://doi.org/10.1177/0734242X211033716.

92 Fulawka, K., "AI Applications in the Telecommunications Industry: Challenging Telecoms with Machine Learning Solutions", Dec. 15, 2021, https://nexocode.com/blog. Accessed Aug. 1, 2022.

93 Levi, D., "6 Common Uses of AI in Telecommunications", Mar. 3, 2018, https://techsee.me. Accessed 1 Aug. 2022.

94 MiPU Staff, "Artificial Intelligence for Telecommunications", 16 Apr. 2020, https://mipu.eu/en/case_study/artificial-intelligence-for-telecommunications. Accessed Aug. 1, 2022

95 M. Dhouibi, A. K. Ben Salem, *et al.*, "Accelerating Deep Neural Networks Implementation: A Survey", *IET Comput. Digit. Tech.* 2021, https://doi.org/10.1049/cdt2.12016 https://ietresearch.onlinelibrary.wiley.com/doi/full/10.1049/cdt2.12016.

96 Ameri, A., Knowledge Currency, U.S. Patent 11,397,897, July 26, 2022. Uncopyrighted material.

97 R. Gandhi, "Introduction to Machine Learning Algorithms: Logistic Regression", May 28, 2018, https://hackernoon.com/introduction-to-machine-learning-algorithms-logistic-regression-cbdd82d81a36. Accessed 2 Jan. 2023.

98 R. Salakhutdinov, G. Hinton, "Deep Boltzmann machines", in 12th International Conference on Artificial Intelligence and Statistics (AISTATS), Clearwater Beach, Florida, USA. Volume 5 of JMLR: W&CP 5, 2009.

99 T. J. Sejnowski, G. E. Hinton, D. H. Ackley, "A Learning Algorithm for Boltzmann Machines", *Cogn. Sci.* 1985; 9: 147–169.

100 S. Abirami, P. Chitra, "The Digital Twin Paradigm for Smarter Systems and Environments: The Industry Use Cases, in *Advances in Computers*, Elsevier, Science Direct, Edited by P. Raj and P. Evangeline Volume 117, Issue 1; Pages 1–368 2020.

101 Q. Sellat, R. Priyadarshini, *et al.*, "Semantic Segmentation for Self-Driving Cars Using Deep Learning", in *Cognitive Big Data Intelligence with a Metaheuristic Approach*, Elsevier Inc.; Academic Press, 2022, https://doi.org/10.1016/C2020-0-02004-9.

102 E. Fenoglio, H. Latapie, *et al*, Deep Fusion Reasoning Engine (DFRE) for Prioritizing Network Monitoring Alerts, U.S. Patent 11,595,268, Feb. 28, 2023. Uncopyrighted material.

103 C. Jose, Y. Mishchenko, *et al*, Wakeword Detection Using a Neural Network, U.S. Patent 1,521,599, Dec. 6, 2022. Uncopyrighted material.

104 A. Dhiman, K. Gupta, D. K. Sharma, "An Introduction to Deep Learning Applications in Biometric Recognition". In *Hybrid Computational Intelligence for Pattern Analysis, Trends in Deep Learning Methodologies*, Editor(s): Piuri P., Raj S., et al, Academic Press, 2021, Pages 1–36,. https://doi.org/10.1016/B978-0-12-822226-3.00001-5. ISBN: 9780128222263. https://www.sciencedirect.com/science/article/pii/B9780128222263000015. Accessed Feb. 1, 2023.

105 IBM Staff, "Where Does RPA Fit Into the Automation Landscape?", https://www.ibm.com/cloud/resources/rpa-buyers-guide-rpa-automation-landscape. Accessed Mar. 15, 2023.

106 T. Hastie, R. Tibshirani, J. Friedman, The Elements of Statistical Learning, Springer Series in Statistics, 2001, with a second edition in 2009. Hardcover ISBN: 978–0–387-84857-0. https://doi.org/10.1007/978-0-387-84858-7.

107 G. James, D. Witten, *et al*, An Introduction to Statistical Learning: With Applications in R, Springer Texts in Statistics, 1st edition, 2013. Corr. 7th printing 2017 Edition. ISBN-10: 1461471370, ISBN-13: 978–1461471370. Accessed Mar. 15, 2023.

2

Current and Evolving Applications to Natural Language Processing

2.1 Scope

Speech synthesis, automatic speech recognition (ASR),[1] and basic text-to-speech (TTS) have been around for decades. Now, at the intersection of computer science, artificial intelligence (AI), and computational linguistics (CL), one finds advanced fields that include natural language processing (NLP), natural language understanding (NLU), and natural language generation (NLG), that combine older capabilities with newer content creation techniques. NLP, NLU, and NLG deal particularly with the generation of answers to questions or with the generation of other prose. Often the output of these systems, also called chatbots, is text, but the output can also be voice; often the input is text, but it can also be voice (for example, via a *wakeword*) – at press time, the input is typically a textual query.

At this juncture, all these language related fields have become heavily dependent on AI. Language processing has existed for some time and has utilized a number of techniques. In recent years, however, these disciplines, and NLG in particular, have seen significant progress because of the application of machine learning (ML) and, more specifically, deep learning (DL) techniques. Neural networks (NNs), in particular recurrent neural networks (RNNs), now comprise the leading approaches to NLP/NLU tasks. The use of DL has resulted in more fluent and coherent generation, also with controllability of style, sentiment, and speed. These advances support high-quality language modeling, query answering, dialogue generation, machine translation, and data-to-text generation – in another example, an AI-capable Internet of Things (IoT) device may be capable of receiving or issuing a command using speech and/or sustaining a conversation.

---

1 ASR is also occasionally known as speech-to-text (STT).

AI Applications to Communications and Information Technologies: The Role of Ultra Deep Neural Networks, First Edition. Daniel Minoli and Benedict Occhiogrosso.
© 2024 The Institute of Electrical and Electronics Engineers, Inc.
Published 2024 by John Wiley & Sons, Inc.

ASR and TTS deal more with basic voice processing itself. Speech synthesis per se from a text form uses traditional vocoding mechanisms, although AI is also being used to enhance sentence fluency and intonation. ASR answers the question "what did he/she say?"; NLP answers the question "what did he/she mean?". NLP plays a critical role in the subdiscipline of "generative AI" that has received considerable attention of late from researchers, technology vendors, the public, politicians, labor unions, employers and employees, and regulators.

This chapter focuses on NLP technologies while the next chapter focuses on ARS/TTS technologies. The topic is consistent with the scope of this text because (i) some of the elements could be implemented or realized across a network (as a cloud service), (ii) some systems entail verbal responses that are in the voice domain, and (iii) the field deals with the broader concept of "communications."

2.2 Introduction

A basic description of these fields follows (synthesized from [1–18]):

- NLP allows machines to understand voice commands or written language. NLP deals with the interactions between human (natural) languages and computerized systems and, in particular, is concerned with programming hosts to effectively process large bodies of natural language material. NLP can be perceived as a subfield of AI, focusing on technologies that interpret natural language inputs. The combination of NLU and NLG represents NLP (see Figure 2.1):
  - NLU is a subfield of NLP concerned with technologies that can be used to draw conclusions on the basis of some input information or structure. It aims at enabling computing devices (for example, but not limited to, digital assistants) to derive meaning from input data (for example, text data and/or audio data) representing natural language. NLU is the process of mapping human language to internal computer representation of information. It entails

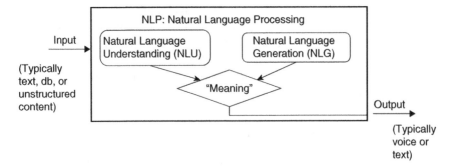

Figure 2.1 Positioning of the subdisciplines.

several tasks such as, but not limited to, textual entailment (natural language inference), question answering, semantic similarity assessment, and document (text) classification.

- NLG is a subfield of NLP that enables the machine to communicate in "spoken conversation" and/to derive meaning from natural language inputs, such as spoken inputs. It refers to the set of tasks that produce fluent text from input data and other contextual information. NLG aims at using computers to process datasets and automatically generate narrative stories about those datasets. More specifically, NLG can be perceived as a subfield (and/or a subtask) of AI concerned with technologies that are able to produce language as an output based on certain input information or structure (e.g. where the input constitutes data about a matter to be analyzed and expressed in natural language). AI and CL are employed by NLG systems to automatically generate human-understandable text or auditory content. NLG is more than just TTS.

- CL is a multidisciplinary field dealing with (theoretical) computational modeling of a natural language. Probabilistic language models are typical, e.g. a unigram model, a bigram model, or an N-gram model. The unigram model assumes that use of words is completely independent of each other; the bigram model is a model that assumes that use of words depends exclusively on one previous word; the N-gram model is a model that assumes that use of words depends on $N - 1$ previous words.

- TTS deals with transforming textual and/or other data into audio data that is synthesized to resemble quality human speech.

- ASR is concerned with transforming audio data associated with speech into a token or other textual representation of that speech. Speech recognition can be seen as a subfield of NLP (and CL), although it is also a distinct field in its own merit. ASR and NLU are often used together as part of a language processing component of a system.

Refer again to Figure 2.1 for a positioning of the NLP/NLU/NLG subdisciplines. Figures 2.2–2.4 provide basic illustrative examples of an NLP/NLG environment. The Appendix to this chapter provides a basic glossary of some key terms and concepts related to this discussion, based on a variety of references (but specifically including [1–18, 20]). Tables 2.1 and 2.2 highlight the key concepts.

Fundamentally, ML approaches applicable to NLP depend on the use of unsupervised or supervised training for language processing. In *unsupervised* learning, new concepts are formed from unannotated examples by clustering the examples into coherent groups; in supervised training, decision rules or case-based classifiers are utilized with a representation of knowledge. This method is preferred and it has been utilized extensively in NLP in recent years. The goal of *supervised* learning is to deduce representations from annotated training examples using

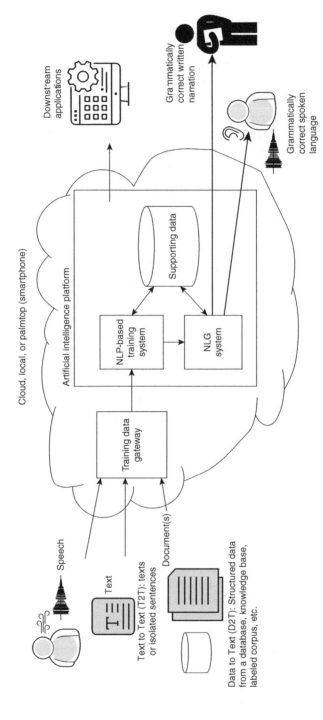

Figure 2.2 Basic generic example of NLP/NLG environment.

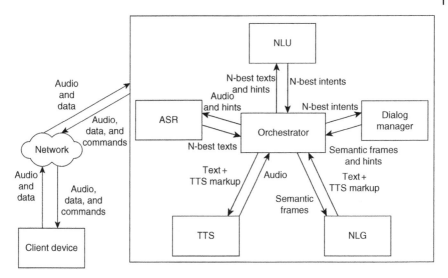

Figure 2.3 Illustrative networked environment with a spoken language processing system that includes ASR, TTS, NLG, and NLU functionality. *Source:* Reference [19]/ U.S. Patent.

basic algorithms; thereafter, the acquired knowledge is used to classify future instances. In this context, DL is the use of NN methods for language processing, including document classification, language modeling (LM), machine translation (MT), and multilingual output, among other language-related computerized activities. NLP problems require modeling long-term dependencies.

In the acoustical domain, TTS mechanisms are used for speech synthesis for delivering intelligible, natural-sounding, and expressive speech – TTS has a long history that spans MIT's formant synthesis and Bell Labs' diphone-based synthesis. Hidden Markov models (HMM) synthesis has been used in the recent past; more recently, DL-based statistical approaches have been shown to advance text analysis and the generation of the waveforms. Recent milestones include Google's WaveNet and the Tacotron (I and II) sequence-to-sequence (S2S) models.

NLP applications that generate new texts from existing (usually human-written) texts include, among many others [17]:

- Simplification of texts, for example, to make the content more accessible for low literacy readers;
- Generation of paraphrases of input sentences;
- Summarization and/or fusion of related texts or sentences to make them more concise;
- MT from one language to another;
- Automatic generation of questions for educational and other purposes;

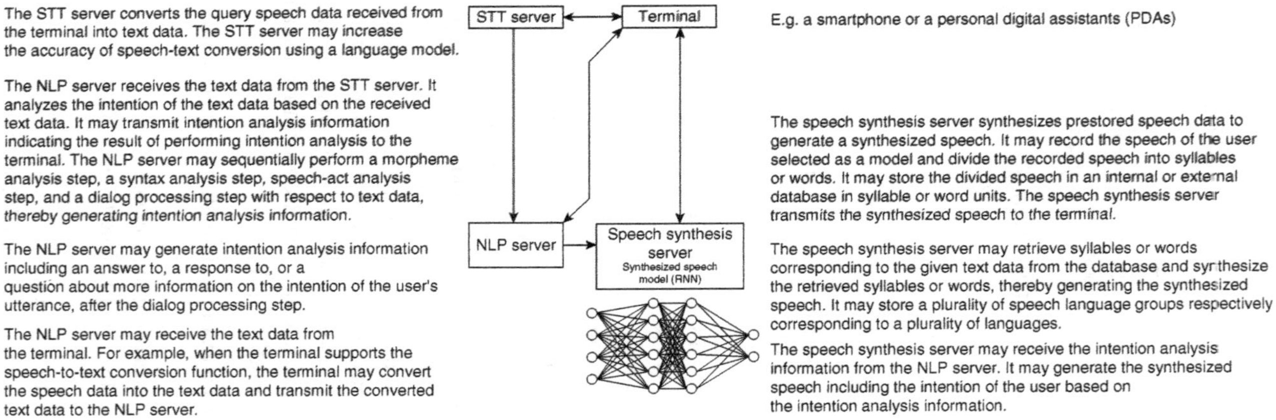
The STT server converts the query speech data received from the terminal into text data. The STT server may increase the accuracy of speech-text conversion using a language model.
The NLP server receives the text data from the STT server. It analyzes the intention of the text data based on the received text data. It may transmit intention analysis information indicating the result of performing intention analysis to the terminal. The NLP server may sequentially perform a morpheme analysis step, a syntax analysis step, speech-act analysis step, and a dialog processing step with respect to text data, thereby generating intention analysis information.
The NLP server may generate intention analysis information including an answer to, a response to, or a question about more information on the intention of the user's utterance, after the dialog processing step.
The NLP server may receive the text data from the terminal. For example, when the terminal supports the speech-to-text conversion function, the terminal may convert the speech data into the text data and transmit the converted text data to the NLP server.
STT server
Terminal
NLP server
Speech synthesis server
Synthesized speech model (RNN)
E.g. a smartphone or a personal digital assistants (PDAs)
The speech synthesis server synthesizes prestored speech data to generate a synthesized speech. It may record the speech of the user selected as a model and divide the recorded speech into syllables or words. It may store the divided speech in an internal or external database in syllable or word units. The speech synthesis server transmits the synthesized speech to the terminal.
The speech synthesis server may retrieve syllables or words corresponding to the given text data from the database and synthesize the retrieved syllables or words, thereby generating the synthesized speech. It may store a plurality of speech language groups respectively corresponding to a plurality of languages.
The speech synthesis server may receive the intention analysis information from the NLP server. It may generate the synthesized speech including the intention of the user based on the intention analysis information.

Figure 2.4 Example of speech synthesizer with STT and NLP mechanisms. *Source:* Adapted from Ref. [15].

Table 2.1 Basic terminology.

Term	Concept
Natural language processing (NLP)	A subfield of AI, focusing on technology that interprets natural language inputs. Makes use of NLU and NLG.
Natural language understanding (NLU)	The process of mapping human language to internal computer representation of information.
Natural language generation (NLG)	NLG is a subfield of NLP that enables the machine to communicate in "spoken conversation" and/to derive meaning from natural language inputs, such as spoken inputs. Mechanisms for generating *text* or *speech* from nonlinguistic input. The set of computer tasks/processes that generate fluent text from input data and other contextual information; it enables the machine to communicate in "spoken conversation." NLG includes both text-to-text generation and data-to-text generation.

Table 2.2 Basic NLP/NLU/NLG methods.

Term	Concept
"Attention" models	Recurrent neural network (RNN) models that iteratively process inputs by selecting relevant data at every step. A process of weighing influence of different parts of input data.
Deep learning (DL) in NLP	Use of neural network (NN) methods for document classification, language modeling, machine translation, multilingual output and larger-context modeling.
	Generative AI: AI systems that generate text, media, or image content, where questions are asked with NLP and the answer, new intelligent content, is generated and presented with NLP-based text, graphics, audio, music or video.
Language model(ing) (LM)	A model of the written letters, words and sentences or phrases. Can be trained using statistical or ML methods. A joint probability distribution over sequences of words: given a sequence of words of length n, a language model assigns a joint probability $P(w_1, w_2, ..., w_n)$ to said sequence. Probabilistic language models such as an N-gram model, a bigram model, or a unigram model. Advanced state of the art LMs are also known as large language models (LLMs).
Long short-term memory (LSTM) networks	A type of RNN that is capable of learning order dependence in sequence prediction problems, such as machine translation and speech recognition. LSTM utilizes feedback connections. The advantages of LSTM-based systems are (i) supporting sequences of varying lengths, (ii) avoiding data sparseness, and (iii) not requiring a large number of parameters; they utilize projection of histories into a low-dimensional space and allow similar histories to share representations.

(Continued)

Table 2.2 (Continued)

Term	Concept
Machine learning (ML) in NLP	The use of supervised or unsupervised training for NLP. Decision trees, rules, or case-based classifiers are utilized in supervised training examples. A representation of knowledge is postulated with basic algorithms, for inducing such representations from annotated training examples and utilizing the acquired knowledge to classify future instances. In unsupervised learning new concepts are formed from unannotated examples by clustering them into coherent groups.
Statistical methods/ statistical learning	LM methods employed in NLP; these include, but are not limited to hidden Markov model (HMM), maximum entropy models, and the expectation maximization method utilized in machine translation (MT).
Surface realization	Determining specific word forms and/or flattening the sentence structure into a string.
Transformer	A DL method that undertakes a small (constant) number of steps. In each step, it applies a self-attention mechanism which directly models relationships between all words in a sentence, regardless of their respective position.

- Weather and financial reports;
- Sport announcing and reports, news stories and virtual "newspapers" from sensor data;
- Data analysis and business intelligence;
- Summaries of patient information in clinical contexts;
- IoT device status and maintenance reporting;
- A specific example includes Gmail's Smart Compose which parses email content and recommends short responses.

NLG, along with NLU, is at the basis of chatbots and voice assistants. Applications for speech recognition technology that started in smartphones are now expected to become more widespread in the home with extensive deployment of smart devices in the IoT realm.

2.3 Overview of Natural Language Processing and Speech Processing

The Information Age has brought to the fore a deluge of information of all types; such deluge can be difficult for people to absorb as a reading exercise. Part of the solution is to employ computers for applying NLP techniques to interpret the data in a given realm and to reduce the volume to manageable patterns of communication. A typical business requirement (or goal) may be to aggregate and then synthesize large volumes of human communications from the many different information modes now routinely available. Solutions may entail computers applying NLP techniques to interpret the many human communications available, seeking to analyze, group, and categorize said human communications into streamlined patterns.

Communication via a natural language requires two baseline functions: (i) producing "text" (written or spoken), which is the purpose of NLG and (ii) understanding it. From a slightly different perspective, NLG functions are typically undertaken in two phases: (i) sentence planning – deciding on the overall sentence structure and (ii) surface realization – determining specific word forms and translating the sentence structure into a string.

DL has gained popularity in NLP and CL in the past decade, being able to improve how narrative stories are generated from datasets while relieving users from the need to directly code and/or re-code the narrative generation system. An advanced NLP system is trainable and is domain-independent. This has opened up the use of the ML-based narrative generation system to a much wider base of users (e.g. including users who do not have specialized programming knowledge) [3, 21].

Summarizing Chapter 1 for a quick review of how NNs work, a neuron unit in an NN receives input value from an external source (or from some other node) and computes an output value. Each input has an associated weight (w) that has been (pre)assigned on the basis of its relative importance to other inputs. The input layer feeds information from the outside world to the NN and comprises input nodes. The hidden layer(s) has no direct connection with the external world, but performs computations and transfers information from the input nodes to the output node(s). The output layer is responsible for final computations and for transferring information from the NN to the downstream entities or systems, being responsible for classification problems, for the final, classified outputs. Each neuron unit in the hidden layer(s) applies a function to the weighted sum of its inputs: each neuron unit receives some number of inputs, performs a dot product, and (optionally) applies a transformation, say with a function $f(n)$. Since many real-world phenomena have nonlinear behaviors, function f – known as the activation function – aims at introducing nonlinearity into the output value of a neuron unit. An activation function takes a single value and performs a predefined mathematical operation on it. Formally, *Neuron n* Output $=f$ (sum(input$_n(i)$*weight$_n(i)$)). There are a number of activation functions commonly used, including but not limited to: (i) sigmoid, which takes a real-valued input and maps it to range between 0 and 1; (ii) tanh, which takes a real-valued input and maps it to the range $[-1, 1]$; and (iii) ReLU (Rectified Linear Unit) which takes an input and thresholds it at zero (replaces negative values with zero).

In recent years, NNs have advanced with increased levels of abstraction for grammatical and semantic generalizations by making use of *backpropagation*. Backpropagation is a process where the weights of the NN connections are iteratively adjusted to minimize the discrepancy (error) between the actual output value of the NN and the desired output value. It entails a training process where the weights of the NN are updated by an algorithm that computes the gradient (direction and rate of increase of the data) of a defined loss function with respect to the weights in the NN for a paired input–output instance. NNs have also achieved successes in sequential modeling using *feed-forward networks* and RNNs, including RNNs with long short-term memory (LSTM) elements.

2.3.1 Feed-forward NNs

A feed-forward network – also sometimes called a "vanilla neural network with fully connected layers," such as the multilayer perceptron – treats all input features as unique (discrete) and independent of one another. For example, one could encode information about university students with features utilized by the NN including gender, high school GPA, ice cream preference, car driven to school, school where diploma was obtained, and number of uncles. Given any feature, one cannot automatically infer something about the proximity feature in the set; proximity has no meaning. For example, knowing the high school GPA provides little inference on the ice cream preference. This dataset may be fine for generic demographic data mining, but is less useful in cases where there is an underlying structure to the set of data under consideration – for example, in a well-formed English sentence, or in an image – in the context of structure in imaging, given a specific green pixel in a panoramic picture of nature, there is a high likelihood that another pixel next to it in several directions will also be green; additionally, with a series of related pictures, the photo may be taken from slightly different angles. A basic feed-forward network, however, encodes features in a way that makes it conclude that elements in the picture, say adjacent pixels of a farm building, *are different, distinct elements.*

2.3.2 Recurrent Neural Networks

RNNs can process not only single data points (such as static images), but also sequences of data (e.g. in speech or video). In RNNs, new sequences are generated by iteratively sampling from the network's output distribution, then feeding in the sample as input at the next step. Thus, in an RNN, the previous state of the network influences the output of the current state of the network. As seen in

Chapter 1, the RNN comprises an input layer to inject the input data vector, hidden layers to support the recurrent function, a feedback mechanism to enable a "memory" of previous states, and an output layer to output a result for downstream consumption or use. The RNN operates based on time steps where the state of the RNN at a given time step is affected by the previous time step by way of the feedback mechanism. For a given time step, the state of the hidden layers is defined by the previous state and the input at the current time step, namely, $s_t = f(Ux_t + Ws_{t-1})$, where W and U are matrices comprising parameters. For example, an initial input (x_1) at a first-time step can be processed by the hidden layer, and a second input (x_2) can be processed by the hidden layer utilizing state information that is determined during the processing of the initial input (x_1). The activation function f is generally a nonlinearity which depends on the application of specific implementation of the RNN.

However, RNNs are unable to store information about past inputs for very long cycle periods, limiting their ability to model long-range environments and also making them subject to instability. The issue is that if the network's predictions are only based on the last few inputs, and these inputs were themselves predicted, the model cannot recover from previous mistakes [22]. Having a longer memory is an improvement because even if the network cannot make sense of its recent history, it can look further back in the past to formulate its predictions. The best remedy is to use LSTMs.

Gated recurrent models (GRMs) are a gating mechanism in RNNs; they are similar to LSTMs described next, but with a "forget gate," being that it lacks an output gate; they utilize fewer parameters than LSTMs [23].

2.3.3 Long Short-Term Memory

An LSTM network is a type of RNN that is able to learn order dependence in sequence prediction problems, such as MT and speech recognition; an LSTM network utilizes feedback mechanisms. Recurrent networks use their feedback connections to store information about recent input events in the form of activations ("short-term memory," in contrast to "long-term memory" captured by slowly changing weights). This capability is important for a variety of applications, including speech processing. Learning to store information over extended time intervals using recurrent backpropagation typically takes a long time, often because of insufficient and decaying error backflow. Older algorithms used for learning what to place in short-term memory take a considerable amount of time or do not work at all, particularly when minimal time lags between inputs and corresponding teacher signals are long. LSTM mechanisms are designed to *overcome error back-flow problems* – in conventional methods, the temporal evolution of the backpropagated error exponentially depends on the size of the weights, and error signals "flowing backward in time" tend to either blow up (have oscillating weights) or vanish (take a long time to learn to bridge long time gaps). LSTM can learn to bridge time intervals of thousands of steps, even in the presence of noisy input sequences, without loss of short time lag capabilities. This can be achieved by an efficient, gradient-based algorithm with an architecture that enforces constant (hence, neither exploding nor vanishing) error flow through internal states of special self-connected units [24]. LSTM is an NN architecture that *allows for constant error flow* by utilizing special units, namely a multiplicative input gate unit (called *in-j*) which is utilized to protect the memory contents stored in cell cj from perturbation by irrelevant inputs, and a multiplicative output gate unit (called *out-j*) which is utilized to protect other units from perturbation by currently unimportant memory contents stored in cell cj. See Figure 2.5.

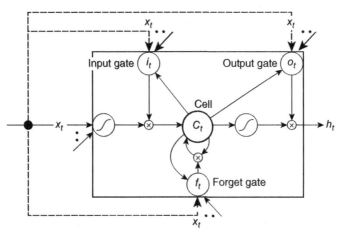

Information is added or removed according to the cell state defined by three different gates:

$i_t = \sigma(W_i[h_{t-1}, x_t + b_i])$ The input gate i_t is responsible for the process of controlling the input

$f_t = \sigma(W_f[h_{t-1}, x_t + b_f])$ The forget gate f_t is responsible for deciding whether the memory cells require to remember or forget its former status.

$o_t = \sigma(W_o[h_{t-1}, x_t + b_o])$ The output gate o_t controls the output activation and determines whether the information from the current cell states needs to be sent or not to the next layer.

Where

W_i, W_f and W_o are the weight matrices, b_i, b_f and b_o are the biases for the input gate, forget gate and output gate, respectively; h_{t-1} is the previous hidden state, and x_t is the input at current time step.

σ is the logistic sigmoid function, used as an activation function in the hidden layer which ranges from 0 to 1: $\sigma(x) = (1 + e^{-x})^{-1}$

Figure 2.5   LSTM memory cell.

2.3.4 Attention

With a convolutional neural network (CNN) model, as discussed in Chapter 1, pixels of an image are processed (read) incrementally, say left to right and top to bottom. The goal is to identify neighboring (local) patterns; for example, darker (black) next to light pixels might indicate an object's boundary. Hence, a CNN takes proximity as a relevant issue; additional layers of processing allow such networks to identify, say, objects, faces, and other physical characteristics.

Textual information and language on the other hand do not carry intrinsic proximity association; nonetheless, they do have content or format sequentiality: text can be seen as existing in a two-dimensional matrix; words, say, left to right and top to bottom. RNNs showed early value in processing two-dimensional sequences of words. RNNs process language sequentially in a left-to-right or right-to-left fashion. Thus, RNNs operate in one direction, utilizing sentence context to predict the next word. RNNs process and absorb a significant amount of information as they proceed along a sentence, enabling them to make reasonable predictions by the time they reach the end of a sentence; this is in contrast to the dearth of context – and predictability – at the beginning of a sentence.

RNNs (and LSTMs) by and large process in one direction: fundamentally RNNs understand words they encounter later in a sentence based on the words they have encountered earlier in the sentence. However, one of the challenges is that in various instances some of the context needed to predict the next word could be closer to the end of the sentence rather than being at the beginning of the sentence. Additionally, the intrinsically small-memory RNNs have a "retention" problem, because there is only so much information they can retain about long-scope dependencies. These challenges impact performance.

"Attention" is a mechanism intended to address these lacunae. Attention refers to a process of weighing the influences of different parts of input data [25–33]. "Attention" – which some have called a breakthrough algorithm, particularly when used in the network called a Transformer – has been shown to produce high-quality results in MT and other NLP tasks, for example in Bidirectional Encoder Representation from Transformers (BERT) and generative pre-training (GPT).[2] Among numerous other applications, the approach is also useful in visual object classification and image caption generation. "Attention" can be combined with neural word embeddings.

As implied, the challenge in NLP is how to provide NNs with the capability to model long-term dependency in sequential data. RNNs and LSTM networks have been used extensively in NLP with good results on various benchmarks. Nonetheless, despite the wide adaption, RNNs are difficult to optimize because of

---

2 Including versions 3, 3.4, and the newest at press time, version 4.

the gradient vanishing and explosion issue; LSTM gating and gradient clipping techniques have not fully addressed the challenges. On average, LSTM language models use 200 context words pointing to the need for additional improvement [34]. "Attention" mechanisms can enable the learning of long-term dependency and improve learning optimization.

While "attention" has been used to complement RNNs or CNNs, it also performs well on its own. "Attention" methods are particularly useful in MT. This mechanism processes two sentences and creates a matrix where the words of one sentence form the rows, and the words of another sentence form the columns; the matrix is used to identify matches, highlighting relevant context. In particular, the mechanism can arrange the same sentence as the columns and the rows to determine how some parts of that sentence relate to other parts of the sentence. An NN using attention mechanisms is able to ignore the noise and accentuate what is relevant, specifically how to connect two related words that in themselves do not carry flags or markers pointing to the other [35]. "Attention" allows the system to assess the totality of a sentence to establish connections between any particular word and its possibly-relevant context.

2.3.5 Transformer

Transformer and NN differ in the manner they process input (which, in turn, contains assumptions about the structure of the data to be processed) and automatically recombine that input into relevant features [35]. The Transformer architecture utilizes an encoder-decoder structure that does not use recurrence and/or convolutions to generate an output. The encoder processes an input to generate encodings that comprise contextual information about which parts of said input are relevant to each other; the decoder processes said encodings and, based on incorporated contextual information, generates an output. Namely, the encoder maps an input sequence to a series of continuous representations; the decoder receives (i) the encoder's output and (ii) the decoder's output at a previous time step, and then generates an output sequence [36, 37]. The architecture starts by converting the input data into an n-dimensional embedding and providing it to an encoder. The encoder and decoder consist of modules stacked on each other several times.

The attention mechanism is parallelized: the multi-head attention mechanism enables the model to pay attention to multiple parts of the key simultaneously (keys are the vector representations of all the words in the sequence). See Figure 2.6 and notice the positional encoding – this gives rise to multiple tokens obtained from segments with the same positional encoding but which, after the attention processing, have a different level of importance. Refer to the source document [38] for full details on this model. Also, see Figure 2.7 for another graphical view of the Transformer model.

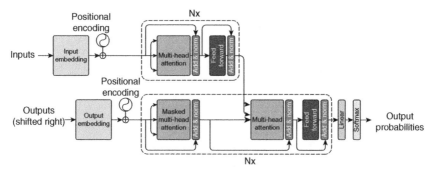

In a scaled dot-product attention, the values V are multiplied and summed with attention-weights a (the Softmax function normalizes the attention value to range between 0 to 1):

$$\text{Attention } (Q, K, V) = \text{Attention-weight } a \times V$$

where

$$a = \text{softmax}\left(\frac{QK^T}{\sqrt{d_k}} \right),$$

with Q representing a matrix that contains the Query, K includes all Keys (the vector representations of all the words in the sequence), and V represents the Values (often these are the same as K). The multi-head attention mechanism consists of multiple attention layers running in parallel, enabling the model to pay attention to multiple parts of the key simultaneously. There is a need to give every word (or part in the sequence) a relative position because a sequence depends on its elements' order; since the Transformer does not include an RNN that can remember how input sequences were fed into the model, the architecture adds these positions to each word's embedded representation.

Figure 2.6 The transformer model. *Source:* Adapted from Ref. [38].

Transformer, introduced by Google in 2017, is a DL model based on a self-attention mechanism that weighs the importance of each part of the input data in some different manner. The "attention" mechanism allows the model to focus on the most relevant parts of the input for each output, and while being similar to RNNs and being designed to process sequential input data such as natural language (perform tasks such as text summarization and translation) they, unlike RNNs, process the entire input at once, not one word at a time. This enables parallel processing, thus reducing training time [36]. As noted, this fostered the development of large pre-trained systems such as GPT and BERT. Transformers utilize "attention" mechanisms to collect information about the relevant context of a given word, and then encode that context in the vector that represents the word. The attention mechanism and the transformer can establish context about a word from distant parts of a sentence, earlier and also later than where the word appears, with the goal of encoding information to understand the word and its role in the sentence. More specifically, the Transformer compares a given word to every other word in the sentence; the output of these comparisons is an attention score for every other word in the sentence. The scores determine how much each of the other words ought to contribute to the next representation of a given word.

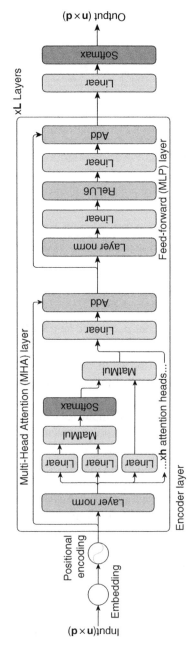

Figure 2.7 Another view of the Transformer model. *Source:* Adapted from Ref. [39].

The Transformer architecture is increasingly becoming a model of choice for NLP and computer vision (CV) problems, replacing RNN models such as LSTM. The Transformer outperforms both RNN and CNN models on English-to-French and English-to-German MT benchmarks; in addition to higher translation quality, the Transformer needs fewer machine cycles (computations) to train and is a better fit for modern ML hardware, speeding up training by up to an order of magnitude [16]. Transformers are able to learn longer-term dependency compared with RNNs/LSTMs, but are still limited by a fixed-length context – even newer mechanism such as Transformer-XL and Compressive Transformer may help (see Glossary for a short description of these more advanced models).

A number of commercially available DL models are based on the Transformer architecture. Google's BERT has been widely used for NLP tasks and since 2020, it has been a key component of the Google search engine (for English language search queries). BERT is principally used for understanding the meaning of text. The "bidirectionality" comes from the fact that it takes the context on both the left and right of a word into account when processing the text (traditional "unidirectional" models only consider the context to the left of a word). Language Model for Dialogue Applications (LaMDA) also uses the Transformer model; it is a large-scale language model for dialogue applications also developed by Google that can be used in chatbots, virtual assistants, and MT systems.

GPT-3 (and newer versions) is a high-quality NLP model developed by OpenAI that also uses the Transformer model aimed at generating coherent human-like text for a given prompt. ChatGPT is a relatively new chatbot built by OpenAI that uses newer versions of the GPT model, namely, GPT-4; the chatbot can interact in a dialogue format and generate textual answers, as discussed in more details in Sections 2.6 and 2.7. GPT-3 was released in 2020, and, along with the 3.5 version, it was used to create the Dall-E image generation tool and the chatbot ChatGPT – two products that generated a lot of industry- and popular-press and spurred other tech companies to pursue AI more aggressively [40]. Reinforcement Learning with Human Feedback (RLHF) is used in ChatGPT, combining reinforcement learning with human input (implemented as a system using reward or punishment signals).

2.4 Natural Language Processing/Natural Language Understanding Basics

Work in NLP/NLG started in the early days of computing (1950s and 1960s); the discipline gathered steam in the 1980s becoming a distinct field of research. The early systems were template-based followed by the use of semantic networks. Basic commercial applications started to emerge in the late 1990s. These early systems utilized a three-step pipeline as shown in Figure 2.8. NN applications started to appear in the late 1990s–early 2000s (e.g. the *DISCERN* system [41]).

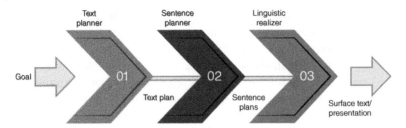

Figure 2.8 Early Reiter/Dale pipeline architecture of an NLG system. *Source:* Reference [18]/with permission of Elsevier.

Figure 2.9 (from [42]) depicts, for illustrative purposes, an example of a context-specific linguistic model to process input text that corresponds to the natural language data being received. Processed input text is thus generated, and the device performs semantic analysis of said processed text utilizing a NLU model to generate a recognition result.

NN usage in NLP has been fostered by increasing research in the field in the past two decades; the recent resurgence of interest in NNs is partially attributable to advances in hardware, enabling the support of resource-intensive learning algorithms. In particular, extensive research has taken place on learning statistical LMs; these models estimate the word and phrase distribution in natural language. Any NLP application needs to computationally process the words in a language before any of the more complex processing is done [7]. The techniques embodied in statistical language involve learning transition of words or phrases in the vocabulary based on the words or phrases encountered in the history: models are trained utilizing an independence assumption (between different segments of a document); the challenge, however, is this assumption may not be valid in all cases.

Labeled data for learning linguistic-specific tasks is scarce, making it challenging for models to train adequately and then perform satisfactorily (on the other hand, large unlabeled text corpora are abundant). In fact, most DL methods need substantial amounts of manually-labeled data; this restricts their applicability in domains that suffer from a dearth of annotated resources. It follows that models that make use of linguistic information from unlabeled data are a useful alternative to the task of gathering more annotation, which can be time-consuming and expensive. Even in cases where considerable supervision is available, learning good representations in an unsupervised fashion can provide a significant performance improvement.

2.4.1 Pre-training

The predicament discussed in the previous paragraph has led to the extensive use of pre-trained word embeddings to improve performance for a gamut of NLP tasks. Pre-trained word embeddings are part-and-parcel of modern NLP systems,

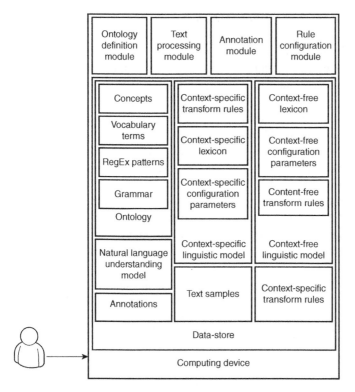

Figure 2.9 NLP processor adapted for context-specific processing of natural language input. *Source:* Reference [42]/U.S. Patent.

offering improvements over embeddings learned (acquired) from scratch. The pre-training is undertaken on a large corpus of text. Pre-training is the process of first training a model on one task or dataset; then utilizing the parameters or model from this training to train another model on a different task or dataset. This furnishes the model a head-start instead of starting from nothing. The typical process is: (i) start with an ML model (MLM) M and datasets A and B; (ii) train M with dataset A. Then, before training the model on dataset B, (i) initialize some of the parameters of M with the values of the model as it was trained on A and (ii) finally train M on dataset B. Pre-training can be used in transfer learning, classification, and feature extraction.[3]

---

3 There are several pre-trained models used in industry and academia to date. Each of these achieves different performance levels and is used for different tasks. Some well-known examples of computer vision (CV) are [43]: VGG-16, ResNet50, Inceptionv3, and EfficientNet, among others. Some popular pre-trained models for NLP tasks: GPT, BERT, ELMo, XLNet, and ALBERT, among others.

In NLP, pre-training entails sentence-level tasks including natural language inference and paraphrasing, which aim to predict the relationships between sentences by analyzing them as a whole. Pre-training also includes token-level tasks such as named entity recognition and question answering, where models are required to produce fine-grained output at the token level [44, 45].

Pre-training of general language representations include unsupervised feature-based approaches and unsupervised fine-tuning approaches (note that the learning of widely-applicable representations of words has utilized NN and non-NN methods in the recent past). Semi-supervised approaches for language understanding tasks using a combination of unsupervised pre-training and supervised fine-tuning have shown promise [46]. However, leveraging more than word-level information from unlabeled text is itself challenging. It is unclear what type of optimization objectives are maximally effective at learning text representations that are useful for transfer; additionally, there is no consensus on the most effective way to transfer these learned representations to the target task.

In the context of fine-tuning approaches, sentence or document encoders which produce contextual token representations have been developed which are pre-trained from unlabeled text and fine-tuned for a supervised downstream task [45]. Normalization and tokenization take place during the processing phase before performing semantic analysis. Tokenization segments input text into various segments, and normalization brings the format of those segments into alignment with some standard forms [42]. See Figure 2.10 for an example.

As illustrated in Figure 2.11, a natural language system may classify each document (in a collection of documents) into specified topics or labels (a "*document-scope task*"), or extract portions of text of the documents related to specified topics or labels (a "*span-scope task*"). Typically, before the performance of feature extraction, a document is partitioned into a number of smaller pieces of text, namely into "*tokens.*"

Feature extraction is used to identify information from textual documents that can be utilized to train natural LMs and then process untested data. Traditionally, the concept of feature extraction involves converting the text of a document into an array of strings (e.g. a plurality of textual characters, referred to as features)

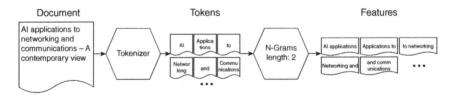

Figure 2.10 Tokenization example.

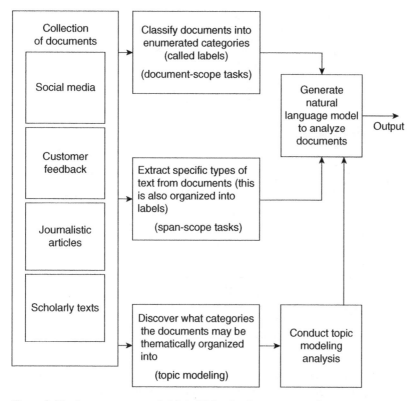

Figure 2.11 Language-agnostic ML in NLP using feature extraction.
Source: Reference [21]/U.S. Patent.

based on a set of rules defined by a particular feature extracting algorithm, referred to as a "*feature type.*" The feature extracting algorithm may be performed on the tokens, depending on the particular feature type selected. For example, a feature type may designate each of the tokens as a feature; as another example, a feature type may designate each pair of adjacent tokens as a feature. More than one feature type may be utilized at the same time; for example, the features may include each of the tokens and each pair of adjacent tokens [21].

2.4.2 Natural Language Processing/Natural Language Generation Architectures

There are three broad approaches to NLP/NLG architectures (e.g. see [17]):

1) *Modular architectures*: an approach where there is a clear division among sub-tasks (for example, see Figures 2.8 and 2.12);

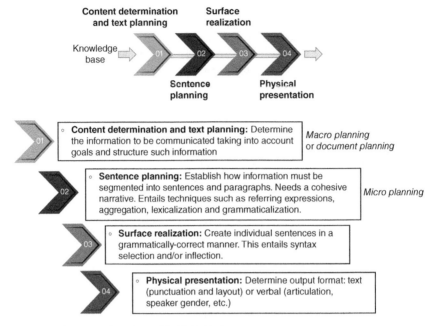

Figure 2.12 Typical modern NLP/NLG pipeline.

2) *Planning perspectives – viewing text generation as planning tasks*: the process of identifying a sequence of one or more actions to satisfy a particular goal – this approach links NLG to AI methodologies and offers a more integrated, less discrete or modular perspective on the various sub-tasks; and,

3) *Integrated or global approaches (now the trend in NLG)*: an architecture that cuts across task divisions, typically by relying on statistical learning of correspondences between (nonlinguistic) inputs and outputs – as noted in Chapter 1 (Glossary), statistical learning is methodology for ML that has foundations in the fields of statistics and functional analysis; it deals with the statistical inference problem of finding a predictive function based on a set of data; problems addressed by statistical learning are problems of regression (when the output takes a continuous range of values) or classification (the output is an element from a discrete set of labels).

The first two approaches are rule-based and perceive the relationship between a planned action and its impact on the context, being as fixed *a priori*. However, in real-life situations, there always is ambiguity, noise, and uncertainty. Thus, a more robust solution to generate communicative outputs is based on *stochastic mechanisms* (specifically, a Markov decision process), also in conjunction with reinforcement learning (RL). In a Markov decision process, there are states associated with possible actions, and each state-action pair is associated with a probability p_t of moving from a state at time *t* to a new state at *t* + 1 via a given action/decision. In the subtending learning algorithm, transitions are associated with a reinforcement signal implemented via a specified reward function that measures the desirability of the generated output. Learning typically involves phases, where distinct generation strategies or outputs become associated with different rewards. One should make note that RL is seen as being better at addressing uncertainty in dynamic environments compared to supervised learning or classification, being that these approaches do not support adaptation in a dynamically-changing context.

Clearly, in its responses an NLP must include information that is relevant to the stakeholders in the particular context and domain; the information must be presented in logical order, conforming to the language's syntax of the language – the information must be organized in sentences, paragraphs, or even sections, also in the domain-specific style, jargon, and tenor.

In the second step (Figure 2.12), there are two basic methodologies for sentence planning [47, 48]:

- Template-based: Texts have pre-defined structure with gaps. Gaps are then filled with pertinent data. These systems include linguistic capabilities to generate grammatically correct text.
- Dynamic creation: At a microscopic level, sentences are generated dynamically from semantic representations and a desired linguistic structure. At a macroscopic level, sentences are organized into a logical narrative.

Markov chain models were initially used for dynamic creation; in the recent past, LM methods have been utilized. LM takes into account recent words encountered in the stream in the prediction process of the next word to use. An N-gram model, cited earlier in this chapter, is a way to construct a language model. In an N-gram model, the predictive distributions are determined by counting exact matches between the recent history and the training set.

More recently, *NNs have been employed*. Specifically, RNN models, LSTM models, gated recurrent NN models, and transformer networks have been utilized problems [38, 49–53]. As noted earlier, a LSTM is a variant of an RNN model and GRMs are a gating mechanism in RNNs with a "forget gate" (being that it lacks an output gate). RNNs make use of the sequential nature of text and "recall" previous words to predict the next word. Ongoing research has endeavored to push the reach of recurrent LMs and encoder-decoder architectures, for example, as discussed in [54–56].

RNNs manage computation along the symbol positions of the input and output sequences; when they align the positions to steps in computation time *t*, they generate a sequence of hidden states based on previous hidden state(s) and the input for position *t*. The challenge is that the sequential process limits, or even precludes, parallelization which would be beneficial, if not critical, at longer sequence lengths [38]. LSTMs offer an improvement over RNNs at recalling longer word sequences.

2.4.3 Encoder-Decoder Methods

In a S2S-based[4] NLP, tasks are supported by the DL models with an encoder-decoder architecture. NNs, for example, as those utilized for MT, often contain an encoder reading the input sentence and generating a representation of it, and a decoder that then generates the output sentence word by word while referencing the representation generated by the encoder. Encoders are trained to represent the contextual information within input sequence utilizing encodings; the decoders are trained to map these encodings to the target sequences [58]. Compared to template-based systems, NN models provide less control but are more flexible.

In the encoder-decoder framework the RNN is used to encode the input into a vector representation, which then serves as the auxiliary input to a decoder RNN [59]. This decoupling between encoding and decoding enables the sharing of the encoding vector across multiple NLP tasks in a multi-task learning setting. Encoder-decoder architectures are particularly well-suited to S2S tasks such as MT – MT entails mapping of variable-length input sequences in the source language, to variable-length sequences in the target [60].

Many simple NLG models are based on RNNs and S2S model; these NLG models generate sentences by jointly optimizing sentence planning and surface realization utilizing a simple cross entropy loss training criterion. The issue remains that the simple encoder-decoder architecture tends to generate complex and long sentences because the decoder has to learn all grammar and diction knowledge [61].

More recently, proposals have been made where the NNs utilize an encoder-decoder architecture in conjunction with an "attention" layer ("attention"

---

4 Sequence-to-sequence (S2S, aka seq2seq) models are a special class of RNN architectures that, among other applications, are used to solve complex Language problems such as Machine Translation (Google Translate) and chatbots. As the name implies, there are two components – an encoder and a decoder. Both encoder and the decoder are LSTM (or sometimes gated recurrent unit [GRU]) models.

The *Encoder* reads the input sequence and summarizes the information in an internal state vectors or context vector (in case of LSTM these are called the hidden state and cell state vectors); one discards the outputs of the encoder and only preserve the internal states. This context vector aims to encapsulate the information for all input elements to help the decoder make accurate predictions. The LSTM reads the data, one sequence after the other (if the input is a sequence of length "*t*," one says that LSTM reads it in "*t*" time steps). (The output is a probability distribution over the entire vocabulary which is generated by using a softmax activation). The *Decoder* is an LSTM whose initial states are initialized to the final states of the Encoder LSTM, i.e. the context vector of the encoder's final cell is input to the first cell of the decoder network. Utilizing these initial states, the decoder starts generating the output sequence, and these outputs are also taken into consideration for future outputs. The decoder comprises a stack of LSTM units where each predicts an output y_t at a time step t. Each recurrent unit accepts a hidden state from the previous unit and produces and output as well as its own hidden state. Softmax is used to create a probability vector that determines the final output (e.g. word in the question-answering problem) [57].

mechanism), cited earlier, providing an end-to-end approach; here the encoder produces a hidden representation of the input text, and the decoder generates the text. During training, the encoder in the encoder-decoder-based approach with the "attention-based" mechanism is forced to weight parts of the input encoding more heavily when predicting certain portions of the output during decoding. This mechanism eliminates the requirement for direct input–output alignment because attention-based models are able to learn input–output correspondences based on loose couplings of input representations and output texts [62].

2.4.4 Application of Transformer

"Attention" mechanisms utilized in conjunction with RNNs have become accepted techniques for sequence modeling and transduction modeling in various tasks; this allows modeling of dependencies without concerns to their distance in the input or output sequences [52, 63, 64]. In particular, Transformer, is a recently-proposed model architecture (introduced in 2017) that deliberately avoids using recurrence and instead relies exclusively on an attention mechanism to establish global dependencies between input and output; in particular, it has been used for transfer learning in NLP tasks [38]. Transformers assess the relationships among words in context. The Transformer method has performed well on various tasks such as MT, document generation, and syntactic parsing. The Transformer allows for significantly more parallel processing. Two well-known Transformer models are BERT and GPT – GPT-3 (and newer versions) is an autoregressive LM with almost 200 billion parameters. Several unsupervised pre-training models based on the Transformer architecture have emerged of late (including the GPT and BERT models, which have been setting records in accuracy in NLP) [45, 46]. Additionally, Transformer architecture has been applied to other S2S tasks, such as image captioning and video description.

In summary, Transformers are models for processing sequential data based on multi-head attention. An important feature of Transformers is that entire sequences may be processed in matrix–matrix products in parallel (instead of one token/input at a time). Transformers consist of two-layer feedforward blocks and multi-head attention operations. Multi-head attention computes relationships between sequence elements by deriving query (Q), key (K), and value (V) sequences and computing dot products with a softmax nonlinearity in between [38, 39] (refer back to Figure 2.6). Transformer topologies work well for large models (in NLP and CV) with billions or even trillions of parameters.

Although large-scale Transformers are now popular, they remain an expensive solution, considering the floating point operations per second (FLOPS) requirements of these advanced DL models. Up to the present, DL hardware has been extensively developed in digital electronics, including graphics processing units

(GPUs) and field programmable gate arrays (FPGAs) [65]. Besides hardware accelerators, optical NNs (ONNs) are now being used or investigated for possible solutions to these heavy computing requirements. References [39, 66–71] note that the rapidly increasing size of DL models has caused renewed and growing interest in alternatives to digital electronic computers; for example, leaning toward the use of ONNs that could, in theory, dramatically reduce the energy cost of running state-of-the-art NNs.

2.4.5 Other Approaches

Two models often used with NLP are (i) the Bag-of-Words (BOW) model and (ii) the Continuous Bag-of-Words (CBOW) model. The BOW is a simplified representation utilized in NLP and information retrieval where a text, such as a sentence or a document, is represented as the bag of its words, ignoring grammar, and even word order, but retaining multiplicity. The BOW is a commonly used method of document classification where the frequency of the occurrence of each word is used as a feature for training a classifier. The CBOW model operates by predicting the probability of a word when provided with a context, e.g. where the context may be a single word or group of words; that is, given a single context word, CBOW predicts a single target word [72].

Generative Adversarial Networks (GANs), also combined with RL, have recently been proposed for NLP. GANs are a class of ML algorithms used in unsupervised learning environments, realized by a system of two NNs contesting with each other in a zero-sum mathematical game framework [73, 74]. Given input variables x (the observed data values) and output variables y (determined values), GAN-based models learn a joint probability distribution for $p(x, y)$. Early applications of adversarial model training involved computer security such as spam filtering, intrusion detection. More recently, they have been applied to image processing; here, GANs have been used to generate images that appear authentic to human observers; this is accomplished by training the generator to generate a synthetic image that fools the discriminator into accepting such image as an actual image, rather than it being a synthetic – while attempting to fool the discriminator, the generator learns the distribution of true images.

In the proposed NLP application, the GAN repurposes the min/max model from mathematical game theory to generate outputs in an unsupervised manner. As noted, the GAN framework entails a generator and a discriminator: the generator acts as an adversary and seeks to fool the discriminator by producing synthetic outputs based on a noise input, and the discriminator seeks to differentiate synthetic outputs from the true input [72]. The GAN model is utilized to generate a bag-of-N-grams, where the N-grams could be characters, words, phrases, or the other portions of natural language content. The bag-of-N-grams is encoded as a

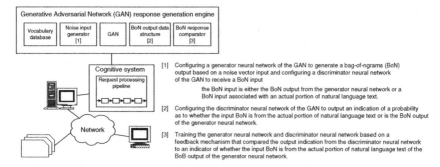

Figure 2.13 GAN-based NLP system: The GAN learns word co-occurrences and their positions jointly in a portion of natural language content. *Source:* Adapted from Ref. [72].

probability distribution over a given vocabulary. N-grams that do not belong to the bag have zero probability, while N-grams in the bag have some probability weight. The bag-of-N-gram representation that is generated from the trained generator G of the GAN provides a large unlabeled set of data that may be used to perform NLP operations. See Figure 2.13 and refer to [72] for a fuller discussion.

2.5 Natural Language Generation Basics

As noted in Section 2.1 and Figure 2.1, NLG is a process where an algorithm endeavors to generate human-understandable text; it is not an easy undertaking, even when AI is employed. NLG is the process or mechanism of generating narratives, descriptions, dialogs, story-telling in a natural language starting from/with structured data. It has been described as the construction of computer systems than can produce understandable texts in human languages (for example, but limited to, English, Italian, French, and German) from some underlying nonlinguistic representation of information; the current definition includes not only data-to-text generation but also text-to-text generation [75–80].[5]

NLG is a subfield of AI dealing with technology that produces human-understandable language as output based on some input information or structure, where that input constitutes data about some situation to be assessed and

---

5 Some observers state that precisely defining NLG is somewhat challenging and there may be blurred boundaries among disciplines. There is agreement on what the output of an NLG system should be (text or voice), but what the exact input is can vary substantially, including flat semantic representations, numerical data and structured knowledge bases, and visual input such as image or video [17].

expressed in natural language.[6] Human audible/linguistic (or text) content is constructed from a nonlinguistic representation of information. NLG has been used in many data-to-text applications enabling users to gain insights from underlaying datasets of all types. One of many examples of NLG is automated journalism and sports journalism in particular; verbal weather reports is another example (however, there still are challenges in developing and text generation systems in generic commercial applications).

As noted earlier in Figure 2.1, NLG is a sub-field of NLP. NLG often operates in conjunction with NLU, also as was seen in Figure 2.3; in cases where the input is unstructured text, NLU assists in the step of producing a structured representation that can then be used by NLG. NLG is often a more complex task than NLU. The output can be in the acoustical domain or can be in the textual domain – in the latter case it is easy to use a TTS module to deliver acoustical (spoken language) signals.

As described earlier, in recent years, the standard approaches NLG tasks have been based on RNNs to model the temporal dynamics in the sequence data, especially in conjunction with LSTM networks and gated recurrent units/gated recurrent models (GRUs/GRMs) [49, 81, 82]. These approaches have shown good performance against many benchmarks. However, although RNN structures work relatively well, they nonetheless suffer from long-term dependency problems, low training-inference speed (the process is also hard to be parallelized because of the consecutive operations), and from gradient vanishing and explosion (backpropagation procedure through time suffers from gradient vanishing and explosion issues); LSTM is also challenged when dealing with longer sentences, e.g. more than 200 context words) [58, 83].

Traditionally, the NLG problem of converting input data into output text was tackled by splitting it up into a number of subproblems. The following subproblems are frequently found in many NLG systems [17, 18]:

1) Content determination: processes for deciding which information to include in the text under construction;
2) Text structuring: processes for determining in which order information will be presented in the text;
3) Sentence aggregation: processes for deciding which information to present in individual sentences;
4) Lexicalization: processes for finding the right words and phrases to express information;

---

6 According to some observers, although NLG has been a part of AI and NLP from the early days going back to the 1970s, as a field it has not been fully embraced by researchers in these communities and has only recently begun to take full advantage of recent advances in data-driven, ML/DL approaches [17].

5) Referring expression generation: processes for selecting the words and phrases to identify domain objects; and

6) Linguistic realization: processes for combining all words and phrases into well-formed sentences.

Figure 2.14 depicts an example process flow for NLP-based training of an NLG system [1].

NLG systems have been classified as standard (also known as real) NLG and template-based NLG. The former systems employ generic linguistic insights, while the latter systems are systems that map the nonlinguistic input directly to the linguistic surface structure [84] (refer again to Figure 2.12).

Many NLG systems use template approaches to translate input data into text. Conventional designs, however, typically suffer from a number of shortcomings such as (i) constraints on how many data-driven ideas can be communicated in any-one sentence, (ii) constraints on variability in word choice, and (iii) limited capabilities of analyzing datasets to determine the content that should be delivered to a user [3]. A system that trains an NLG system to produce style-specific natural language outputs has to combine techniques of NLG and NLP/NLU in such a manner that the system not only understands the meanings and styles that constitutes the training data, but also is able to translate these stylistic understandings and meanings into a configuration that is usable by the NLG system.

Figure 2.14 Example of a process flow for NLP-based training of an NLG system. *Source:* Reference [1]/U.S. Patent.

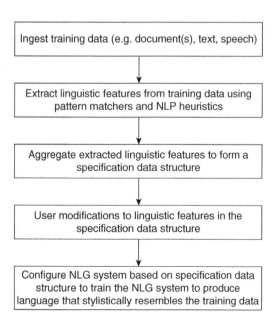

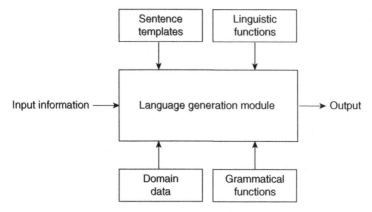

Figure 2.15 Typical elements of a template-based systems NLG.

Computational phonology can also be utilized in NLG [8]. Phonology is the study of the sounds used in a language, their internal structure, and their composition into syllables, words, and phrases.

Template-based systems are NLG systems that map nonlinguistic input directly to linguistic surface structure. See Figure 2.15. One of the advantages of these systems is that they can be used when the linguistic rules are not available, or where constructions have unpredictable meaning. There are several template-based NLG systems.

There is a number of high-level architectures for NLG systems, one of which is a sequential pipeline architecture. A pipeline architecture can typically include a document planning module (for content determination and document structuring), a microplanning module (for lexicalization, aggregation, and referring expression generation), and surface realization module (for linguistic realization and structure realization) [75, 85–87]. See Figure 2.16. Commercially available system at press time included Quill (Narrative Science), that indeed uses DL, and other systems such as Arria NLG, AX Semantics, NaturalOWL (open source, multilingual), SimpleNLG (open source), Yseop, and Wordsmith (Automated Insights, an early entrant).

NLP/NLG evaluation (to compare two systems or to establish if new mechanism offers improvements) is a subdiscipline that emerged in the 1990s. Aspects such as fluency/readability, accuracy, clarity, correctness, and relevance, and thus are of interest in these evaluations. Two general approaches have emerged: black box evaluation and glass box evaluation. In the former, one evaluates the entire system; in the latter, one evaluates various system's components, for example aggregation or document structuring.

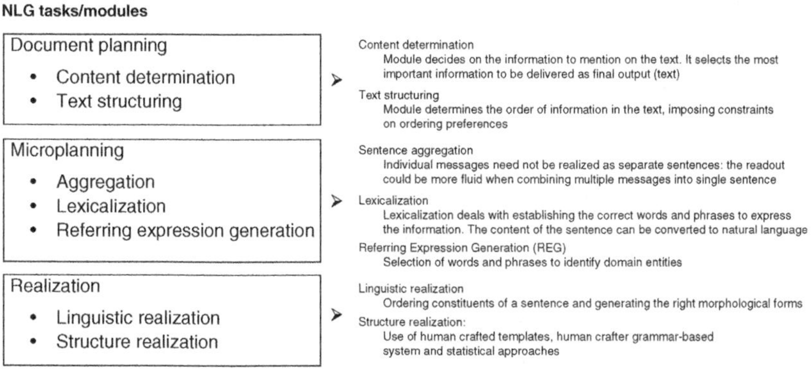

NLG tasks/modules

Document planning
• Content determination
• Text structuring

➤ Content determination
  Module decides on the information to mention on the text. It selects the most
  important information to be delivered as final output (text)

Text structuring
  Module determines the order of information in the text, imposing constraints
  on ordering preferences

Microplanning
• Aggregation
• Lexicalization
• Referring expression generation

➤ Sentence aggregation
  Individual messages need not be realized as separate sentences: the readout
  could be more fluid when combining multiple messages into single sentence

Lexicalization
  Lexicalization deals with establishing the correct words and phrases to express
  the information. The content of the sentence can be converted to natural language

Referring Expression Generation (REG)
  Selection of words and phrases to identify domain entities

Realization
• Linguistic realization
• Structure realization

➤ Linguistic realization
  Ordering constituents of a sentence and generating the right morphological forms

Structure realization:
  Use of human crafted templates, human crafter grammar-based
  system and statistical approaches

Figure 2.16 Typical NLG tasks for converting inputs. *Source:* Adapted from Ref. [85].

2.6 Chatbots

Chatbots are specific systems that implement NLP/NLU/NLG techniques, inter-acting with users in a very conversational way:

- People may engage in human-to-computer *dialogs* with interactive software applications typically known as "chatbots" (also referred to as "bots," "auto-mated assistants," "intelligent personal assistants," "interactive personal assis-tants," "personal voice assistants," or "conversational agents").
- A chatbot may be also configured to *cause actions* to be performed in response to natural language user inputs, such as text-based natural language inputs or verbal inputs, for example, triggered by a wakeword.

Using chatbots humans may issue commands, queries, and/or requests using free-form natural language input, as discussed in earlier sections the input may be verbal utterances that are converted into text and then processed, and/or may be typed free-form natural language text .

Automated assistants are typically invoked using predetermined vocal utter-ances (e.g. "Hi Bixby" for a Samsung smartphone). It should be noted that these assistants are adept at communicating with users in some common languages, such as English, but are less able to communicate in other languages. Figure 2.17 shows a typical system.

A chatbot can be implemented utilizing *software only* (e.g. the chatbot is a digi-tal entity implemented with coded programs or instructions executable by one or more processors) running on various hardware; or, *utilizing a combination of hardware and software.* A chatbot can be implemented in various physical sys-tems or devices, such as, but not limited to, a computer, a mobile phone, a watch,

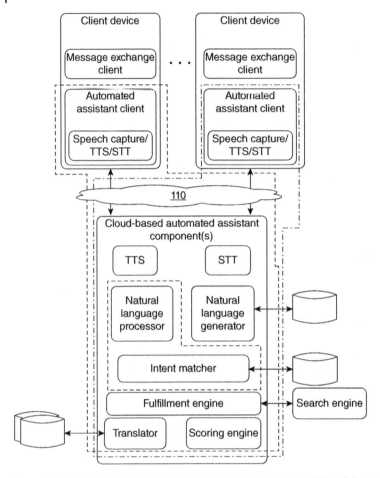

Figure 2.17 Typical Chatbot-based system. *Source:* Reference [88])/U.S. Patent.

an appliance, an automated personal assistant [89]. The processing performed by the chatbot system is typically implemented by a pipeline of components including an ASR subsystem and an NLU subsystem that may also include a semantic parser and an intent classifier.

One press time example included OpenAI's ChatGPT (Chat Generative Pre-trained Transformer) Version 4 (https://chat.openai.com), which is a chatbot that is able to provide lengthy, apparently thoughtful and thorough responses to questions and prompts. ChatGPT can, for example, compose complex essays, generate volumes of convincing text, draft marketing pitches, create poems and jokes, compose books, suggest interior decorating schemes, and can create computer code.

ChatGPT is a large LM trained on a very large set of online information to create responses. The chatbot uses RLHF – an algorithm that relies on human responses to enable intuitive chats which also embody a degree of memory. OpenAI has made the claim that the "format makes it possible for the tool to answer follow-up questions, admit its mistakes, challenge incorrect premises, and reject inappropriate requests" [90–92].

In the Spring of 2023 OpenAI unveiled the successor to GPT-3/GPT-3.5 that spawned viral services ChatGPT and Dall-E and set off an intense competition among technology companies. OpenAI stated that the new version of the technology, GPT-4, is more accurate, creative and collaborative; OpenAI has made the claim that the tool is "40% more likely to produce factual responses than GPT-3.5 on our internal evaluations" [40]. GPT-4 can also handle text and image queries – thus, a user can submit a picture with a related question and ask GPT-4 to describe it or answer questions. GPT-4 is available to OpenAI's paid ChatGPT Plus subscribers; developers can sign up to build applications with it.

ChatGPT's capabilities (or follow-on systems or developments) could eventually replace standard online search engines such as Google. [93]. Since its release in late 2022, ChatGPT has set off a tech craze, prompting rivals to launch similar products and companies to integrate it or similar technologies into their apps and products [94]. According to some observers "ChatGPT has shaken Google out of its routine," with the company reportedly jump-starting AI development intended to create a cadre of new products and demonstrate a version of its search engine with chatbot features; in early 2023 it launched its own large language model (LLM), Bard, also making financial investment in OpenAI rival Anthropic. Microsoft Corp has been quick to add the generative AI models to its own products: in early 2023, Microsoft reportedly announced it was "investing billions of dollars into OpenAI" (apparently, more than US$13 billion), and announced it was planning to incorporate AI tools such as ChatGPT into all its products (with GPT-4 powering its Bing search engine), making these products available as platforms for other businesses to build on [95, 96]. Amazon also announced around the same time its own family of LLMs, called Titan. Amazon also introduced generative AI services such as Bedrock to assist developers enhance software and Amazon Web Services (AWS) HealthScribe to help doctors create patient visit summaries. Additionally, Amazon was reportedly designing two types of microchips for training and accelerating generative AI. These custom chips, Inferentia and Trainium, offer AWS customers an alternative to training their LLM on Nvidia GPUs.

However, there were still potential issues (even acknowledged by the software provider), such as spreading incorrect information, issue unsafe content, give plausible-sounding but incorrect or nonsensical answers, perpetuate biases based on the pool of data on which it is trained, and also threaten some professions.[7,8,9,10,11,12] The growth in patent filings for chatbots and NLP technologies, combined with a competitive atmosphere of greater investment and interest in AI from Big Tech, implies that big intellectual property battles may ensue in the near future. Some call generative AI "the next $100 billion tech revolution" [102].

There is a major concern in the field of Education that students will use generative AI to prepare their term papers and essays. While there may be some form of watermarking included in the future, it is also expected that tools to test whether a text is generated by AI will soon emerge. Such a tool would, for example, take

---

7 A press time quote, similar in theme to many others, stated [97]: "In the next five years, it is likely that AI will begin to reduce employment for college-educated workers. As the technology continues to advance, it will be able to perform tasks that were previously thought to require a high level of education and skill. This could lead to a displacement of workers in certain industries, as companies look to cut costs by automating processes ... There you have it, I guess: ChatGPT is coming for my job and yours, according to ChatGPT itself ... The technology is, put simply, amazing. It generated that first paragraph instantly, working with this prompt: 'Write a five-sentence paragraph in the style of The Atlantic about whether AI will begin to reduce employment for college-educated workers in the next five years'."

8 Consistent with the phrase 'we live in interesting times', the following press time press blurb illustrates the new dynamics: "Google's ... new artificial intelligence-powered chatbot [Bard] gave a wrong answer in a promotional video, as investors wiped more than $100bn off the value of the search engine's parent company, Alphabet . ... Alphabet stock slid by 9% during regular trading in the US ..... Experts pointed out that promotional material for Bard, Google's competitor to Microsoft-backed ChatGPT, *contained an error in [a] response by the chatbot*" [98].

9 Example of quote: "The increasing sophistication of programs like ChatGPT has led to unease over the future of film-making . . . Can an AI program really write a good movie? Writers Guild of America proposed that ChatGPT would absolutely be allowed to write scripts in the future, provided that the credit (and the money) goes to the human writer who came up with the prompts in the first place" [99]. An extensive U.S. screenwriters and actors strike in the Summer of 2023 focused, among other issues, on the role of AI in the creation of entertainment content.

10 In the Spring of 2023 Elon Musk, Apple cofounder Steve Wozniak, and thousands of AI experts and industry executives called for a six-month pause in developing systems more powerful than OpenAI's GPT-4, in an open letter (https://futureoflife.org/open-letter/pause-giant-ai-experiments) citing potential risks to society [94].

11 According to Nobel Prize-wining labor economist, Christopher Pissarides, the ChatGPT revolution opens the door to a four-day week by providing a major productivity boost for a large number of jobs; this comes with possible employment implications, given that productivity gains would require people to work fewer hours [100].

12 Chatbots typically collect text, voice and device information as well as data that can reveal the location, such as the IP address; chatbots also gather data such as social media activity, which can be linked to your email address and phone number. Concerns about the growing abilities of chatbots trained on large language models, such as OpenAI's GPT-4, Google's Bard and Microsoft's Bing Chat, are surfacing. People are aware of the privacy risks posed by search engines, but experts are of the opinion that chatbots could be even more data-hungry: their conversational nature can catch people off guard and encourage them to give away more information than they would have entered into a search engine. Each time one asks an AI chatbot for help, micro-calculations feed the algorithm to profile individuals [101].

consecutive streams of, say, four words iteratively from the essay in question and then invoke well-known AI tool 1, 2, 3, and so on, and individually compare what the AI tool generated with those four words against the essay text in question, under some established similarity measure. After repeating the process for the entire essay using the several well-known AI tools, a combined "similarity" or "genesis" or "origin" score is computed providing an aggregated probability measure of AI composition usage.

In addition to a textual or verbal (synthesized speech) response, or information retrieval processing to determine an answer to a question contained in a natural language input, as noted, a chatbot can also cause actions to be performed in response to natural language user inputs (e.g. spoken and/or text-based natural language inputs). For example, for the natural language input "play workout music," a system may output music from a user's playlist; or for the natural language input "turn on the lights," a system may turn on "smart" lights associated with a user's profile. Furthermore, a system may be configured to perform NLU processing to determine an *intent* representing a natural language input (that is, intent classification processing), and one or more portions of the natural language input that enable the intent to be carried out. For example, for the natural language input "play [song name] by [artist name]," the system may determine the natural language input corresponds to a "Play Music" intent, may determine "[song name]" is a song title, and may determine "[artist name]" is an artist name. Based on such determinations, the system may identify audio data corresponding to the song title "[song name]" and the artist's name "[artist name]," and may output the audio data as audio to the user [103].

An illustrative NLP system, a chatbot, was depicted in Figure 1.2; it included the following elements:

- ASR
- natural language input NLU
- A knowledge base (KB)
- Skills components
- TTS
- User recognition
- Orchestrator

A simpler system is depicted in Figure 2.18. Here the NLP/chatbot obtains an audio input and determines a query associated with the audio input based on the audio input and the context of the interface. As described in [104], the process includes feeding the audio input to a voice recognition engine to determine raw texts corresponding to the audio input, and feeding the raw texts and the context of the current interface to an NLP engine to determine the query associated with the audio input. The process may also include pre-processing the raw texts based on at least one of: lemmatizing, spell-checking, singularizing, or sentiment

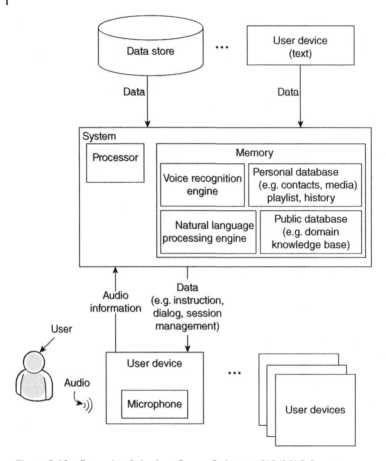

Figure 2.18 Example of chatbot. *Source:* Reference [104]/U.S. Patent.

analysis to obtain pre-processed texts; then matching the pre-processed texts against preset patterns – or if not detecting any preset pattern matching the pre-processed texts, tokenizing the texts; and then vectorizing the tokenized texts to obtain vectorized texts.

There are a number of controversies associated with chatbots because machine-generated text may be "non-factual" or "toxic." There is work underway to mitigate some of these matters. One may endeavor to address these problems by applying discriminators over the output to enforce appropriate properties after the fact; for example, one can apply a secondary model as a re-ranker over a small collection of outputs. With a large set of candidates, a secondary model could more easily find an acceptable output without having to undertake more extreme steps such as re-training the initial generation model. Diversity in outputs has been advocated as a desirable goal for applications such as story generation and/or dialogue. See, for example, reference [105] for a discussion of the topic.

2.7   Generative AI

Chatbots discussed in the previous section are examples of generative AI systems. Generative AI has received considerable research-, market-, and popular-press attention of late, being called 'disruptive technology' and/or 'revolutionary' by many observers, interested parties, and agencies. Some see generative AI as being able to approximate human behavior in the content-generation space. It can be seen as assistive technology for nontechnical users and it is expected, by some, to have significant job loss implications.

Generative AI refers to AI systems that are able to generate (that is, create) NLP text, media, or image content, or other types of outputs, where questions are asked (with NLP) and answers are generated and presented (with NLP, or graphics, or other outputs). Generative AI text generators rely on the NLP technologies described earlier; generative AI image generators utilize CNNs among other methods and models (some described in Chapter 4).

Clearly, Generative AI is a subset of AI; it is focused on creating new content, such as text, literature, graphics, or music, that uses ML techniques to generate new material that is similar to the training data; massive, largescale training datasets are utilized. Generative AI systems (models) are designed and trained to generate text and graphics such as reports, news stories, movie scripts, blog posts, program code, poetry, and artwork. Hence, generative AI applications span social media, education, search engines, business proposals, movie scripts, literature creation, answering queries, image creation, song writing, deep fakes, voice, and every area of information processing. Generative AI addresses the world of content production, and it is expected to have major implications in the marketing-, software-, design-, entertainment-fields, among other fields of human endeavor.

ChatGPT is a key example of a chatbot with *NLP-in, NLP-out text*. DALL-E 2 is an image-generation system text-to-image (*NLP-in, image-out*) DL model; specifically, it can be used (i) to generate detailed images based on text descriptions, (ii) to create images from partial images, and (iii) to generate image-to-image translations subtended by a text prompt. Midjourney is another text-to-image model that turns text-based prompts into images.

To develop new material, the technology uses its inputs (the data it has consumed and a user prompt) and experiences (exchanges with users that help it "learn" new knowledge and what is correct/incorrect). Affiliated algorithms are used to generate new content in the form of audio, code, images, text, simulations, and movies. Generative AI systems utilize complex ML models to predict, for example, the word that follows based on previous word sequences, or the subsequent image based on words describing previous photos [2.x]. Generative AI employs GANs mentioned earlier. As discussed, a GAN is comprised of two NNs: a generator for creating new data and a discriminator for evaluating the data. The generator and discriminator collaborate, with the generator refining its outputs depending on the discriminator's feedback until such time as it generates content that is practically indistinguishable from real-world information.

As noted in the previous subsection, generative models have been advanced by leading technology companies including Google (BERT and LaMDA), Facebook (OPT-175B, BlenderBot), and OpenAI (GPT-4 for text). Once developed, a generative model can also be specialized or sub-setted for a specific content domain that entail fewer resources; for example, BERT models can be specialized for biomedical material (BioBERT), legal content (Legal-BERT), and French text (CamemBERT).

Generative AI examples include:

- *Text Generators* that can create, for example, articles, real-time chats, product descriptions, or content summarization. Press time examples include but are not limited to: ChatGPT-4.
- *Image Generators/Video Generators* that can create, for example, videos, sales presentations, corporate communications videos, employee training clips, and YouTube content; these AI video generators are designed and trained to convert text into good-quality videos and provide intuitive video editing tools. Text-conditioned models, such as DALL-E 2 and Stable Diffusion, enable novice users to generate detailed imagery given only a text prompt as input [2.y]. Other press time examples include but are not limited to: Gen-2 by Runway (runwayml.com), Synthesia, Text2Live, SinFusion, Tunea-Video.
- *Voice Generators* that employ speech synthesis to create quality verbal content also typically emulating some specific known personality; they can replicate any style of voice using text-to-voiceover techniques. Press time examples include but are not limited to: Microsoft's VALL-E (Voice Agnostic Lifelike Language model) and Synthesis.
- *Music Generators* that can be used to create music, modify songs, customize melodies by learning from large amounts of input content. Press time examples include but are not limited to: Soundraw.io and Amper AI.
- *Graphic Design Generators* that enhance picture editing and 3D architecture development, producing designs based on the context. Currently designers spend a large amount of time on activities such as cropping, scaling, and color correction; AI systems automate these processes. These systems also support adaptive design technologies that alter the look of websites or online marketing materials based on who is browsing.Text generators have attracted a large amount of research-, vendor-, and public interest in recent times. Ominous concerns have been raised in many quarters, including the concern that many jobs will ultimately be eliminated, being replaced by AI-based systems. Thus, one then asks "what is all the rage with generative AI (at writing time)?"

A. Basic Glossary of Key AI Terms and Concepts Related to Natural Language Processing

(Based on a variety of references but specifically including [1] – [20, 34, 36, 57, 106, 107]).

Term	Definition/explanation/concepts
Attention models	Recurrent neural network (RNN) models that iteratively process inputs by selecting relevant data at every step. This approach extends the applicability range of end-to-end training methods by enabling the construction of networks with external memory.
Chatbot	(aka ChatterBot or digital assistant) An application that simulates human conversations with users to answer their questions, typically via chat or an instant message. Software application(s) used to support an on-line chat conversation utilizing text or text-to-speech (TTS), replacing direct interaction with a live human agent. Chatbots are designed to simulate the manner a person would behave as a conversational counterpart.
Compressive Transformer	An extension of the Transformer model that maps past hidden activations (known as short-term and/or granular memories) into a smaller set of compressed representations (known as long-term and/or coarse memories). The model learns to query both long-term memory and short-term memory, utilizing the same attention mechanism both for granular and coarse memories. It embodies Transformer-XL's approach of keeping a memory of past activations in each layer to preserve longer contextual records [36, 107].
Computational linguistics (CL)	A multidisciplinary field dealing with (theoretical) computational modeling of a natural language. Currently, the term is perceived as being a near-synonym of natural language processing (NLP) and/or language technology. Also see *language model(ing)*, below.
Deep learning (DL) in natural language processing (NLP)	Use of neural network (NN) methods for document classification, language modeling, machine translation, multilingual output, and larger-context modeling. Deep learning (DL) can be used in combination with composable communication goal statements and with an ontology to facilitate a user's ability to quickly develop story outlines that can be used by a narrative generation system, without any need to directly generate computer code.
Dialog processing	A processing step for establishing whether to answer the user's utterance, respond to the utterance or inquire about additional information.
Discourse	The area of linguistics that deals with the aspects of language usage that go beyond the sentence – and, in particular, deals with the study of coherence and salience.
Finite-state machines	(Automata and transducers) fundamental logical mechanism used in NLP and CL. Can be probabilistic and non-probabilistic.

(Continued)

(Continued)

Term	Definition/explanation/concepts
Gated recurrent models (GRMs)	(aka gated recurrent units – GRUs) are gating mechanisms used in/ with RNNs. GRMs/GRUs are similar to LSTMs, but with a "forget gate," being that they lack an output gate, they utilize fewer parameters than LSTMs. These NNs deal well with speech signal modeling and NLP especially with smaller datasets.
Gated recurrent unit (GRU)	GRU is a simplified version of the LSTM RNN model. A GRU uses only one state vector and two gate vectors (the reset gate and the update gate). In GRU, the input and forget gates are combined and controlled by one gate, making the GRU simpler than LSTM, and the GRU model has been found to outperform LSTM when dealing with smaller datasets.
Generative AI systems	AI systems that are able to generate NLP text, media, or image content, among other types of outputs, where questions are asked (with NLP) and answers are generated and presented (with NLP or graphics or other outputs). Generative AI aims at creating new content, such as text, graphics, or music, that uses ML techniques to generate new material. Also see Glossary of Chapter 1.
Intention analysis	Determining the purpose or intention of a sentence, for example, whether the user asks a question, makes a request, or expresses some basic emotion.
Language Model(ing) (LM)	A joint probability distribution over sequences of words: given a sequence of words of length n, a language model assigns a joint probability $P(w_1, w_2, ..., w_n)$ to said sequence. Language models are used to address a variety of problems in CL, particularly speech recognition. Probabilistic language models such as an N-gram model, a bigram model, or a unigram model. The N-gram model is a language model that assumes that use of words depends on $(n-1)$ previous words. The bigram model is a language model that assumes that use of words depends on only one previous word. The unigram model refers to a language model that assumes that use of all words is completely independent of each other and calculates the probability of a word string by a product of the probabilities of words. Recurrent neural network (RNN), long short-term memory (LSTM) network, and gated recurrent models (GRMs) are well-established approaches for language modeling.
Lemmatization	The process that aims at removing inflectional endings and aims to return the base or dictionary form of a word, which is known as the lemma; properly using a vocabulary and morphological analysis of words.
Lexis/lexicon	The basic raw material of natural language, e.g. words, tokens, types, lemmas, phrasemes.
Linguistic realization	A task that deals with ordering constituents of a sentence; it also deals with generating the right morphological forms (including, where needed) verb conjugations and agreement. Various approaches exist, including human-developed templates, human-developed grammar-based systems, and stochastic/statistical approaches.

Term	Definition/explanation/concepts
Long short-term memory (LSTM) networks	A type of RNN that is capable of learning order dependence in sequence prediction problems, such as machine translation and speech recognition. LSTM utilizes feedback connections. This type of RNN can process not only single data points (such as static images), but also complete sequences of data (e.g. speech or video). LSTM-based systems were the dominant type of RNN for language modeling tasks of late. The advantages of LSTM-based systems are (i) supporting sequences of varying lengths, (ii) avoiding data sparseness, and (iii) not requiring a large number of parameters; they utilize projection of histories into a low-dimensional space and allow similar histories to share representations. Also see Glossary of Chapter 1.
Machine learning (ML) algorithms	Data analysis algorithms. These algorithms include but are not limited to NNs, k-nearest neighbor systems, fuzzy logic/possibility theory, Boltzmann machines, vector quantization, support vector machines (SVM), maximum margin classifiers, inductive logic system Bayesian networks, Petri nets (e.g. finite state machines), classifier trees (e.g. perceptron trees, support vector trees, Markov trees, decision tree forests, random forests), betting models and systems, artificial fusion, sensor fusion, image fusion, reinforcement learning, pattern recognition, augmented reality, and automated planning.
Machine learning (ML) in NLP	The use of supervised or unsupervised training for NLP. Decision trees, rules, or case-based classifiers are utilized in supervised training examples. A representation of knowledge is postulated with basic algorithms for inducing such representations from annotated training examples and utilizing the acquired knowledge to classify future instances. In unsupervised learning, new concepts are formed from unannotated examples by clustering them into coherent groups.
Machine translation (MT)	Conversion of a text (or speech) from one language to another, with the goal of fluency and accuracy.
Morpheme analysis	Morphemes are the smallest units having a meaning. Morpheme analysis consists of processing steps (i) for classifying the text data corresponding to the speech uttered by the user into morphemes and (ii) for determining the part of speech of each of the classified morphemes.
Morphology	The study of the structure of words and how words are formed by combining morphemes, that is, combining the smaller units of linguistic information.
Natural language generation (NLG)	The task of generating *text* or *speech* from nonlinguistic input. The set of computer tasks/processes that generate fluent text from input data and other contextual information; it enables the machine to communicate in "spoken conversation." NLG includes both text-to-text generation and data-to-text generation. Communication utilizing a natural language entails two basic skills: producing 'text' (written or spoken) and understanding it; NLG refers to the former. Example: a system that collects publicly available information (e.g. in human-written texts) and compiles these data into a book.

(Continued)

(Continued)

Term	Definition/explanation/concepts
Natural language processing (NLP)	A subfield of AI focusing on technology that interprets natural language inputs.
Natural language understanding (NLU)	The process of mapping human language to internal computer representation of information
Ontologies	Knowledge models of a domain of interest; a set of concept definitions with the aim of showing the properties of a given technical area and how these areas are related. It encompasses the (or a) representation, naming, and definition of (i) the categories, (ii) the properties, and (iii) the relations between the concepts, data, and entities that substantiate the given technical area. An ontology may be represented as a hierarchical structure. In this context, the structure contains a (large) number of nodes, where each node represents either (i) an "actionable intent" or (ii) a "property" (a parameter) relevant to one or more of the "actionable intents." The linkage between an "actionable intent" node and a "property" node defines how a parameter (represented by the property node) relates to the task represented by the "actionable intent" node. An actionable intent node with its linked concept nodes is described as a "domain." For example, for an "Hotel Reservation Domain" one can construct an ontology that includes a "hotel reservation" node (i.e. an actionable intent node). Property nodes "hotel," "date/time" (for the reservation), and "party size" are each directly linked to the actionable intent node (i.e. the "hotel reservation" node). In addition, property nodes "bed type," "price range," "phone number," and "location" are sub-nodes of the property node "hotel," and are each linked to the "hotel reservation" node (i.e. the actionable intent node) through the intermediate property node "hotel."
Phonology	The application of computational techniques to the representation and processing of phonological information.
Pre-training	The process of first training a model on one task or dataset, then utilizing the parameters or model from this training to train another model on a different task or dataset. This furnishes the model a head-start instead of starting from nothing.
Recurrent neural network (RNN)	See Glossary of Chapter 1.
Referring Expression Generation (REG)	The task of selecting words or phrases to identify domain entities; enables a system to communicate adequate information to distinguish one domain entity from other domain entities.

Term	Definition/explanation/concepts
Self-attention	A technique that enables the model to consider the entire input sequence when processing each element.
Semantics	The study of linguistic meaning utilizing mathematical characterizations; for example, logical representations of (English) sentences, and how meanings are composed using a grammar.
Sentence planning	Deciding on the overall sentence structure.
Sequence-to-sequence (S2S) models	S2S (aka seq2seq) models are a special class of RNN architectures that utilize an encoder and a decoder. Both encoder and the decoder are LSTM models. The *Encoder* reads the input sequence and summarizes the information in an internal state vectors or context; this context vector aims to encapsulate the information for all input elements to help the decoder make accurate predictions. The *Decoder* is an LSTM whose initial states are initialized to the final states of the Encoder LSTM, i.e. the context vector of the encoder's final cell is input to the first cell of the decoder network. Utilizing these initial states, the decoder starts generating the output sequence, and these outputs are also taken into consideration for future outputs. The decoder comprises a stack of LSTM units where each predicts an output y_t at a time step t. Each recurrent unit accepts a hidden state from the previous unit and produces and output as well as its own hidden state. Softmax is used to create a probability vector that determines the final output [57].
Speech recognition	(also known as automatic speech recognition) Methods for converting the speech waveform, an acoustic signal, into a sequence of words, typically using statistical modelization of the speech signal. Techniques entail acoustic-phonetic modeling, lexical representation, language modeling, decoding, and model adaptation. Speaker-independent, large-vocabulary continuous speech recognition is a typical desideratum. Initially, applications included dictation and interactive systems for limited domain information access; more recently one sees a broader coverage of languages with interest in transcription systems for information archival and retrieval, media monitoring, machine-based translation, automatic subtitling and speech analytics.
Speech-act analysis	A processing step for analyzing the intention of the speech uttered by the user utilizing the result of the syntax analysis step; determining the intention of a sentence (e.g. whether the user asks a question, makes a request, or expresses simple emotion).

(Continued)

(Continued)

Term	Definition/explanation/concepts
Statistical methods	Language modeling methods employed in NLP; these include, but are not limited to hidden Markov model (HMM), maximum entropy models, and the expectation maximization method utilized in machine translation (MT). Applications include: (i) recognition applications based on Shannon's Noisy Channel, e.g. Optical Character Recognition (OCR), speech recognition, spelling correction, part-of-speech tagging, and MT; and (ii) discrimination/ranking applications (e.g. sentiment analysis, information retrieval, spam email filtering, and author identification). (also see Glossary of Chapter 1)
Surface realization	Determining specific word forms and/or flattening the sentence structure into a string.
Syntax	The arrangement of words and phrases to generate well-formed sentences in a language. How linguistic elements (such as words) are joined together to form phrases or clauses. The portion of the grammar dealing with this.
Syntax analysis	A processing step for classifying the text data into a noun phrase, a verb phrase, an adjective phrase, and so on, utilizing the result of the morpheme analysis step and then determining a relation between the classified phrases.
Text-to-speech (TTS) Synthesis	Mechanisms for delivering intelligible, natural-sounding, and expressive speech. Speech synthesis has a long history that spans MIT's formant synthesis and Bell Labs' diphone-based concatenative synthesis. Hidden Markov models (HMM) synthesis has used in the recent past. More recently, DL-based statistical approaches have been shown to advance text analysis well as the generation of the waveforms. Recent examples include Google's Wavenet and the Tacotron (I and II) sequence-to-sequence (S2S) models.
Transformer	A DL method that undertakes a small (constant) number of steps. In each step, it applies a self-attention mechanism which directly models relationships between all words in a sentence, regardless of their respective position.
Transformers XL (extra long)	A new variation on the transformer model that uses relative positional encoding and a recurrence mechanism. The model retains the previously-learned segment in a hidden state, and utilizes it for the current segment rather than calculating each segment's hidden state from scratch. It processes the first segment just as a regular Transformer and retains the output of the hidden layer while processing the next segment. Recurrence speeds up the evaluation process [34, 36].
Word representation	A fundamental step in NLP is to represent words as mathematical entities so that the words can be read and manipulated by computational models. Specifically, one can represent words as vectors in a nth-dimensional R^n space in such a way that similarities in the vector space correlate with semantic similarities between words.

References

1 D. J. Platt, N. D. Nichols, *et al.* US11042713B1, applied artificial intelligence technology for using natural language processing to train a natural language generation system, June 22, 2021. Uncopyrighted material.

2 K. Cho, "Deep learning", in *The Oxford Handbook of Computational Linguistics*, 2nd Edition, R. Mitkov, (Editor), 2017. https://doi.org/10.1093/oxfordhb/ 9780199573691.013.55.

3 A. R. Paley, N. D. Nichols, *et al,* Applied artificial intelligence technology for performing natural language generation (NLG) using composable communication goals and ontologies to generate narrative stories. U.S. Patent 20200401770, Dec. 24, 2020. Uncopyrighted material.

4 M. Poesio, "Discourse", in *The Oxford Handbook of Computational Linguistics*, 2nd Edition, R. Mitkov, (Editor), 2017. https://doi.org/10.1093/oxfordhb/ 9780199573691.013.32.

5 P. Hanks, "Lexis", in *The Oxford Handbook of Computational Linguistics*, 2nd Edition, R. Mitkov, (Editor), 2017. https://doi.org/10.1093/oxfordhb/ 9780199573691.013.017.

6 R. Mooney, "Machine learning", in *The Oxford Handbook of Computational Linguistics*, 2nd Edition, R. Mitkov, (Editor), 2017. https://doi.org/10.1093/ oxfordhb/9780199573691.013.016.

7 K. Oflazer, "Morphology", in *The Oxford Handbook of Computational Linguistics*, 2nd Edition, R. Mitkov, (Editor), 2017. https://doi.org/10.1093/oxfordhb/ 9780199573691.013.006.

8 S. Bird, J. Heinz, "Phonology", in *The Oxford Handbook of Computational Linguistics*, 2nd Edition, R. Mitkov, (Editor), 2018. https://doi.org/10.1093/ oxfordhb/9780199573691.013.30.

9 D. Beaver, "Semantics", in *The Oxford Handbook of Computational Linguistics*, 2nd Edition, R. Mitkov, (Editor), 2017. https://doi.org/10.1093/oxfordhb/ 9780199573691.013.29.

10 L. Lamel, J.-L. Gauvain, "Speech recognition", in *The Oxford Handbook of Computational Linguistics*, 2nd Edition, R. Mitkov, (Editor), 2017. https://doi.org/10.1093/oxfordhb/9780199573691.013.37.

11 C. Samuelsson, S. Štajner, "Statistical Methods: Fundamentals", in *The Oxford Handbook of Computational Linguistics*, 2nd Edition, R. Mitkov, (Editor), 2017. https://doi.org/10.1093/oxfordhb/9780199573691.013.40.

12 K. W. Church, "Statistical models for natural language processing", in *The Oxford Handbook of Computational Linguistics*, 2nd Edition, R. Mitkov, (Editor), 2017. https://doi.org/10.1093/oxfordhb/9780199573691.013.54.

13 T. Dutoit, Y. Stylianou, "Text-to-speech synthesis", in *The Oxford Handbook of Computational Linguistics*, 2nd Edition, R. Mitkov, (Editor), 2017. https://doi.org/10.1093/oxfordhb/9780199573691.013.38.

14 O. Levy, "Word Representation", in *The Oxford Handbook of Computational Linguistics*, 2nd Edition, R. Mitkov, (Editor), 2017. https://doi.org/10.1093/oxfordhb/9780199573691.013.57.

15 J. Chae, S. Han, Speech synthesizer using artificial intelligence, method of operating speech synthesizer and computer-readable recording medium. U.S. Patent 11, 417,313, Aug. 16, 2022. Uncopyrighted material.

16 J. Uszkoreit, "Transformer: A Novel Neural Network Architecture for Language Understanding", Aug. 31, 2017, https://ai.googleblog.com/2017/08/transformer-novel-neural-network.html. Accessed Sept. 1, 2022.

17 A. Gatt, E. Krahmer, "Survey of the State of the Art in Natural Language Generation: Core Tasks, Applications and Evaluation", *J. Artif. Intell. Res.* 2018; 61: 65–170.

18 E. Reiter, R. Dale, *Building Natural Language Generation Systems*, Cambridge University Press, Cambridge, UK, 2000.

19 L. Mathias, Y. Shi *et al,* Architecture for Multi-Domain Natural Language Processing, U.S. Patent 11,176,936, Nov. 16, 2021. Uncopyrighted material.

20 Y. Kim, J. Bridle, *et al*, Detecting a Trigger of a Digital Assistant. U.S. Patent 11,532,306, Dec. 20, 2022. Uncopyrighted material.

21 R. J. Munro, S. D. Erle *et al,* Methods and Systems for Language-Agnostic Machine Learning in Natural Language Processing Using Feature Extraction. U.S. Patent 20210081611, Mar. 18, 2021. Uncopyrighted material.

22 A. Graves, "Generating Sequences with Recurrent Neural Networks", Aug. 2013, arXiv:1308.0850.

23 K. Cho, B. van Merrienboer, *et al*, "On the Properties of Neural Machine Translation: Encoder-Decoder Approaches", 2014, arXiv:1409.1259.

24 S. Hochreiter, J. Schmidhuber, "Long short-term memory", *Neural Comput.* 1997; 9(8): 1735–1780.

25 S. Agarwal, G. Sastry, *et al*, "Learning Transferable Visual Models from Natural Language Supervision", in Meila, M. and Zhang, T., (Editors), Proceedings of the 38th International Conference on Machine Learning, volume 139 of Proceedings of Machine Learning Research, pp. 8748–8763. PMLR, 18–24 Jul. 2021.

26 Ramesh, A. and Pavlov, M., *et al*, "Zero-Shot Text-To-Image Generation", in Meila, M. and Zhang, T. (eds.), Proceedings of the 38th International Conference on Machine Learning, volume 139 of Proceedings of Machine Learning Research, pp. 8821–8831. PMLR, 18–24 Jul. 2021.

27 J. Yu, Z. Wang, *et al*, "CoCa: Contrastive Captioners are Image-Text Foundation Models", in Transactions on Machine Learning Research, 2022. ISSN: 2835-8856.

28 S. Reed, K. Zolna, *et al*, "A Generalist Agent", in Transactions on Machine Learning Research, 2022. ISSN: 2835-8856.

29 A. Radford, K. Narasimhan, "Improving Language Understanding By Generative Pre-Training", 2018, https://openai.com/blog/language-unsupervised. Accessed Feb. 1, 2023.

30 J. Devlin, M. W. Chang, *et al*, "BERT: Pre-Training of Deep Bidirectional Transformers for Language Understanding", in Proceedings of the 2019 Conference of the North American Chapter of the Association for Computational Linguistics: Human Language Technologies, Volume 1 (Long and Short Papers), pp. 4171–4186, Minneapolis, Minnesota, June 2019. Association for Computational Linguistics. https://doi.org/10.18653/v1/N19-1423.

31 T. B. Brown, B. Mann *et al*, "Language Models are Few-Shot Learners", 2020, https://arxiv.org/abs/2005.14165.

32 K. Lu, A. Grover *et al*, "Pretrained Transformers As Universal Computation Engines", 2021, arXiv:2103.05247.

33 A. Lewkowycz, A. Andreassen, *et al*, "Solving Quantitative Reasoning Problems with Language Models", 2022, https://arxiv.org/abs/2206.14858.

34 Z. Dai, Z. Yang, *et al*, "Transformer-XL: Attentive Language Models Beyond a Fixed-Length Context", June 2019, https://arxiv.org/pdf/1901.02860.pdf.

35 C. Nicholson, "A Beginner's Guide to Attention Mechanisms and Memory Networks", 2022, https://wiki.pathmind.com/attention-mechanism-memory-network. Accessed Sept. 1, 2022.

36 Datagen Staff. "What is the transformer architecture and how does it work?", 2022, https://datagen.tech/guides/computer-vision/transformer-architecture. Accessed Feb. 3, 2023.

37 X. Wang, Z. Xu, *et al*, Pretraining Framework for Neural Networks, U.S. Patent 2023/0019211, Jan. 19, 2023. Uncopyrighted material.

38 A. Vaswani, N. Shazeer, *et al*, "Attention is All You Need", in 31st Conference on Neural Information Processing Systems (NIPS 2017), Long Beach, CA, USA, 2017.

39 M. G. Anderson, S.-Y. Ma, *et al*, "Optical Transformers", Feb. 20, 2023, https://arxiv.org/pdf/2302.10360.pdf.

40 D. Bass, R. Metz, R., "ChatGPT Creator openAI Debuts New GPT-4 AI System", Mar. 14, 2023, https://www.bloomberg.com/news/articles/2023-03-14/openai-unveils-next-version-of-the-ai-tool-that-birthed-chatgpt?srnd=premium. Accessed Mar 14, 2023.

41 R. Miikkulainen, "Text and Discourse Understanding: The DISCERN System", in *A Handbook of Natural Language Processing: Techniques and Applications for the Processing of Language as Text*, R. Dale, H. Moisl, H. Somers, (Editors), Marcel Dekker Inc., 2002.

42 J.-F. Lavallee, K. W. D. Smith, Processing Natural Language Text with Context-Specific Linguistic Model. U.S. Patent 10,789,426, Sept. 29, 2020. Uncopyrighted material.

43 E. Budu, "What Does Pre-Training a Neural Network Mean?", March 16, 2023, www.baeldung.com, https://www.baeldung.com/cs/neural-network-pre-training. Accessed Mar. 18, 2023.

44 T. Brown, B. Mann, *et al,* Language Models are Few-Shot Learners, Part of Advances in Neural Information Processing Systems 33, NeurIPS 2020.

45 J. Devlin, M.-W. Chang, *et al,* "BERT: Pre-Training of Deep Bidirectional Transformers for Language Understanding", in Proceedings of NAACL-HLT 2019, pp. 4171–4186, Minneapolis, Minnesota, June 2–June 7, 2019.

46 A. Radford, K. Narasimhan, K., *et al,* "Improving Language Understanding by Generative Pre-Training", Technical report, OpenAI, 2018, https://s3-us-west-2. amazonaws.com/openai-assets/research-covers/language-unsupervised/ language_understanding_paper.pdf. Accessed Feb. 5, 2023.

47 Devopedia, "Natural Language Generation", Version 3, Feb. 20, 2020, https:// devopedia.org/natural-language-generation. Accessed Aug. 15, 2022.

48 A. Gatt, K. Emiel, "Survey of the State of the Art in Natural Language Generation: Core Tasks, Applications and Evaluation", *J. Artif. Intell. Res.* 2018; 61: 65–170.

49 S. Hochreiter, J. Schmidhuber, "Long Short-Term Memory", *Neural Comput.* 1997; 9(8): 1735–1780.

50 J. Chung, Ç. Gülçehre, *et al,* "Empirical Evaluation of Gated Recurrent Neural Networks on Sequence Modeling", 2014, CoRR, abs/1412.3555.

51 I. Sutskever, O. Vinyals, Q. V. Le, "Sequence to Sequence Learning with Neural Networks", *Adv. Neural Inf. Proces. Syst.* 2014; 2: 3104–3112.

52 D. Bahdanau, K. Cho, Y. Bengio, "Neural Machine Translation by Jointly Learning to Align and Translate", 2014, CoRR, abs/1409.0473.

53 K. Cho, B. van Merrienboer, *et al,* "Learning Phrase Representations Using RNN Encoder-Decoder for Statistical Machine Translation", 2014, CoRR, abs/1406.1078.

54 Y. Wu, M. Schuster, *et al,* "Google's Neural Machine Translation System: Bridging the Gap Between Human and Machine Translation", 2016, arXiv:1609.08144.

55 M.-T. Luong, H. Pham, C. D. Manning, "Effective Approaches to Attention-Based Neural Machine Translation", 2015, arXiv:1508.04025.

56 R. Jozefowicz, O. Vinyals, *et al,* "Exploring the Limits of Language Modeling", 2016, arXiv:1602.02410.

57 Singh, P., "A Simple Introduction to Sequence to Sequence Models", Nov. 9th, 2020, https://www.analyticsvidhya.com/blog/2020/08/a-simple-introduction-to-sequence-to-sequence-models. Accessed Apr. 1, 2023.

58 H. Li, Y.C. Wang, *et al,* "An Augmented Transformer Architecture for Natural Language Generation Tasks", in 2019 International Conference on Data Mining Workshops (ICDMW), 2019, https://doi.org/10.1109/ ICDMW48858.2019.9024754.

59 I. Sutskever, O. Vinyals, Q. V. Le, "Sequence to Sequence Learning with Neural Networks", in Proceedings of NIPS'14, 2014.

60 N. Kalchbrenner, P. Blunsom, "Recurrent Continuous Translation Models", in Proceedings of EMNLP'13, 2013.

61 S. Y. Su, K. L. Lo, *et al*, "Natural Language Generation by Hierarchical Decoding with Linguistic Patterns", in NAACL-HLT 2018, arXiv:1808.02747v2, 2018, https://doi.org/10.48550/arXiv.1808.02747.

62 D. Bahdanau, K. Cho, Y. Bengio, "Neural Machine Translation by Jointly Learning to Align and Translate", in Proceedings of ICLR'15, 2015.

63 Y. Kim, C. Denton, *et al*, "Structured Attention Networks", in International Conference on Learning Representations, 2017.

64 A. Parikh, O. Täckström, *et al*, "A Decomposable Attention Model for Natural Language Inference", in Proceedings of the 2016 Conference on Empirical Methods in Natural Language Processing, pp. 2249–2255, Austin, Texas. Association for Computational Linguistics, 2016.

65 A. Reuther, P. Michaleas, *et al*, "Survey of Machine Learning Accelerators", 2020, arXiv:2009.00993, https://arxiv.org/abs/2009.00993.

66 G. Wetzstein, A. Ozcan, *et al.*, "Inference in Artificial Intelligence with Deep Optics and Photonics", *Nature*. 2020; 588(7836): 39–47, https://doi.org/10.1038/s41586-020-2973-6. ISSN: 1476-4687.

67 A. Sebastian, M. L. Gallo, *et al.*, "Memory Devices and Applications for In-Memory Computing", *Nat. Nanotechnol.* 2020; 15(7): 529–544, https://doi.org/10.1038/s41565-020-0655-z.

68 M. A. Nahmias, T. F. De Lima, *et al.*, "Photonic Multiply-Accumulate Operations for Neural Networks", *IEEE J. Select. Top. Quant. Electron.* 2020; 26: 1–18, https://doi.org/10.1109/JSTQE.2019.2941485.

69 P. Stark, F. Horst, *et al.*, "Opportunities for Integrated Photonic Neural Networks", *Nanophotonics.* 2020; 9(13): 4221–4232, https://doi.org/10.1515/nanoph-2020-0297.

70 C. Huang, V. J. Sorger, *et al.*, "Prospects and Applications of Photonic Neural Networks", *Adv. Phys. X.* 2021; 7(1), https://doi.org/10.1080/23746149.2021.1981155.

71 B. J. Shastri, A. N. Tait, *et al.*, "Photonics for Artificial Intelligence and Neuromorphic Computing", *Nature Photon.* 2021; 15(2): 102–114, https://doi.org/10.1038/s41566-020-00754-y.

72 D. Dua, C. N. Dos Santos, C. N. Zhou, Generative Adversarial Network Based Modeling of Text for Natural Language Processing, U.S. Patent 11,281,976, Mar. 22, 2022. Uncopyrighted material.

73 I. J. Goodfellow, J. P. Abadie, *et al*, "Generative Adversarial Nets", in Advances In Neural Information Processing Systems (NIPS 2014), Dec. 8–13, 2014.

74 A. Radford, L. Metz, S. Chintala, "Unsupervised Representation Learning with Deep Convolutional Generative Adversarial Networks", Cornell University, Nov. 19, 2015, arXiv: 1511.06434v1. Accessed Dec. 30, 2017.

75 E. Reiter, R. Dale, "Building Applied Natural Language Generation Systems", *Nat. Lang. Eng.* 1997; 3(1): 57–87.

76 J. A. Bateman, M. Zock, "Natural Language Generation", in *The Oxford Handbook of Computational Linguistics*, R. Mitkov, (Editor), Oxford University Press, Oxford, UK, 2005.

77 L. Wanner, "Report Generation", in *Handbook of Natural Language Processing*, 2nd Edition, N. Indurkhya, F. Damerau, (Editors), Chapman and Hall/CRC, London, 2010.

78 B. Di Eugenio, N. Green, "Emerging Applications of Natural Language Generation in Information Visualization, Education, and Health-Care", in *Handbook of Natural Language Processing*, 2nd Edition, N. Indurkhya, F. Damerau, (Editors), Chapman and Hall/CRC, London, 2010.

79 E. Krahmer, M. Theune, *Empirical Methods in Natural Language Generation*, Springer, Berlin & Heidelberg, 2010.

80 S. Bangalore, A. Stent, *Natural Language Generation in Interactive Systems*, Cambridge University Press, 2014.

81 R. J. Williams, D. Zipser, "A Learning Algorithm for Continually Running Fully Recurrent Neural Networks", *Neural Comput.* 1989; 1(2): 270–280.

82 K. Cho, B. van Merrienboer, *et al*, "Learning Phrase Representations Using RNN Encoder-Decoder for Statistical Machine Translation", in Proceedings of the 2014 Conference on Empirical Methods in Natural Language Processing (EMNLP), 2014, pp. 1724–1734, Doha, Qatar. Association for Computational Linguistics.

83 S. Hochreiter, Y. Bengio, *et al.*, "Gradient Flow in Recurrent Nets: The Difficulty of Learning Long-Term Dependencies", in *A Field Guide to Dynamical Recurrent Neural Networks*, S. C. Kremer, J. F. Kolen, (Editors), IEEE Press, 2001.

84 K. Deemter, M. Theune, *et al.*, "Real Versus Template-Based Natural Language Generation: A False Opposition?", *Comput. Linguist.* 2005; 31(1): 15–24, https://doi.org/10.1162/0891201053630291.

85 M. H. D. Y. Gunasiri, "Automated Cricket News Generation in Sri Lankan Style Using Natural Language Generation", Thesis at University of Colombo School of Computing, 2019.

86 C. Mellish, D. Scott, *et al.*, "A Reference Architecture for Natural Language Generation Systems", *Nat. Lang. Eng.* 2006; 12(1): 1.

87 A. Gatt, E. Krahmer, "Survey of the State of the Art in Natural Language Generation: Core Tasks, Applications and Evaluation", *J. Artif. Intell. Res.* 2018; 61: 65–170.

88 J. Kuczmarski, V. Jain, *et al*, Facilitating Communications with Automated Assistants in Multiple Languages, U.S. Patent 11,354,521, June 7, 2022. Uncopyrighted material.

89 S. A. Teserra, Using Semantic Frames for Intent Classification. U.S. Patent 11,538,468, Dec. 27, 2022. Uncopyrighted material.

90 S. M. Kelly, "This AI Chatbot is Dominating Social Media with Its Frighteningly Good Essays", CNN Business, Dec. 5, 2022, https://www.cnn.com/2022/12/05/tech/chatgpt-trnd/index.html. Accessed Feb. 8, 2023.

91 A. Sharma, "OpenAI's New ChatGPT Bot: 10 Dangerous Things It's Capable Of", Dec. 6, 2022, https://www.bleepingcomputer.com/news/technology/openais-new-chatgpt-bot-10-dangerous-things-its-capable-of. Accessed Feb. 10, 2023.

92 B. Cost, "Rise of the Bots: 'Scary' AI ChatGPT Could Eliminate Google Within 2 Years", New York Post, Dec. 6, 2022, https://nypost.com/2022/12/06/scary-chatgpt-could-render-google-obsolete-in-two-years. Accessed Feb. 12, 2023.

93 N. Grant, "Google Calls in Help from Larry Page and Sergey Brin for A.I. Fight", New York Times, Jan. 20, 2023.

94 Reuters Staff, "Italy Curbs ChatGPT, Starts Probe Over Privacy Concerns", Apr. 1 2023, https://www.cnbc.com/2023/04/01/italy-curbs-chatgpt-starts-probe-over-privacy-concerns.html. Accessed Apr. 2, 2023.

95 S. Schechner, "Microsoft Plans to Build OpenAI, ChatGPT Features Into All Products", Wall Street Journal, Jan. 17, 2023.

96 R. Klar, "Microsoft Investing Billions in ChatGPT Maker", The Hill, 23 Jan. 2023, https://thehill.com/policy/technology/3826573-microsoft-investing-billions-in-chatgpt-maker. Accessed Feb. 14, 2023.

97 A. Lowrey, "How ChatGPT Will Destabilize White-Collar Work", The Atlantic, Jan. 20, 2023, https://www.theatlantic.com/ideas/archive/2023/01/chatgpt-ai-economy-automation-jobs/672767. Accessed Feb. 16, 2023.

98 D. Milmo, "Google AI Chatbot Bard Sends Shares Plummeting After It Gives Wrong Answer", The Guardian, Feb. 8, 2023, www.theguardian.com. Accessed Feb. 18, 2023.

99 S. Heritage "Can an AI Program Really Write a Good Movie? Here's a Test", The Guardian, Mar. 24, 2023, https://www.theguardian.com/film/2023/mar/24/chapgpt-movie-script-ai. Accessed Feb. 22, 2023.

100 T. Rees, "ChatGPT Opens Door to Four-Day Week, Says Nobel Prize Winner", Financial Review, Apr. 5, 2023, https://www.afr.com/technology/chatgpt-opens-door-to-four-day-week-says-nobel-prize-winner-20230406-p5cyki. Accessed Apr. 12, 2023.

101 O'Flaherty, K., "Cybercrime: Be Careful What You Tell Your Chatbot Helper...", The Guardian, Apr. 9, 2023, https://www.theguardian.com/technology/2023/apr/09/cybercrime-chatbot-privacy-security-helper-chatgpt-google-bard-microsoft-bing-chat. Accessed Apr. 10, 2023.

102 Subin, S., "Piper Sandler Calls Generative A.I. the Next $100 Billion Tech Revolution, Names Stocks to Play It", CNBC, Apr. 14, 2023, https://www.cnbc.com/2023/04/14/piper-sandler-calls-generative-ai-the-next-100-billion-tech-revolution.html. Accessed Apr. 14, 2023.

103 H. Wang, T. Wang, *et al*, Complex Natural Language Processing. U.S. Patent 11,398,226, July 26, 2022. Uncopyrighted material.

104 H. E. Cheng, J. Jian, System and Method for Natural Language Processing, U.S. Patent 10,719,507, July 2020. Uncopyrighted material.

105 J. Xu, S. R. Jonnalagadda, G. Durrett, "Massive-Scale Decoding for Text Generation Using Lattices", in Proceedings of the 2022 Conference of the North American Chapter of the Association for Computational Linguistics: Human Language Technologies, pp. 4659–4676, July 10–15, 2022.

106 J. Chorowski, D. Bahdanau, *et al*, "Attention-Based Models for Speech Recognition", 2015, arXiv: 1506.07503.

107 J. W. Rae, A. Potapenko, *et al*, "Compressive Transformers for Long-Range Sequence Modelling", June 13, 2019, https://arxiv.org/pdf/1911.05507v1.pdf.

108 R. Shevde, "How Generative AI is affecting Graphic Design in 2023", Mar 01 2023, available online at https://designwizard.com/blog/how-generative-ai-is-affecting-graphic-design/, accessed June 1, 2023.

109 P. Esser, J. Chiu, *et al*, "Structure and Content-Guided Video Synthesis with Diffusion Models", https://research.runwayml.com/gen1, arXiv:2302.03011v1 [cs.CV] 6 Feb 2023, https://doi.org/10.48550/arXiv.2302.03011.

3

Current and Evolving Applications to Speech Processing

3.1 Scope

Automatic speech recognition (ASR) and text-to-speech (TTS) synthesis have been around for several decades. ASR maps a speech signal to the corresponding sequence of textual words; it deals with the generation of the most probable linguistic sequence given an input acoustic sequence. As discussed in Chapter 2, TTS is often used in conjunction with natural language processing (NLP); however, it can exist on its own.

Although ASR and TTS deal with basic voice processing, including traditional vocoding mechanisms, artificial intelligence (AI), especially machine learning (ML)/deep learning (DL) techniques, is increasingly being used to enhance the accuracy of recognition in ASR, and to enhance sentence fluency and intonation in TTS. ASR performance has improved significantly over time – e.g. as measured by the word error rate (WER) – especially with the use of DL techniques and in close-field environments (where the microphone is in the immediate proximity of the speaker). The last decade has also seen major breakthroughs in speech synthesis by applying neural network (NN) methods and end-to-end modeling.

In an NN, neurons are chained together in a defined logical structure. A dense neural network (DNN) uses a hierarchical cascade of several layers of processing units; each successive layer uses the output from the previous layer as input. There are several different ways of using DNNs to generate robust speech features in ASR. As noted in Chapter 1, a neuron (cell) combines the inputs it receives with a set of coefficients, or weights, that either dampen or amplify the inputs. These weighted input signals are summed and the resulting value is passed through the node's activation function to determine if, and to what extent, that signal should progress further through the network to affect the ultimate outcome (this being

AI Applications to Communications and Information Technologies: The Role of Ultra Deep Neural Networks, First Edition. Daniel Minoli and Benedict Occhiogrosso.
© 2024 The Institute of Electrical and Electronics Engineers, Inc.
Published 2024 by John Wiley & Sons, Inc.

Table 3.1 Cascaded TTS operation.

	Current/typical cascaded TTS systems
Intermediate representation	Mel spectrogram
Objective function	Continuous signal regression
Training data	Hundreds of hours, e.g. ~600 hours

similar to neurons in the human brain that fire when they register sufficient stimuli). To analyze the incoming data and generate assessments (e.g. recognize units of speech), NNs/DNNs utilize extracted features; a feature is an individual measurable property of a (physical) process being observed – the concept of feature is similar or related to the concept of a variable used in linear regression (LR).

ASR classically entails an architecture that has a number of cascaded sequential layers, for example preprocessing, feature extraction, optimization, prediction, and decision making; for each layer, many distinct algorithms are typically used. *End-to-end (E2E) learning*, on the other hand, encompasses training a complex learning system represented by a single model (specifically by a DNN), rather than a *cascaded arrangement*, that encompasses the complete system under consideration and bypasses the intermediate layers typically present in traditional pipeline designs.

Currently, TTS systems also typically utilize a cascaded pipeline with an acoustic model (AM) and a vocoder utilizing mel spectrograms as the intermediate steps. Table 3.1 depicts this arrangement. Traditional TTS systems do not support in-context learning. In-context learning is an emergent behavior in language models (LMs), where the LM performs a task by conditioning exclusively on input-output examples without attempting to optimize any parameters. It is used in the "few-shot task learning" environment. In-context learning was part of the original GPT-3 design (highlighted in Chapter 2) as a process to use LMs to learn tasks when given only a few examples. During in-context learning, one gives the LM a prompt that consists of a list of input-output pairs that demonstrate a task; at the end of the prompt, one appends a test input and allows the LM to make a prediction just by conditioning on the prompt and predicting the next tokens [1]. Although advanced TTS systems can synthesize high-quality speech from single or multiple speakers, it still requires high-quality clean acoustic data (e.g. from the recording studio) [2]. Given that the training data is relatively small, current TTS systems suffer from poor generalization: speaker similarity and speech naturalness decline significantly *in the zero-shot scenario*.

The ASR/TTS topic is consistent with the scope of this text because (i) some of the elements could be implemented or realized across a network (as a cloud service), (ii) some systems entail verbal responses that are in the voice domain, and (iii) the field deals with the broader concept of "communications." There is a large body of research on this topic and new methods or improvement of existing methods are routinely brought forth by researchers. This chapter aims at covering only the broad technologies in play and not be a highly technical review of the "breaking news" advancements and perennially novel proposals in the field.

The Appendix at the end of this chapter provides a basic (non-exhaustive) glossary of key AI terms and concepts covered in this chapter.

3.2 Overview

ASR entails methods for converting the acoustic speech waveform into a sequence of textual words. This is traditionally done using statistical modeling of the speech signal. Techniques entail acoustic-phonetic modeling, lexical representation, language modeling (using LMs), decoding, and model adaptation. Speaker independence, large vocabulary, and continuous speech recognition are typical desiderata. ASR can be used in a wakeword context or in a more general context of continuous (real-time or post-storage) content processing. A well-known example of a wakeword/digital assistant is "Alexa."

3.2.1 Traditional Approaches

Typically, in an ASR context, some appropriate microphone captures the audio. The next step is to determine whether speech is detected. This is done using various mechanisms such as linear classification, support vector machines (SVMs), and decision trees. Thereafter, Gaussian mixture model (GMM) or hidden Markov model (HMM) techniques are (have been) utilized to compare the audio data to one or more AMs in storage – the AMs may include models corresponding to silence, noise, or speech (noise may include environmental noise or background noise). Punctuating this point, ASRs traditionally utilized the pattern matching approach, with the classifier commonly being based on HMM techniques. In training, a sequence of observations is provided and the objective is to find a model that provides the best fit of the observation data. In a wakeword environment, a wakeword detection component may be utilized to determine when the user intends to actually speak an input stream to the system.

Much of the large-vocabulary speech recognition is still based on models such as HMMs, tree lexicons, or N-gram LMs that can be represented by weighted finite-state transducers (WFSTs).[1] The HMM has been one of the most commonly

---

1 A finite-state transducer (FST) is a finite automaton whose state transitions are labeled with both input and output symbols. A path through the transducer encodes a mapping from an input symbol sequence to an output symbol sequence. A weighted finite-state transducer (WFST) places weights on transitions in addition to the input and output symbols. Weights may encode probabilities, durations, penalties, or any other quantity that accumulates along paths to compute the overall weight of mapping an input sequence to an output sequence. Weighted transducers are a good choice to represent the probabilistic finite-state models prevalent in speech processing [3].

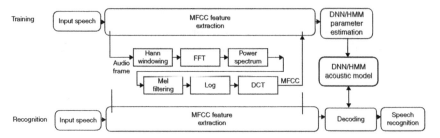

Figure 3.1   Basic ASR system. MFCC, Mel Frequency Cepstral Coefficients; FFT, Fast Fourier Transform; DTC, Discrete Cosine Transform; HMM, Hidden Markov Model.

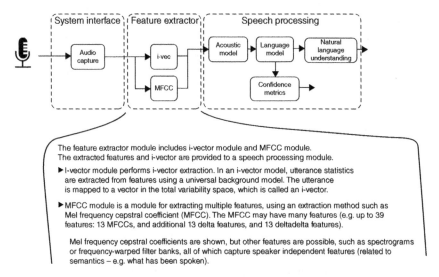

Figure 3.2   Voice processing system example. *Source:* Adapted from Ref. [4].

used approaches for ASR; however, accelerated ML techniques using DNNs have enabled, in the recent past, the gradual replacement of HMMs and GMMs which, as stated, were heretofore widely used for ASR.

Figure 3.1 depicts such a basic ASR system. In this figure, one can see an extraction module followed by an HMM/DNN module. Figure 3.2 depicts a typical example of a feature extraction module.

Feature vectors extracted from a timeframe of the audio recording can be represented as a fixed-dimension vector (e.g. one-second timeframe of audio can have 1,000 fixed-dimension vectors). Such vectors can be represented using mel frequency cepstral coefficients (MFCCs). MFCCs provide a representation of speech samples based on frequency, for example, from a cepstral representation of an audio clip such that the frequency bands are equally spaced on the mel scale.[2] The mel frequency cepstrum (MFC) is a representation of the short-term power spectrum of an acoustic clip based on a linear cosine transform of a log power spectrum on a nonlinear scale of frequency. The coefficients, the MFCCs of an acoustic clip, define a small set of features (usually about 10–20) that concisely describe the shape of a spectral envelope. The MFCC set makes up an MFC in the aggregate. The cepstral representation of the acoustic clip can be perceived as a "spectrum of a spectrum."

HMM is a model where the system under consideration is assumed to be a Markov process X with hidden, but unobservable, states. The assumptions embedded in HMM are: (i) there is an observable process Y whose outcomes are "influenced" or "impacted " by the outcomes of X in some known manner; (ii) the outcome of Y at time t_i must be "influenced" only by the outcome of X at time t_i; and (iii) the outcomes of X and Y at $t < t_i$ must not affect the outcome of Y at t_i. Since X cannot be observed directly, the goal is to learn about behaviors in X by observing behaviors in Y. Notice that while the model states are hidden, the parameters of the model are not hidden; namely, while the model state may be hidden, the state-dependent output of the model is visible. Information about the state of the model can be inferred from the probability distribution over possible output tokens given the fact that each model state creates a different distribution.

Thus, the HMM is a statistical Markov model where the system being modeled is assumed to be a Markov process. The Markov process is characterized by a grouped sequence of internal states. The transition among these states occurs based on a given probability of a transition; the aggregate of all the probabilities can be represented as a transition matrix. HMMs are a kind of Bayesian network constructed by combining a hidden Markov layer and a second layer of outputs, where the second layer depends probabilistically on the hidden states of the first layer [4]. For example, there could be just two hidden states supporting the basic division among vowels and consonants. When more states are included, the HMM can model (discriminate) the initial and final letters of a word, vowel followers and preceders, phones and phonemes, and so on.

Recall that ASR answers the question "what did he/she say?"; NLP answers the question "what did he/she mean?". In some cases, the ASR function suffices; in other cases, the ASR needs to be complemented by NLP functions. Comprehensive speech recognition is implemented by using ASR to process speech (e.g. one or more utterances) and then using an LM and/or NLP to interpret and/or understand the essence of the processed speech.

In the context of the former (ASR only), a typical ASR system includes a front-end speech pre-processor that extracts representative features from the speech

---

2 The mel scale approximates the human auditory system's response.

input. Typically, the *front-end speech pre-processor* performs a Fourier transform on the speech input to extract spectral features that characterize the speech input as a sequence of representative multi-dimensional vectors. An ASR system may then include one or more *speech recognition models* (e.g. AMs and/or LMs) and implement one or more speech recognition engines (e.g. HMMs, N-gram LMs, and the newer GMMs, DNNs). The speech recognition engines are used to process the extracted representative features to generate preliminary recognitions results (e.g. phonemes, sub-words), and then the final text recognition results (e.g. words, word strings, or sequence of tokens). In the context of the latter (ASR + NLP) the recognized text string result can be passed to NLP module for intent deduction as a sequence of words or tokens corresponding to the speech input. A speech recognition confidence score may be generated enabling the system to rank the candidate text outputs.

Just focusing here initially on wakewords, wakeword detection is typically undertaken without performing linguistic or semantic analysis. Instead, the audio data is analyzed to determine if specific characteristics of the audio data match preconfigured acoustic waveforms or audio signatures; namely, the audio data is analyzed to determine if the audio data "matches" pre-stored data that corresponds to a wakeword. One approach is to utilize a general large vocabulary and continuous speech recognition systems to decode audio signals. Another approach utilizes one or more HMMs for each wakeword and non-wakeword speech signals, respectively (non-wakeword speech includes other spoken words, background noise, and so on). Viterbi decoding can be used to search the best path in the decoding graph, and the decoding output is then processed to make a decision regarding wakeword presence or lack thereof. This approach can be extended to include discriminative information by including a hybrid DNN-HMM decoding framework. In some implementations, the wakeword detection component may utilize DNN/recurrent neural network (RNN) structures directly without HMM mechanisms; such an architecture may estimate the posterior threshold of wakewords with context data, either by stacking frames within a context window for the DNN or using the RNN; follow-on posterior threshold tuning or smoothing can be applied for decision making [5]. Once the wakeword detection component detects a wakeword, the device may "wake up" and begin transmitting audio data to full ASR system (either locally-based or in the cloud) and/or to an NLP system (e.g. as one discussed in Chapter 2). The ASR component transcribes the audio data into ASR output data; such output data may include one or more ASR hypotheses.[3] However, the keywork recognition is typically limited to one vocal

---

3 The ASR output data may include one or more ASR hypotheses in the form of one or more textual interpretations or one or more tokens. Each ASR hypothesis could be associated with a score representing a confidence of ASR processing performed to generate the ASR hypothesis with which the score is associated.

word or small phrase, e.g. it may only recognize the word "Hi Bixby." See Figure 3.3 for an example of a wakeword-based system.

More generally, Figure 3.4, synthetized from [7], depicts an NN-based example of a vocal command ASR system that enables users to interact directly with a device, such as a smartphone, to translate voice commands into actionable textual commands. As seen in the figure, the typical functional elements of an ASR system are *feature extraction* (in a front end) and *inference* (in a back end).

In an ASR process a series (set) of acoustic features are extracted from the acoustic speech signal. In the front end, one of the first steps aims at extracting a short-time representation of the spectral envelope for each speech frame. As illustrated in Figure 3.4, to generate the MFCC matrix the audio frames may be processed through the Hann windowing operation, the fast Fourier transform (FFT) operation, the power spectrum operation, the mel filtering operation, the log operation, and the discrete cosine transform (DCT) operation.

3.2.2 DNN-based Feature Extraction

DNN models have come to the forefront of ASR in recent years and now constitute the state-of-the-art in the field; in particular, one has seen in the recent past a growing popularity of DNNs in conjunction with HMMs; ML-only (E2E) systems are also emerging.

Until recently, most ASR systems in the market utilized a frame-based model where an input waveform was converted into a sequence of frames of distilled features. The goal of feature extraction is to represent a window of speech samples with a *feature vector* that captures (to the degree possible) the intrinsic phonetic content of the speech. Conventional preprocessed methods such as MFCC, perceptual linear predictive (PLP)[4] and mel filter bank are common, being perceived to be fairly robust and being based on physiological models of the human auditory system. As implied above, most conventional ASR systems to date utilize traditional features such as log mel filter banks; these features are low dimensional representations of the speech signal and they preserve the information required to achieve high recognition performance. The mel features are derived by element-wise multiplication of the magnitude spectrum with positive mel filter weights followed by pooling [9].

In recent years, various studies have proposed different methods for DNN-based feature extraction and feature learning from a raw waveform for large vocabulary

---

4 The PLP technique uses three concepts from the psychophysics of hearing to derive an estimate of the auditory spectrum: (i) the critical-band spectral resolution, (ii) the equal-loudness curve, and (iii) the intensity-loudness power law. The auditory spectrum is then approximated by an autoregressive all-pole model [8].

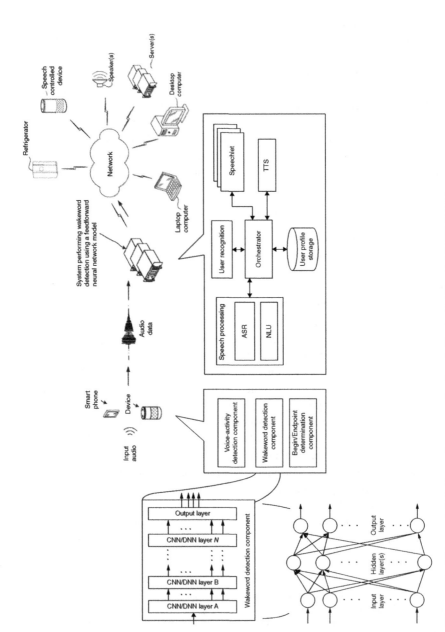

Figure 3.3 Wakeword detection using a neural network. *Source:* Reference [6]/U.S. Patent.

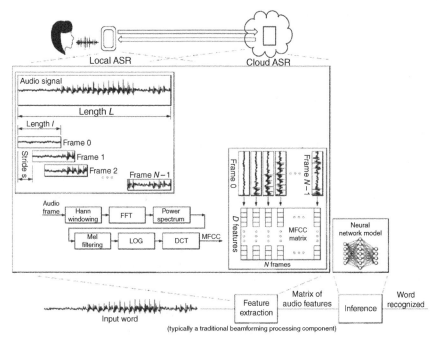

Processing includes dividing audio data into frames extracting a mel frequency cepstral coefficient (MFCC) from the frames. A feature extraction module segments an audio data into a plurality of *N* audio frames (labelled as frames 0, 1, 2, ... , and *N* − 1). An audio frame may have an overlapping portion with its immediate succeeding audio frame. The non-overlapping portion of the audio frame is called a stride. From the *N* audio frames, the feature extraction module generates a DxN mel frequency cepstral coefficient (MFCC) matrix. Each element in the MFCC matrix is a coefficient representing an audio feature. A different column of the MFCC matrix corresponds to a different audio frame. A different row of the MFCC matrix representing a different type of audio features. For example, the first row of the MFCC matrix includes coefficients representing the power features of the audio frames; thus, the first element of the first column of the MFCC matrix (MFCC[0][0]) is a coefficient representing the power feature of the first audio frame (frame 0). The extracted features (i.e. the MFCC matrix) are the input to the neural network model of the inference unit, which comprises a neural network. The inference unit outputs the word recognized.

Figure 3.4 ASR elements: feature extraction and inference. *Source:* Adapted from Ref. [7].

speech recognition, with raw time signal directly as input to a DNN; however, the raw waveform methods do not *per se* outperform the conventional methods.

As described in [7], the listening device (i) receives audio data containing speech and (ii) performs a keyword-spotting operation to find a keyword or wake-word in a frame of audio data. If/when a keyword is detected, the listening device may then connect to a cloud network, such as the Internet or other commercial cloud, to perform more extensive ASR functionality on the audio data to detect commands or other material. By connecting to a cloud, the listening device can take advantage of the cloud's resources to perform more extensive ASR functionality, given that ASR requires fast, complex processing and large memory resources. Detecting a keyword from the audio data can be done with a local NN with adequate, but somewhat limited, edge-based processing capabilities; after the keyword is identified by the local NN, the cloud-based NN can be invoked to bring to bear its more powerful level of processing.

The audio data needs to be processed before it is usable by the local NN (and possibly later by the cloud-based NN). Local processing includes segmenting audio data into frames to generate a MFCC matrix from the frames [10]. The MFCC matrix is used by the feature extraction module to extract pertinent audio features; the extracted features are then sent to the local NN. The local NN is trained to identify the first syllable of a keyword (say "Hi") and the second syllable of the keyword (say "Bixby"). For each frame of audio data, the NN produces outputs, by inference, related to the probability that the respective syllables are present in the audio frame. After a user vocalizes the keyword, the listening device will typically initiate an information exchange with a cloud to recognize commands or other linguistic material from audio data.

The local NN – say, an RNN such as a long short-term memory (LSTM), for example, a 11-node, 1-layer LTSM network followed by a multilayer perceptron (MLP) – may be trained to search the first part of an audio frame for the syllable "Hi" and then search for the second part of an audio frame to detect the syllable "Bixby." Supervised training may be used to train the NN.

The NN may output a probability that an audio sample includes a keyword. The probability is determined based on the inputs to the NN and weighting relationships between internal nodes of the NN.[5] During training, the probability is fed back to the recognition device which adjusts the internal weighting of the NN depending on the accuracy of the outputs in view of the identified audio samples. This training process may continue until the NN is able to satisfactorily identify the desired sounds. After the keyword is identified, the audio data that follows may be sent to the cloud where feature extraction capabilities are also employed to compute appropriate MFCC matrices, which are then processed by a more sophisticated cloud-resident NN (say, for example, a 64-node, 6-layer LTSM NN plus a 3-node MLP). (Refer to [7], if desired, for additional details.)

3.3 Noise Cancellation

In practical terms, noisy and reverberant environments are typical in business settings (and home applications). Figure 3.5 depicts this situation graphically. Although, as noted, recent advances in DL have led to performance improvement

---

5 In addition, a confidence score may be calculated based on the outputs of the local NN: probabilities output by the local NN may be compared with thresholds to determine whether the probability is considered to be high or low – in some cases, the confidence score may just entail a pass or fail scheme, where a pass indicates that the audio data meets a minimum threshold to conclude that a keyword has been uttered.

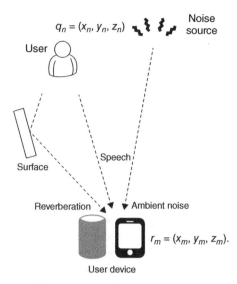

An array of M sensing microphones may be located positions $r_m = (x_m, y_m, z_m)$.

The sound field is modeled as the superposition of the wave fields generated by N acoustic sources located at $q_n = (x_n, y_n, z_n)$.

The time-domain samples of each microphone are segmented into frames of K samples, and each frame is converted to the frequency domain using the Discrete Fourier Transform (DFT).

When additive noise is present, the $M \times 1$ output vector for a single frequency ω_k on a single frame can be modeled as $x(\omega_k) = V(\omega_k)y(\omega_k) + \eta(\omega_k)$, here $y(\omega_k) = [y_0(\omega_k)\, y_1(\omega_k), \dots , y_{N-1}(\omega_k)]^T$ represents the source signals in the frequency domain and $\eta(\omega_k)$ represents additive noise also in frequency-domain.

The $M \times N$ array matrix $V(\omega_k) = [v(q_0,\omega_k)\, v(q_1,\omega_k), \dots , v(q_{N-1},\omega_k)]$ defines the transfer function between source n and sensor m at frequency ω_k.

The information on the propagation medium and the characteristics of the propagating wave (i.e., free-field or reverberant conditions, etc.) is encoded in the vector $v(q_n,\omega_k)$ (also called the steering vector).

Figure 3.5   Distant-talking speech recognition environment.

in ASR, ASR in noisy and reverberant environments remains somewhat challenging. The performance in noisy environments and in far-field scenarios (where the microphone is not in the immediate proximity of the speaker) (also known as distant-talking speech recognition) is still significantly inferior to the performance in clean and close-field speech situations because of unwanted noise and signal reflections. Basic ASR techniques are degraded or unable to accurately extract text data corresponding to speech utterances in the presence of ambient noise (e.g. heating systems, fans, automotive traffic, background sound systems such as TVs), or reflection noise (reverberations). One can model the observed signal $\mathbf{X}$, the Signal of Interest (SOI), in the short-time Fourier transform (STFT) domain as the comingling of the target image $\mathbf{Y}$ and a distortion $\mathbf{N}$ overlaid by noise sources or by reverberation effects: $X(t, f) = Y(t, f) + N(t, f)$, where t and f denote time frame indexes and the frequency, respectively. If the reverberation and noise are not cancelled in front of the ASR-capable devices, then speech analysis will be degraded.

3.3.1 Approaches

Noise cancellation mechanisms are typically used to deal with noise. Signal impairments and distortions due to noise can be addressed with a dereverberating and/or de-noising *front-end*, although there are also a number of *back-end* approaches, including back-ends relying on DNNs. Long-established front-end techniques are based on spatial speech processing and *separation techniques* such as, but not limited to, Non-negative Matrix Factorization (NMF).

Two types of filters can be used for speech enhancement in ASR: *data-independent multichannel filters* and *data-dependent multichannel filters*; the latter relies on the estimation of the statistics of the noisy signal. The multichannel Wiener filter (MWF) is an example of a data-dependent multichannel filter, which aims at optimizing the mean squared error (MSE); an enhancement, the speech distortion weighted multichannel Wiener filter (SDW-MWF), offers a trade-off between the noise reduction and the speech distortion. Typically, multichannel filters have been developed for environments with constrained microphone arrays: the positions and number of microphones are fixed, and a so-called *fusion center* gathers all the signals of the microphone array [12]. The centralized MWF aims at estimating the speech component s_i of the i-th sensor of the microphone array. Computing the solution matrix requires the knowledge of noise-only periods and speech-plus-noise periods. This is typically obtained with a voice activity detector (VAD)[6] or a time-frequency (TF) mask.

Various front-end methods include *denoising autoencoders* and *beamforming* [13, 14].

- *Traditional auto-encoders* consist of an encoder and a decoder; the encoder is a deterministic mapping that transforms an nth dimensional input vector x of the observed signal into a hidden representation w. The resulting hidden representation w is then mapped back to a reconstructed vector z in the input space.[7]

---

6 A VAD utilizes raw audio data values and features such as, but not limited to frequency, energy, zero-crossing rate, to assess threshold values to determine if speech is present.

7 z is not an exact reconstruction of x, but a set of parameters of a probability distribution that may generate x with high probability.

A denoising autoencoder (DA) is a variant of the traditional autoencoder. Here, the DA is trained to reconstruct a clean input x from a corrupted version of it; the corrupted input $x\sim$ is mapped, as is the case with the basic autoencoder, to a hidden representation w from which one reconstructs a new vector z (the DA is trained to minimize the average reconstruction error over a clean training set to have z as close as possible to the uncorrupted input x) [15]. By adjoining multiple layers of encoders and decoders, the DA becomes a deep denoising autoencoder (DDA).

- The *beamformer* method can be used to dereverberate and denoise the SOI. This method utilizes spatial diversity to its advantage to extend the classic TF filtering. To improve performance of actual field speech, microphone arrays can be utilized to enhance the speech signal and eliminate (or minimize) unwanted noise and signal reflections. A generic, common approach for beamforming design is to seek to maximize the signal-to-noise ratio (SNR) by selecting parameters that maximize the receive signal power, while minimizing the interference noise; other approaches include, but are not limited to, the least mean squares (LMS) error method and maximum likelihood method.

Acoustic beamforming has been proposed and used as a front-end processing technique for ASR for a number of years. Beamformers entail multichannel signal processing: Beamforming techniques are commonly applied to signals captured by sensor arrays to enhance signals received from desired directions while reducing background noise and localized interference. Usually, beamforming is applied to digital discrete-time signals. In situations where the directions of the desired source signal and interfering source signals are known, this knowledge, combined with assumptions on the background noise characteristics, can be used to derive the beamformer coefficients for each sensor [16–18]. An acoustic beamformer can be perceived as a multiple-input single-output linear time-invariant filter. Acoustic beamforming has progressed in recent years, including the use of new objective functions, such as the MWF, and the consideration of arbitrary acoustic transfer functions (ATFs) from the speech source to the microphones [19].

3.3.1.1 Delay-and-Sum Beamforming (DSB)
A common multichannel signal processing method to deal with noise is *delay-and-sum beamforming* (DSB). Here, signals from different sources are aligned in time to adjust for the propagation delay from the speaker to each individual microphone; these signals are then mixed to a single channel: this mixing has the effect of enhancing the signal from the desired direction and attenuating noise coming from other directions. The challenge, however, is that it is difficult to precisely estimate the time delay of arrival in environments where the signal encounters many reflections, leading to reverberations; furthermore, DSB does not take

into account the effect of spatially-correlated noise. At low acoustical frequencies, signals arriving from neighboring directions exhibit very small phase differences across the microphones; as a consequence, at lower frequencies, the DSB has poor angle discrimination. Fortunately, this improves as the frequency increases, resulting in a narrower and sharper mainlobe.

Performance can be improved by using *filter-and-sum* (FS) techniques, where a linear filter is applied to each channel before summing [20]. (In geometrically-based microphone array configurations, the key difference between the signals of the individual channels is their phases, much less in their amplitudes – however, in real-life situations such arrangements do not really exist.) This solution entails processing the signal in the frequency domain, which is what a "conventional beamformer" does (this beamformer is sometimes also known as the Bartlett beamformer). The solution works reasonably well for situations with compact microphone arrays. In a DSB implementation, the delays are discrete and, thus, it is challenging to create a perfect alignment between the channels resulting in quantization errors; the conventional beamformer solves the quantization problem because shifting the signal in time is no longer restricted to discrete steps; however, this beamformer suffers from spatial aliasing given that at certain frequencies the signals from different directions (microphones) have the same phase differences [16].

3.3.1.2 Minimum Variance Distortionless Response (MVDR) Beamformer

Instead of maximizing the signal from the target direction, the Minimum Variance Distortionless Response (MVDR) uses the statistics of the noise field to minimize the energy from directions outside of the target direction; this is done by approximating or modeling the background noise information. The *MVDR beamformer* requires and uses information about the background noise. An MVDR beamformer is able to reduce the levels of the sidelobes in comparison to DSB, while also narrowing down the mainlobe. Nonetheless, the MVDR still exhibits a broader mainlobe at low frequencies (but which, again, narrows with increasing frequency). Some improvements have been proposed (e.g. but limited to [21, 22]).

As just noted, a well-known beamformer approach is to base it on the MVDR principles (this method is also used for radar antennas and ultrasound applications). The MVDR beamformer (with MVDR filters as a front end) is a data-adaptive beamforming approach whose goal is to minimize the variance of the beamformer output. If the noise and the underlying desired signal are uncorrelated, as is typically the case, the variance of the captured signals is the sum of the variances of the desired signal and the noise; therefore, the MVDR approach seeks to minimize this sum, thus mitigating the effect of the noise. The MVDR beampattern, however, is strongly dependent on the number of microphones and the length of the attached finite impulse response (*FIR*) filters [23]. See Figure 3.6 for an example. Using only

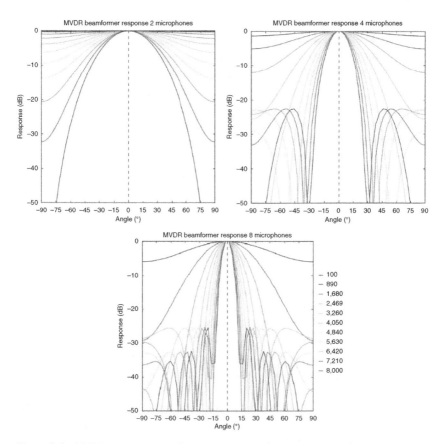

Figure 3.6 MVDR beampatterns. Patterns are strongly dependent on the number of microphones and the length of the attached FIR filters. This beamformer is a uniform linear array with omnidirectional elements, the sampling rate is set at 16 kHz, and the length of the attached type II linear phase FIR filters is set to 64. Multiple sources and multiple sensing devices may also be present. *Source:* Courtesy Vocal.com.

two microphones has almost no value-added; using four microphones provides considerable improvements with the sidelobes having 20 dB of suppression. Using eight microphones allows the mainlobe to be 15° as the Nyquist frequency is approached, and the sidelobes offer about 25 dB of suppression.

3.3.1.3 Non-adaptive Beamformer

As discussed, conventional *adaptive beamformers* are able to attenuate and reject signals from unwanted directions. The adaptation process, however, can be disturbed and the output signal may become distorted if the beamformer is exposed to even mildly reverberant conditions, because if there were reflected components of the target signal that arrive from the same direction as noise, the adaptation process would also try to cancel the desired signal. *Non-adaptive systems*, such as FS FIR beamformers, can provide data independence and robust performance. The filter coefficients can be calculated as a constrained optimization problem and the robustness control can thus be added there, taking into account the statistical properties of the gain and phase errors in the array sensor system [24, 25].

In addition to the MVDR-based beamformer just discussed, microphone array-based methods such as the Linear-Constrained Minimum Variance (LCMV) beamformer, the Generalized Sidelobe Canceller (GSC), and post-filters implemented with the Wiener filter, are often used for speech enhancement to eliminate or minimize reverberation and noise (for example, LCMV can be used for the dereverberation in the case of multiple *speech* sources). Additionally, the generalized eigenvalue (GEV) beamforming, which maximizes the SNR of the beamformer output in each frequency, is a popular signal processing technique that can improve speech recognition performance.

3.3.1.4 Multichannel Linear Prediction (MCLP)

At press time, the most effective dereverberation method was based on multichannel linear prediction (MCLP) deconvolution [26–31]. The MCLP-based methods model the reverberation components as a delay of the speech signal in the time domain and the reverberation is subtracted from the speech signal itself – this method can be applied to the situation where the room impulse response (RIR) cannot be reliably estimated. The Kalman filter-based dereverberation algorithm that combines the MCLP and the GSC, known as the Integrated Sidelobe Cancellation and Linear Prediction (ISCLP), has also been proposed. See [32] for other noise reduction methods.

Conventional ASR systems achieve their best WER accuracy under the condition in which the in-field received data matches the data used to train the models (known as "matched condition training"). For example, as noted in [33], an ASR system trained with *close-talk data* (e.g. talk into a headphone used by a speaker) performs better on close-talk test data and worse on far-talk test data (e.g. talk captured by microphone array mounted on the walls or ceiling of a room in which the speaker is present); the reverse case where the training is done with *far-talk data*, also achieves the same WER performance; finally, an ASR system trained with *mixed close-talk data and far-talk* data achieves a similar WER accuracy for both close-talk test data and far-talk test data, which is lower than the case of matched condition training.

3.3.1.5 ML-based Approaches

Beamforming that utilizes ML techniques has emerged of late. Figure 3.7 illustrates the architecture of an ASR system synthetized from [33] that uses beamforming and ML methods. Figure 3.8 from the same source illustrates the

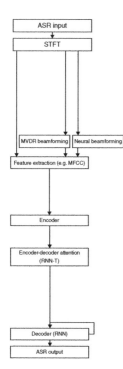

The speech is processed framewise typically utilizing a temporal window duration of 20–40 ms. The STFT is used for the signal analysis of each frame and its output is fed to the feature extraction blocks. The processing of a feature extraction block can include:

1) conversion of STFT into power spectrum;
2) computing filter bank features by applying triangular filters on a Mel-spectrum to the power spectrum to extract frequency bands; alternatively (or in addition) a discrete cosine transform (DCT) can be applied to the filter bank features to generate a compressed representation of the filter bank features, such outputs being the MFCCs.

In some cases the STFT output is fed to the neural beamforming block, particularly when far-talk input is multichannel input from a microphone array, in which case the neural network of the neural beamforming block learns how to map the noisy multichannel input to an enhanced single-channel signal (if the far-talk input is a single-channel input from a single microphone, the neural beamforming block can be skipped). In other implementation the neural beamforming block can be replaced by a conventional beamformer (e.g. MVDR beamformer) that transforms a multichannel signal into a single-channel signal using a data-adaptive beamforming solution, i.e., an optimal beamforming solution is found analytically based on signal processing knowledge.

The encoder recognizes the speech segment and produces the ASR output.

The output are fed to an encoderdecoder attention module (e.g. attention-based encoder-decoder system or recurrent neural network transducer (RNN-T). The attention-based encoderdecoder model learns a direct mapping of input features to outputs based on joint learning of the language model, the acoustic model, and the pronunciation model. The encoder maps the input features (e.g., filter bank or MFCC) into high-level representations. The attention mechanism goes through the whole input sentence (or segment) to calculate the weights which reflect the frames that should be focused for the current output unit (e.g. phoneme, wordpiece, or word), then feeds the weighted features into the decoder.

The decoder serves as a language model that utilizes attention to summarize the encoder's representations to produce the ASR output. The decoder is a RNN that takes the encoder output *and* the last output of the decoder at $t-1$ as inputs to produce the output at time t

Figure 3.7 Example of ASR system that makes heavy use of NN mechanisms. *Source:* Adapted from Ref. [33].

architecture of an (end-to-end[8]) ASR system. Here, one has an encoder configured for close-talk data input and a second encoder configured for multichannel far-talk input. Speech features (e.g. filter bank, MFCCs, and so on) are extracted from the close-talk input (e.g. a headphone-based input) and from the far-talk input (e.g. one or more microphones in an array); the extracted speech features are used to decide which type of encoder (i.e. short-talk encoder or far-talk encoder) to use to generating the ASR output.

---

8 In this narrow context an *end-to-end ASR system* is a system that is designed from the get-go to deal simultaneously with both close-talk data/signals and far-talk data/signals. However, although in a multichannel end-to-end ASR system, the NN arrangement contains a *neural beamforming* front-end that replaces the *traditional signal-processing-based beamforming component*, the acoustic and language models are both typically located in a single back-end NN model. Where the input of an ASR system could be a combination of close-talk and far-talk data, the best case scenario achievable with typical end-to-end ASR system is obtained by training the NN model with mixed close-talk and far-talk data; however, this results in a lower level of ASR accuracy than the case of matched condition training. See [33] for a proposed extension to the traditional end-to-end ASR system.

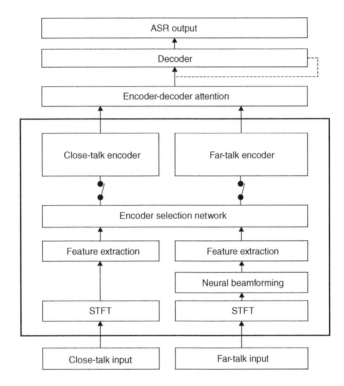

The multichannel far-talk input is processed by a neural beamforming block where the NN of the neural beamforming block learns how to map the noisy multichannel signal to an enhanced single-channel signal.

The neural beamforming block can be replaced by a traditional beamformer (e.g. MVDR beamformer, or delay-and-sum beamformer) that transforms a multichannel signal into a single-channel signal using a data-adaptive beamforming solution; namely, an optimal beamforming solution is derived analytically based on signal processing knowledge.

Figure 3.8 Neural beamforming use in ASR where one encoder configured for close-talk data input and a second encoder configured for multichannel far-talk input. *Source:* Reference [33]/U.S. Patent Application.

There are a number of proposed "experimental" multichannel approaches that directly employ DNNs. One approach is to utilize logarithmic Mel filter bank features of multiple acoustic channels as a parallel input to a convolutional neural network (CNN). Another approach is to use multiple input channels to take advantage of temporal difference information between channels by directly working on the raw waveform, that is, feeding the time domain signals into the DNN. Yet other proposed approach entails jointly training an MVDR beamformer and the AM, where the DNN estimates the beamforming weights for the MVDR beamformer given the time differences of arrivals, performs the beamforming operation, extracts the features and then uses these features to train an AM [19]. Another approach is to first condense multiple input channels to a single enhanced output signal to be fed to the ASR back-end, but utilizing a DNN component in the estimation of the beamformer coefficients.

3.3.1.6 Neural Network Beamforming

Modern ASR systems advantageously use MVDR beamformer with DNN back-ends; this beamforming arrangement is called *neural network beamforming*, also known as Neural Network Adaptive Beamforming (NAB), or simply *neural beamforming*. As noted, classical DSBs estimate the source signal from the microphone signals based on spatial filtering, which is implemented as a weighted sum of the signals captured by the sensors. The neural beamformer uses the complete information available from a microphone array of channels as input to generate a beamforming vector that changes according to the mixed signal characteristics. The data-centric model characteristic of an NN allows it to be trained to filter signals that arrive from a range of directions, resulting in a beampattern that has uniform width across a wide frequency range while and has sharp attenuation outside of the target range [16]. The NN consists as a group of MLPs adjoined with a traditional beamforming filter block, where the aggregate group of MLPs generates a beamforming vector that is specific to each frame of the input information; the beamforming filter block outputs the target signal approximation.

Thus, the basic, newly-advanced approach to handle multichannel signals is with neural beamforming, (i) first utilizing conventional beamforming techniques to condense the multiple signals into one combined signal and (ii) then feeding the signal to a DNN ASR back-end.

One can estimate the necessary statistics for acoustic beamforming by masking the observed signals. DNNs can be utilized to robustly estimate masks that are then used to calculate speech and noise covariance matrices for beamforming processing. This "neural beamforming" combines a DNN-based mask estimator with an analytic formulation to obtain a beamforming vector for speech enhancement.[9] Here, a mask estimation NN is trained on single channel magnitude spectra to achieve independence from the microphone array configuration – the training

---

9 The Time-Frequency (*T–F*) masking-based speech enhancement using supervised DNN learning algorithms has outperformed the traditional techniques. However, to deal with the often-perceptible difference between the oracle mask and the predicted mask in practical settings, proposals have been made to use a CNN-based Generative Adversarial Network (GAN) for inherent mask estimation [34]. This approach utilizes adversarial optimization as an alternative to the other maximum likelihood optimization-based architectures.

minimizes a cross entropy loss with an oracle speech and distortion mask. In the denoising application, the mask estimator is trained to differentiate between speech and noise [11, 19]. Furthermore, when given the correct training inputs, a neural beamformer is also able to perform denoising and dereverberation simultaneously.

A key consideration is how to estimate the MVDR filter coefficients to extract the target signal while mitigating interferences by utilizing the different spatial and spectral properties of the target and the distortions. For the DSB, the filter coefficients can be derived from an estimate of the Direction-of-Arrival (DoA), if the geometry of the microphone array is known. The (relative) ATFs between source and sensors are estimated; additionally, an estimate of the Cross-Power Spectral Density (PSD) matrix of the noise signal is sought. These parameters can be secured by estimating spectral masks for speech and noise, which are *typically obtained by model-based methods* (e.g. but not limited to [35, 36]). Another technique is that instead of using a model-based approach, one would *use a DNN to estimate the masks* [19]. DNNs for mask estimation have already been used in single channels speech enhancement for a while.

With the emergence of embedded microphones in wireless portable devices in the home, office, or car, *ad-hoc* (unconstrained) microphone arrays require attention. These environments benefit from a solution that *distributes the signal processing* over the array instead of utilizing a central processing point. Various methods have been proposed which rely on the knowledge either of the ATF or knowledge of the target signals covariance matrices, or both. DNN-based solutions have advanced to accurately estimate these parameters, typically by predicting TF masks from a *single-channel* input [37–39]. It is also possible to utilize the *multichannel* information to derive better estimates for these parameters; these methods are also DNN-based but they typically are centralized solutions, e.g. a centralized MVDR. Reference [12] discusses a distributed DNN-based mask estimation that can exploit the multichannel data to better predict the masks. Multi-node DNNs can predict a TF mask based on signals coming from several nodes; better TF mask estimates increase the speech enhancement performance.

Neural beamforming has shown to be able to achieve improved overall performance [40]; these models re-estimate a set of spatial filter coefficients at each input frame using an NN. Here, raw multichannel waveform signals are passed into a filter prediction (FP) LSTM[10] (a DNN) whose outputs are used as spatial

---

10 As discussed in previous chapters, an LSTM node includes several gates to handle input vectors (e.g. phonemes from an utterance), a memory cell, and an output vector (e.g., contextual representation). The input gate and output gate control the information flowing into and out of the memory cell, respectively. Forget gates remove information from the memory cell, as/if needed, based on the inputs from linked cells earlier in the NN [41]. Weights for the various gates are adjusted during the training phase.

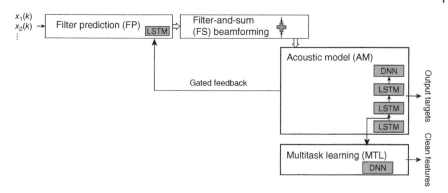

Figure 3.9 Example of NAB model (simplified).

filter coefficients; these spatial filters for each channel are then convolved with the corresponding waveform input, and the outputs are summed together to form a single channel output waveform containing the enhanced speech signal. The resulting single channel signal is passed to a raw waveform AM, which is trained jointly with the FP LSTM layers. This approach requires estimation of a small(er) set of filter coefficients, and results in better WER compared to frequency domain FP.

The NAB model architecture typically consists of cascading set of functional blocks: FP, followed by FS beamforming, followed by AM and followed by multitask learning (MTL). The NAB model entails explicitly feeding activations of the upper layers of the AM from the previous time step, which capture high-level information about the acoustic states as an additional input to the FP layers. Additionally, a gating mechanism can be used to attenuate the potential errors in these predictions. See Figure 3.9. In the AM section, the feature vector generated by the time convolution layer is passed to three LSTM layers followed by a fully connected DNN layer. The outputs of the AM block are output targets and the output of the MTL block are clean features.

As a final observation about deverberation, the weighted prediction error (WPE) (and DNN extensions, DNN-WPE) is a well-known signal processing method that can be used to dereverberate acoustic signals based on long-term linear prediction [11, 28, 42–44]. The basic approach of WPE is to estimate the reverberation tail of the signal and subtract it from the observation to obtain a maximum likelihood estimate of early arriving speech. The approach is based on maximizing the likelihood given a Gaussian with time-varying variances as a source signal model (however, the selection of a proper source model in advance of dereverberation is still a challenging problem, for which a number of solutions have been proposed recently [45]). WPE is a "model-based dereverberation technique" that is utilized in the Google Home speech assistant hardware in online conditions. DNN-based mask estimation for beamforming has shown good noise suppression (e.g. in the CHiME 4 Speech Separation and Recognition Challenge[11]); recently, it has been shown that this estimator can also be trained to perform dereverberation and denoising jointly [11]. DL-based beamforming benefits from a model-based dereverberation technique (e.g. WPE) and vice versa; thus, the integration of WPE and a neural beamformer outperforms stand-alone systems.

3.3.2 Specific Example of a System Supporting Noise Cancellation

This section provides a specific example of a system supporting noise cancellation. As discussed in the previous section, noise cancellation is typically required in an ASR system. As noted, an ML-based noise-cancellation component in the user device (and/or remote cloud system) is a trained model that processes audio data received from at least one microphone and/or other sources and reduces noise in the audio data to generate processed audio data that represents the utterance but does not include the noise. In the example of this section, the audio processing and noise-cancellation component includes [46]:

i) The first ML component may be configured as a *DNN encoder* – the *encoder* processes the audio data to determine one or more high-level features of the audio data, such as tone, pitch, and/or speech rate, as represented by first encoded data.

ii) A second ML component includes at least one *RNN layer* (may be used to store or "remember" a number of seconds of previously received audio data, such as 5–10 seconds of audio data, which may represent an average duration of time of an utterance); for example a gated recurrent unit (GRU)[12] layers and/or LSTM layers, that process the first encoded data to determine a second encoded data.

iii) A third ML component configured as a *DNN decoder* (may be used to process one or more outputs of the second component to determine mask data used to estimate output audio data that includes a representation of an utterance present in the input audio data).

---

11 The CHiME Speech Separation and Recognition Challenge is a challenge that aims at addressing speech separation and recognition in typical everyday listening conditions e.g. family living room. Speech commands are intermingled with the noise. The challenge is to separate the speech and recognize the commands utilizing systems that have been trained on noise-free commands and room noise recordings.

12 As noted in Chapter 2, a GRU is a simplified version of the LSTM RNN model [47–49]. GRU uses only one state vector and two gate vectors (the reset gate and the update gate). In GRU, the input and forget gates are combined and controlled by one gate making the GRU simpler than LSTM, and the GRU model has been found to outperform LSTM when dealing with smaller datasets.

More specifically, in one implementation such as the one shown in Figure 3.10 [46], the user device and/or remote system will typically,

- Process the first and second audio data using a *first component*, which may include at least one DNN layer that may perform a convolution operation on the frequency-domain audio data. The convolution operation may be a function that processes a number of subsets of each item of frequency-domain audio data (e.g. each frame of data) in accordance with a function.
- Process, using a *second component* comprising at least one RNN layer, the first encoded data to determine second encoded data corresponding to at least a second feature of the first and second audio data. The RNN layer(s) may include one or more RNN cells that receive an input that includes a portion of

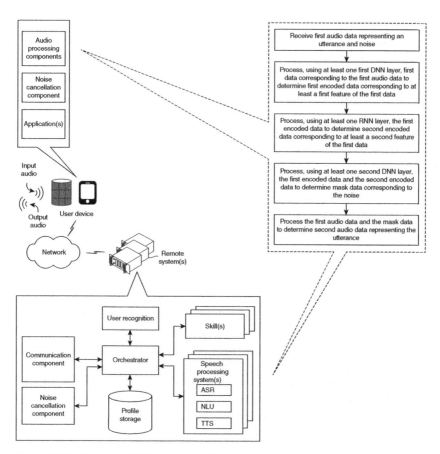

Figure 3.10 Example of ASR system with noise cancellation. *Source:* Adapted from Ref. [46].

an output of that same cell and/or an output of a cell in a subsequent layer. The RNN layer(s) include at least one connection between cells defining a feedback loop, thus permitting the RNN layer(s) to retain information received from previously received input data. The RNN layer(s) may include, for example, one or more LSTM cells, one or more GRU cells, or any other type of recurrent cell. In various implementations, the RNN layer(s) is/are configured to retain information corresponding to 5–10 seconds of previously received audio data, which may correspond to the average duration of time of an utterance.

- Process, using a *third component* comprising at least one second DNN layer, the second encoded data to determine mask data corresponding to the noise. The third component may be a decoder. Similar to the encoder, the decoder may be a network that produces an output for each item of input data (e.g. the second encoded data) received. The device may then process the audio data and the mask data (using, for example, a complex multiplication component) to determine second audio data representing the utterance (e.g. without the representation of the noise).

LSTMs/RNNs can use their feedback connections to store information of recent input events in form of activations; this is a form of "short-term memory" (as compared to "long-term memory" embodied by slowly changing weights). Learning to retain information over extended time intervals via recurrent backpropagation typically requires a long time, often because of decaying and/or insufficient error back flow. As hinted in the previous two chapters, in conventional methods (i) the temporal evolution of the backpropagated error depends exponentially on the size of the weights and (ii) error signals "flowing backwards in time" tend to either blow up (have oscillating weights) or vanish (taking a long time to learn to bridge long time gaps). LSTM is designed to overcome error back-flow problems: an LSTM can learn to bridge time intervals of thousands of steps even in the presence of noisy input sequences, without loss of short time lag capabilities. This is achieved by an efficient, gradient-based algorithm with an architecture enforcing constant (thus, neither exploding nor vanishing) error flow through internal states of special (neural) units [50]. As previously discussed, LSTM is an NN architecture that allows for constant error flow by utilizing special, self-connected units: a multiplicative input gate unit (called in-j) is utilized to protect the memory contents stored in cell cj from perturbation by irrelevant inputs, and a multiplicative output gate unit (called out-j) is utilize to protect other units from perturbation by currently-irrelevant memory contents stored in cell cj. Many ASR systems make use of LSTMs at various stages, as noted just above.

3.4 Training

In training of a DNN architecture used in ASR, a regression mechanism often includes a minimization of a pre-defined cost function. If the cost function value is not within a pre-determined range, based on the known training utterances, backpropagation is used. As discussed in Chapter 1, backpropagation is a common method of training NNs that are used with an optimization method such as a stochastic gradient descent (SGD) method. Use of backpropagation can include propagation and weight update. In the training process, when an input is presented to the NN the derived output of the NN is compared to the desired output and an error value is calculated for each output layer node [41]. The error values are sent backward, propagated backwards until each node in the NN has an associated error value. The backpropagation algorithm utilizes these error values to compute the gradient of the cost function with respect to the weights in the NN. In summary, SGD[13] is used by the selected optimization method to update the weights (parameters) of the network in an effort to minimize the cost function and thus produce a usable final output [41].

There is also a need for adaptation to test conditions. DNNs are able to achieve – with learning – stable operational invariances by utilizing many layers of nonlinear transformations. The invariance of the internal representations with respect to variabilities in the input space increases with depth (the number of layers). However, to improve the results, there is a need to explicitly compensate for unseen variabilities in the acoustics because if the site conditions (e.g. noise, reflections) and the training of the AM are not appropriately matched, then the runtime results (the runtime data distribution) may well differ from the training distribution, causing a degradation in accuracy.

This challenge can be addressed through *explicit adaptation to the test conditions*. Feature-space normalization is one approach; model-based approach is another (other approaches are available, see [52] for an extensive discussion). *Feature-space normalization* increases the invariance to unseen data by transforming the data such that it better matches the training data. In this approach, the DNN learns an additional transform of the input features conditioned on the speaker or the environment. The transform is parameterized by an additional set of adaptation parameters. An effective form of feature-space

---

13 SGD is an effective way of training DNNs, and SGD variants such as momentum and Adagrad have been used to achieve high-end performance – however, while SGD methods are simple and effective, they require careful tuning of the model hyper-parameters, specifically the learning rate used in optimization and the initial values for the model parameters [51].

normalization is constrained (feature-space) maximum-likelihood linear regression (MLLR) where the linear transform parameters are estimated by maximizing the likelihood of the adaptation data under a GMM/HMM AM. To utilize the MLLR with a DNN AM, it is necessary to estimate a single input transform per speaker (using a trained GMM), utilizing the resultant transformed data to train a DNN in a speaker-adaptive training manner. At runtime, another set of MLLR parameters is estimated for each speaker and the data is transformed accordingly. This technique has consistently and significantly reduced the WER across several different benchmarks. One can also estimate the linear transform as an input layer of the DNN. *Model-based approaches* adapt the DNN parameters using data from the target speaker; the goal is to alter the learned speaker-independent representation to improve the classification accuracy for data from a possibly mismatched test distribution. However, directly adapting all the weights of a large DNN is computationally- and data-intensive, and results in large speaker-dependent parameter sets. Refer to [52] for an extensive discussion on this topic.

3.5 Applications to Voice Interfaces Used to Control Home Devices and Digital Assistant Applications

This section discusses some examples of voice interfaces used to control home devices and digital assistant applications, for which the literature abounds. A user device such as a smartphone, a standalone assistant, or laptop computer may be configured to capture sounds and convert those sounds into an audio signal and, after performing speech processing, into a textual/command stream. In particular, intelligent automated assistants support natural, speech-based interfaces between human users and electronic devices or systems, allowing users to interact with devices and/or systems with their own natural language. Automated assistants utilize one or more microphones to sample audio signals from the surrounding environment to detect when a trigger word (e.g. "Hi Bixby"), for initiating a digital assistant session, is verbalized by a user. Voice interfaces can used, for example, to control smart home devices.

Thus, ASR systems allow humans to interact with computing devices using their own voices by identifying spoken words. An application capable of understanding and responding to verbal commands allows users to more easily interact with and use the application. When ASR is combined with natural language understanding (NLU)/NLP processing techniques, it enables speech-based user control of a smart or computing device to perform automated tasks based on the user's spoken commands.

At this juncture, more often than not, a distributed computing environment is used to perform the speech processing, giving rise to a hybrid system. A command represented in the audio data may then be executed by a combination of the remote system and/or the user device. Commands include, but are not limited to performing actions, downloading media, and obtaining information.

The voice interface process typically has three parts (e.g. as discussed in [4], which is a typical system):

1) *Wakeword detection*: elements for high accuracy detection, minimizing the chances of spurious activation of the system – low power consumption is another requirement.

2) *Speech processing activation, once the wakeword is detected*: elements for mapping the audio captured from the microphone to an internal representation that can then be processed by the rest of the system. In many systems, this representation is the text corresponding to the utterance of the user. Also in many systems, an ASR system is utilized in this step, with a combination of HMMs and ML-based DNNs models (the models usually work with three HMM states per phoneme, although new architectures can use a single HMM state per phoneme). In some current systems, the speech processing is performed entirely in a cloud backend rather than the device itself, given that typically the processing power of the embedded systems are much more limited compared to the cloud backend and, thus, one has to settle with limited vocabulary size and limited complexity of the voice commands. Hybrid approaches have been advanced (e.g. [4]), where the speech recognition is performed both on a local device and on a remote server in the cloud; hybrid arbitration is used to select the best ASR result from device and cloud outputs.

3) *Action*: elements mapping the input representation of the voice command (e.g. the text) to the required action.

The user device may include its own ASR system for processing audio data to determine corresponding text data; the ASR system of the user device is typically designed to understand only a subset of all possible words that may be understood by an ASR system operating on one or more remote system(s), that is configured to understand a greater subset or all possible words or parts of words or "acoustic units" [6]. For example, an ASR system of the user device may be capable of detecting only 3,000–4,000 acoustic units while an ASR system operating on the remote device(s) may be capable of detecting 10,000 different sounds.

Hybrid systems are common. In hybrid systems, there is a desire to determine a confidence measure to determine where to perform ASR locally on the device or remotely in the cloud. In a typical implementation, digitized audio command is decoded to generate audio features. An in-domain confidence score is calculated corresponding to the confidence that the plurality of audio features is better decoded by a model trained by a limited set of peripheral device commands. An out-domain confidence score is calculated corresponding to the confidence that the plurality of audio features is better decoded by a model trained by a broader set excluding the peripheral device commands (e.g. a server model). Based on the in-domain and out-domain confidence scores, an audio command is obtained by either (i) processing the audio features with the local processor using a local speech recognition module or (ii) sending the audio features to a remote server for processing. An action is then performed in the peripheral device in accordance with the audio command. A likelihood ratio (LR) is calculated of the in-domain and out-domain confidence scores. Based on the LR, a locally decoded audio command is performed, or the audio features are sent to a remote server for processing to determine the audio command. The LR is calculated using an ASR process combining statistical and ML models.

When a system utilizes a hybrid method, speech processing is segmented into two parts: (i) processing on the local system at the device and (ii) a cloud backend computer or server that is accessible over a network interface. Both the local device and the cloud backend systems include ML-based speech processing capabilities; however, as noted, the local system may have a more limited speech processing capability compared to speech processing capability of the cloud backend system because of the hardware and other computational limitations. These limited speech processing capabilities, compared to the cloud backend system, could include, e.g. smaller vocabulary, simple grammar, lack of contextual information, and so on. In comparison, the cloud backend invariably has larger processing power and can utilize a large vocabulary, is able to process complex natural language commands, and has access to the databases that hold contextual information (the cloud system may be used as a fallback mechanism, for example, in case that the local system is unable to process a given voice command on its own). In a hybrid system, a controller determines where a query should be processed. An example is depicted in Figure 3.11.

For the training modules, an AM employing DNN is used with dynamic models and LMs. In- and-Out filters model the In- and Out-domains. A WFST[14] search

---

14 Weighted finite-state transducer (WFST) has a set of states with one distinguished start state; each state has a final cost (or ∞ for non-final states); and there is a set of arcs between the states, where each arc has an input label, an output label, and a weight (cost). It can be perceived to be a decoding graph $HCLG = \min(\det(H \circ C \circ L \circ G))$, where H, C, L, and G represent the HMM structure, phonetic context-dependency, lexicon, and grammar respectively, and $\circ$ is WFST composition. More specifically [53].

H contains the HMM definitions; its output symbols represent context-dependent phones and its input symbols are transition-ids, which encode the pdf-id and other information.

C represents the context-dependency: its output symbols are phones and its input symbols represent context-dependent phones, that is, windows of N phones;

L is the lexicon; its output symbols are words and its input symbols are phones.

G is an acceptor (that is, its input and output symbols are the same) that encodes the grammar or language model.

See [54] for a discussion of the WFST decoding operation.

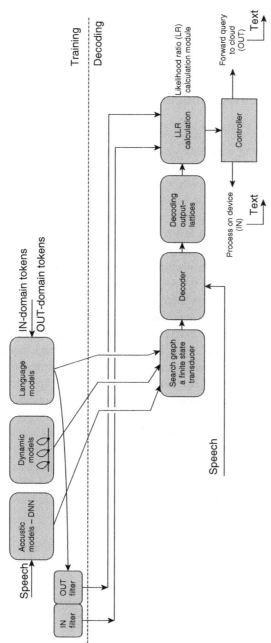

Figure 3.11 Diagram of the likelihood ratio (LR) calculation in an ASR system. All the steps are performed on the device. The training phase is done offline before using the controller. *Source:* Adapted from Ref. [4].

graph can be used by the decoder to decode input speech, and the output is provided as decoding output lattices[15]. The lattices represent the most probable word sequence(s). The decoding lattices, together with the filters, are processed to calculate a LR. The in-domain and out-domain tokens are the training text. For the in-domain, the tokens are a set of in-domain commands, such as volume-up, volume-down.

3.6 Attention-based Models

As noted, conventional ASR solutions implement several subsystems back to back to process speech using a pipeline architecture. In particular, traditional ASR solutions utilize a trained AM as the back-to-back ASR solution along with an LM (e.g. a decoding graph) used to interpret output(s) from the AM. The AM can be created for sub-phonetic units associated with speech; these sub-phonetic units can be represented using HMM and a decoding graph can then be used to interpret them in order to produce a text transcript. Conventional back-to-back ASR solutions have limitations; for example, when endeavoring to adapt such solutions for a new domain with the ensuing need to individually adapt most of the systems or components to make the ASR solutions work well in the new domain.

As discussed in Chapter 2, "attention" models are RNNs that iteratively process their input by selecting relevant content at every step. In the recent past, attention-based models have been applied to a variety of problems, such as, but not limited to, machine translation, visual object classification, and image caption generation. Learning to recognize speech can be viewed as learning to generate a sequence (transcription) given another sequence (speech). Speech recognition, however, requires long input sequences (thousands of frames instead of dozens of words); this introduces a challenge of distinguishing similar speech fragments in a single utterance. In addition, the input sequence can be noisy and may not have as clear structure. E2E trainable speech recognition systems are still based on hybrid systems consisting of a DNN AM, a triphone HMM model, and an N-gram LM. This requires dictionaries of hand-crafted pronunciation and phoneme lexicons, and a multi-stage training procedure to make the components work together.

---

15 There is no uniquely accepted single definition of a lattice. A lattice, in the ASR context, is a representation of the alternative word-sequences that are "sufficiently likely" for a given speech utterance. A lattice can be defined as a labeled, weighted, directed acyclic graph; time information can also be included; phone-level time alignments are also supported (with separate language model, acoustic and pronunciation-probability scores), and HMM-state-level alignments are also produced [54]. In mathematical language, a lattice Λ is a discrete additive subgroup of R^n. For a rank-n lattice in R^n, the rows of a $n \times n$ generator matrix G constitute a basis of Λ and any lattice point x is obtained as $x = zG$, where $z \in Z^n$ [55].

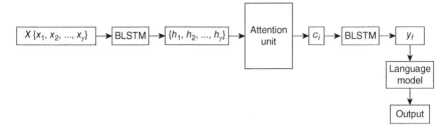

Figure 3.12 BLSTM/Attention example. *Source:* Reference [57]/U.S. Patent.

Reference [56] discusses some attention-based models for phoneme recognition tasks. At each time step in generating an output sequence (phonemes), an attention mechanism selects or weighs the signals produced by a trained feature extraction mechanism at potentially all of the time steps in the input sequence (speech frames). The weighted feature vector then helps to condition the generation of the next element of the output sequence.

Figure 3.12 provides an example from reference [57]. To train the ASR NN, an encoder-decoder architecture with an attention unit can be employed. The encoder (for example, a bi-directional long short-term memory [BLSTM][16] model) is used in this case to learn feature representations that capture correlations between sub-phonetic units and the output of the system. The attention unit can be used to estimate the relative importance of each feature in determining the correct output; it receives the hidden representation vector from the encoder (in some implementations the attention unit can be used to align the NN in such a way that given a sequence of audio, the NN can identify where a character starts and ends). The decoder can be used to construct an output using learned representations; it can use the context vector to predict a character unit. The decoder can be comprised of BLSTM layers and can also have a softmax function. The feature vectors that are indicative of a timeframe of the audio recording (e.g. representing an utterance) can be input into the encoder of the ASR NN. The ASR can then use an LM NN to process the predicted outputs from the ASR NN to determine the likelihood of occurrence of a word from the predicted output. The LM NN predicts a next item given previous items in a sequence. The LM NN can be initially trained using the generic dataset and further adapted using the target dataset. Also see Section 3.11.

---

16 A BLSTM is a method of allowing an NN to make use of data in both backward and forward directions: input flows in two directions [58–60]. BLSTM utilizes most of the data by going the time-step in both directions to preserve the future and the past information. In an LSTM the input flow in one direction, either backward or forward.

3.7 Sentiment Extraction

There are techniques and systems for training and, thereafter, process audio data of a user's speech to determine sentiment information *indicative of emotional state*. However, the assessment of sentiment associated with a sample of audio is complex; this is particularly true if there are more than a small set of mutually exclusive sentiment descriptors used. Raw audio as acquired from one or more microphones is processed to provide audio data that is associated with the enrolled user. Reference [61] provides one example. The system uses one or more ML NN systems during operation to accept audio feature data as input and provide as output sentiment data comprising one or more sentiment descriptors. A VAD module may be used to process the raw audio data and determine if speech is present. The VAD module may generate data that is indicative of individual utterances within the raw audio data. A feature analysis module uses the audio feature data to determine sentiment data; this can be done by determining the MFCC and using the MFCC to determine an emotional class associated with the portion. The emotional class may include one or more of the following conditions: angry, happy, sad, or neutral.

To produce accurate results, the feature analysis module is trained using training data obtained from one or more raters. For example, a rater may be a human who uses a computing device to present sample audio and then provides input indicative of one or more sentiment descriptors. Accurate training data is needed to train the feature analysis module to provide reliable and useful output. The raters generate annotated data by reviewing the presentation of sample audio data and assigning one or more sentiment descriptors from a list of predetermined sentiment descriptors. See Figure 3.13 for an example.

3.8 End-to-End Learning

The last decade has seen major breakthroughs in speech synthesis by applying NNs methods and E2E modeling. E2E learning refers to training a complex learning system represented by a *single model* (specifically, by a DNN) that encompasses the complete system under consideration, bypassing the intermediate layers typically present in traditional pipeline designs – ASR classically entails an architecture that has a number of sequential layers, for example preprocessing, feature extraction, optimization, prediction, decision making; for each layer, many distinct algorithms may be used. To achieve improved results, modifications in the inner layers and the subtending algorithms have to be applied over time, as the science advances. However, since each layer is responsible to solve specific tasks, it is challenging to determine how such changes will affect the system as a whole.

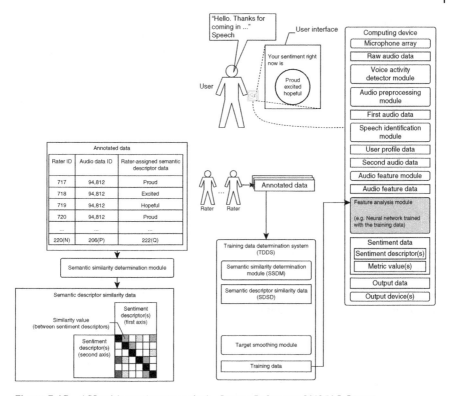

Figure 3.13 ASR with sentiment analysis. *Source:* Reference [61]/U.S. Patent.

ASR systems that utilize an E2E approach have endeavored to overcome the challenges encountered by back-to-back solutions. E2E learning seeks to represent the entire system by a single (DNN) model.

As stated, the traditional approach design for a spoken language understanding system is a pipeline structure with several different components, exemplified by the following sequence [62]:

$$\text{Audio (input)} \rightarrow \text{feature extraction} \rightarrow \text{phoneme detection}$$
$$\rightarrow \text{word composition} \rightarrow \text{text transcript (output)}.$$

A limitation of this pipelined architecture is that each module needs to be optimized separately under different criteria. The E2E approach consists in replacing the long chain with a single NN, allowing the use of a single optimization criterion for enhancing the system:

$$\text{Audio (input)} \rightarrow (\text{NN}) \rightarrow \text{transcript (output)}$$

E2E approaches use a trained ASR model based on a direct sequence-to-sequence (S2S) mapping approach (for example, mapping acoustic features to text-transcripts) (S2S concepts were discussed in Chapter 2). With an E2E solution, a trained ASR model can be optimized utilizing output characters rather than sub-phonetic units; the E2E system can then be applied to predict a sequence of words from the output characters using an external LM [57]. E2E trainable ASRs, however, are still generally based on hybrid systems consisting of a DNN AM, a triphone HMM model, and an N-gram LM [56]. This requires large dictionaries of hand-crafted pronunciation and phoneme lexicons, and a multi-stage training procedure to enable the components interoperate properly.

3.9 Speech Synthesis

Speech synthesis absorbs text and produces a sequence of audio samples that represent the corresponding speech audio. Speech synthesis can be [63]:

- Concatenative – requiring a large set of recorded data samples and storage;
- Parametric – requires little data and is adaptable in more dimensions and over greater ranges;
- Hybrid methods – methods that fit between concatenative synthesis and parametric synthesis;
- Articulatory – parameters apply to models of the human vocal tract;
- ML-based – ML techniques to estimate improved TTS voice parameters. ML techniques train with the goal of reducing a cost function to approximate optimal parameters. Various ML implementations use multivariate LR, SVMs, NNs, and decision tree induction. Some ML systems are supervised and apply past learning to new data; some ML systems are unsupervised, and draw inferences from large datasets. Figure 3.14 depicts pictorially an example.

The recent advancement of E2E DL models has enabled high-quality TTS with excellent natural-sounding results; however, most of these models are trained on large datasets (20–60 hours) recorded with a single speaker in a professional environment. Zero-shot TTS are TTS systems that aim at synthetizing speech into the voice of a target speaker using just a few seconds of speech; adapting the TTS model to synthesize the speech of a novel speaker without re-training the model. Some systems emerging at press time (e.g. [2, 64]) were able to synthesize voices in multiple languages and reduce data requirements significantly by transferring knowledge among languages in the training set (can even introduce new speakers by zero-shot learning on the fly).

An example of an RNN for speech synthesis is described in [65]. Here a hidden layer is connected to a directional edge to form a recurrent structure;

Figure 3.14 Modifying parameters for parametric speech synthesis responsive to listener behavior. *Source:* Reference [63]/ U.S. Patent.

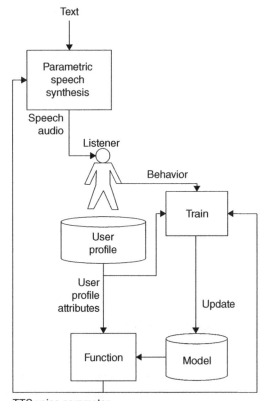

see Figure 3.15. *Xt* denotes input data, *Ht* denotes current hidden data, *H*(*t* − 1) denotes previous hidden data, and *Yt* denotes output data. The input data, the hidden data and the output data can be expressed by feature vectors. Parameters learned by the RNN include a first parameter *W*1 for converting the previous hidden data into the current hidden data, a second parameter *W*2 for converting the input data into the hidden data and a third parameter *W*3 for converting the current hidden data into the output data. The first, second, and third parameters *W*1, *W*2, and *W*3 can be expressed by a matrix. The input data can be a feature vector indicating a word, and the output data can be a feature vector indicating a first probability that an input word belongs to a word phrase (WP) class, a second probability that the input word belongs to an accentual phrase (AP) class and a third probability that the input word belongs to an intonation phrase (IP) class. The previous hidden data may be hidden data of a previously input word, and the current hidden data may be data generated using the hidden data of the previously input word and a feature vector of a currently input word.

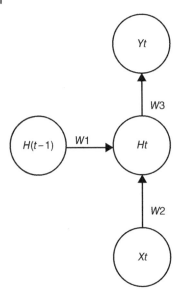

Figure 3.15 RNN model for speech synthesis. *Source:* Adapted from Ref. [65].

3.10 Zero-shot TTS

As mentioned earlier, zero-shot TTS are TTS systems that aim at synthetizing speech into the voice of a target speaker using *just a few seconds of speech*. [64] describes one such system (YourTTS). The system uses the VITS (*variational infer-ence with adversarial learning for end-to-end TTS*) model as the backbone archi-tecture and builds on top of it; the system employs a separately trained speaker encoder model to compute the speaker embedding vectors (*d*-vectors) to pass speaker information to the rest of the model. It uses the half attentive statistics pooling (H/ASP) model as the speaker encoder architecture described in [66] (also see [67]). See Figures 3.16 and 3.17. VITS employs different DL techniques together, including adversarial learning, normalizing flows, variational auto-encoders, transformers, to achieve high-quality natural-sounding output.

3.11 VALL-E: Unseen Speaker as an Acoustic Prompt

Continuing the discussion on zero-shot TTS systems, there have been a number of efforts underway to develop TTS systems that can synthesize high-quality person-alized speech with only a small sample enrolled recording of an unseen speaker as an acoustic prompt. The Lyrebird algorithm developed by the University of Montreal in 2017, requires a full minute of speech to analyze. As a new develop-ment, Voice Agnostic Lifelike Language (VALL-E) model was a press time

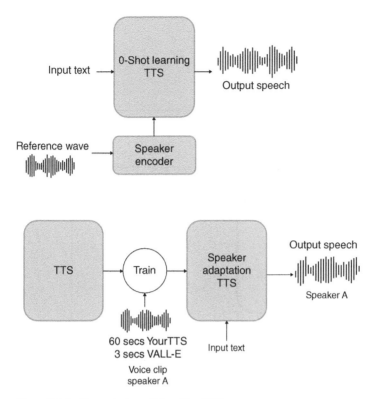

Figure 3.16 General view of Zero-Shot TTS.

announced system developed by Microsoft Corporation that can sample a voice for just three seconds, then mimic that voice – including the emotional tone and acoustics – and say anything one wants[17] (in American, British, and a few European-sounding accents) [69]. VALL-E uses ML techniques such as self-attention (Transformer-based architecture) and MLT. Self-attention is a technique that lets the model consider the entire input sequence when processing

---

17 The potential for misuse, for example, by scam artists exists and is an ethics issue. Microsoft stated that "The experiments in this work were carried out under the assumption that the user of the model is the target speaker and has been approved by the speaker." Unfortunately, AI has already been used for financial fraud: headlines such as the following were appearing at press time: "It sounds like science fiction but it's not: AI can financially destroy your business." In 2022 scammers already stole over US$11 million from unsuspecting consumers by fabricating the voices of loved ones, doctors, and attorneys requesting money. So much so that the US Federal Trade Commission (FTC) felt the need to issue a warning about deepfake AI technology, noting that all (the scammer) needs is a short audio clip of a family member's voice – which he could get from content posted online – and a voice-cloning program. These incidents are not limited to just consumers: businesses of all sizes are quickly falling victim to this new type of fraud [68].

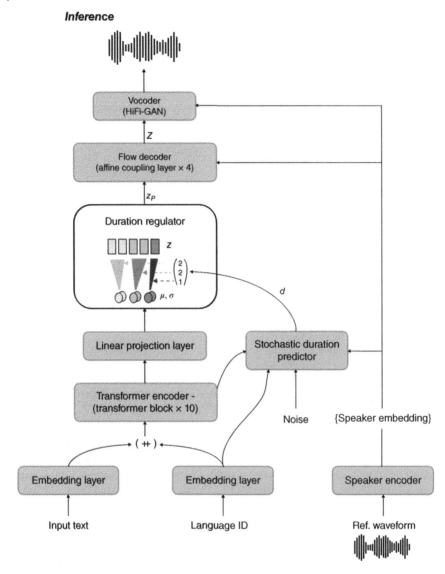

Figure 3.17 YourTSS. *Source:* Courtesy Coqui.

each element; MLT is a technique that allows the model to learn multiple related tasks simultaneously. VALL-E was trained on some 60,000 h of English speech.

VALL-E is a language modeling approach for TTS that entails training a neural codec LM using discrete codes derived from an off-the-shelf neural audio codec model, and treats TTS as a conditional language modeling task rather than

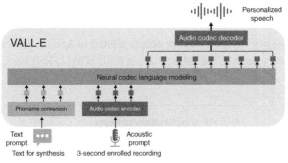

VALL-E replaces the extant pipeline
(e.g. phoneme → mel-spectrogram → waveform)
with the new pipeline
phoneme → discrete code → waveform

VALL-E generates the discrete audio codec codes based on phoneme and acoustic code prompts, corresponding to the target content and the speaker's voice. VALL-E directly enables various speech synthesis applications, such as zero-shot TTS, speech editing, and content creation combined with other generative AI models such as GPT-3.

Figure 3.18 The overview of VALL-E. *Source:* Courtesy Microsoft.

continuous signal regression. VALL-E can be used to synthesize high-quality personalized speech with only a three-seconds enrolled recording of an unseen speaker as an acoustic prompt [2]. See Figure 3.18.

Table 3.2 from [2] summarizes the innovation of VALL-E, an LM approach for TTS, using audio codec codes as intermediate representations, leveraging large and diverse data, leading to strong in-context learning capabilities. In summary, VALL-E was an innovative press time TTS framework with strong in-context learning capabilities, that treats TTS as an LM task with audio codec codes as an intermediate representation to replace the traditional mel spectrogram. VALL-E has in-context learning capability and enables prompt-based approaches for zero-shot TTS that does not require additional structure engineering, pre-designed acoustic features, and fine-tuning as in existing systems.

Table 3.2 A comparison between VALL-E and current cascaded TTS systems.

	Current systems	**VALL-E**
Intermediate representation	Mel spectrogram	Audio codec code
Objective function	Continuous signal regression	Language model
Training data	≤600 hours	60K hours
In-context learning	✗	✓

Source: Courtesy Microsoft.

A. Basic Glossary of Key AI Terms and Concepts

Basic Glossary of key AI terms and concepts related to Speech Processing (based on various industry sources including but not limited to [1, 3, 4, 16, 52–54, 57, 58, 62, 70–72])

Term	Definition/explanation/concepts
Acoustic model (AM)	A model of the spoken sounds. Can be trained using statistical or machine learning methods.
Acoustic transfer function (ATF)	The relationship between the sound level of a source and the sound level at some remote point known as the receiver. The ATF represents the relationship between the sound at its source location and how it is detected by an individual or by a microphone array or. A microphone array may have a number of associated acoustic transfer functions for several different source placements in the area around the microphone array.
Acoustic units	In the context of linguistics phones, phonemes, diphones, triphones, and/or senones.
Adaptive beamformer	An (acoustic) beamformer that performs adaptive spatial signal processing with an array of receivers. The signals are combined in a manner that increases the signal strength from a chosen direction. Signals from other directions are combined in a constructive or destructive manner, resulting in attenuation of the signal from the unwanted direction. An adaptive beamforming system relies on the acoustic (or electromagnetic) principles of phase relationships and wave propagation.
Backpropagation	A common method of training artificial neural networks that are used with an optimization method, such as a stochastic gradient descent (SGD) method.
Beamforming	An approach for obtaining a target signal in the presence of noise. It entails simultaneously sampling the sound field at various locations, with a microphone array, making use of spatial diversity. Captured signals with spatial diversity can be exploited to localize the various sources contributing to the sound field and, thus, enhancing signals coming from the spot containing the desired source, while at the same time attenuating signals from other locations. Beamforming algorithms can be classified into three major categories: data independent, statistically optimum, and adaptive.

Term	Definition/explanation/concepts
Bi-directional long-short term memory (BLSTM)	A method of allowing an NN to make use of data in both backward and forward directions: input flows in two directions. A BLSTM utilizes most of the data by going the time-step in both directions to preserve the future and the past information. In an LSTM the input flow in one direction, either backward or forward.
Confidence scores	Scores indicate a degree of confidence in a correct-recognition of a command. Examples of metrics include, but are not limited to, equal error rate or fixing false positive rates/false negative rates (FPR/FNR) at a certain rate.
Data-independent filter	A filter where the coefficients of the filter are independent of the data. The filter is non-adaptive, but they tend to be computationally efficient.
Delay-and-sum beamformer (DSB)	A beamformer that applies different delays to the signals received at the different microphones in a microphone array, to temporally align all signals coming from a key direction (while misaligning signals coming from other directions); the aligned signals are mixed and averaged across all channels to generate a single processable channel.
	It is a method where signals from different sources (e.g. original source, reflections, echoes, and so on) are aligned in time to adjust for the propagation delay from the speaker to each individual microphone. This classical method estimates the source signal from the microphone signals based on spatial filtering, where the source signal is taken to be the weighted sum of the signals captured by the microphones (or sensors).
Distant-talking speech communication systems	Environments where the speaker is at some distance from the receiving equipment, either in a large room or at some remote location via a telecommunications network.
End-to-end (E2E) ASR system	A system that deals simultaneously with both close-talk data/ signals and far-talk data/signals
End-to-end (E2E) learning	The training of a complex learning system represented by a single model (specifically by a DNN) that encompasses the complete system under consideration, bypassing the intermediate layers typically present in traditional pipeline designs.
Finite impulse response (FIR) filter	A filter where the response to any finite length input is of finite duration, i.e. it settles to zero in finite time; they have applications in signal processing.

(Continued)

(Continued)

Term	Definition/explanation/concepts
Finite-state transducer (FST)	A finite automaton whose state transitions are labeled with both input and output symbols. A path through the transducer encodes a mapping from an input symbol sequence to an output symbol sequence.
Gaussian mixture model (GMM)	A probabilistic model that assumes all the data points are generated from a mixture of a (finite) number of processes with Gaussian probability distribution functions (PDFs), with unknown parameters. A *Gaussian Mixture* is a function that is comprised of several Gaussians PDFs, each identified by $k \in \{1, \ldots, K\}$, with K the total number of clusters of the dataset. Each Gaussian distribution in the mixture has (i) a mean μ that defines its center and a covariance s that defines its width; and (ii) a mixing probability π that defines how small or big the Gaussian function is $\sum_{k=1}^{K}\pi_k = 1$.
	GMM is a clustering algorithm that is more versatile than the k-means hard clustering method, being that the latter associates each point to only one cluster and there is no uncertainty mechanism (probability) covering the case that a data point might be associated, in a given instance, with a different cluster. (Also see the Glossary definition in Chapter 1.)
Gradient	Slope or slant of a surface. Gradient descent means descending a slope to reach the lowest point on that curve or surface.
Gradient descent	A method used for updating the LR parameters a_0 and a_1 over time to reduce the cost as measured by the cost function. The approach starts with some values for a_0 and a_1 and then vary these values iteratively to reduce the cost. In the gradient descent algorithm, the number of steps is the learning rate, which determines how fast the algorithm converges to the minimum point– a smaller learning rate could get one closer to the minimum but takes more time to reach the said minima, while a larger learning rate converges sooner but there is a chance that one could overshoot the minimum.
Hidden Markov model (HMM)	A model where the system being modeled is assumed to be a Markov process X with hidden, unobservable, states. The assumptions embedded in the model are: (i) there is an observable process Y whose outcomes are "influenced" by the outcomes of X in some known way; (ii) the outcome of Y at time t_i must be "influenced" only by the outcome of X at time t_i; and (iii) that the outcomes of X and Y at $t < t_i$ must not affect the outcome of Y at t_i. Since X cannot be observed directly, the goal is to learn about behaviors in X by observing behaviors in Y.

Term	Definition/explanation/concepts
	The model states are hidden but the parameters of the model are not hidden, namely, while the model state may be hidden, the state-dependent output of the model is visible. Information about the state of the model can be inferred from the probability distribution over possible output tokens given that each model state creates a different distribution.
	For example, consider two individuals: Alice and Bob. Bob likes to eat pizza, sandwiches, and hamburgers, and he generally tends to pick which food to eat depending on his feelings. Alice has a general understanding of the likelihood that Bob is happy or upset and his tendency to pick food at lunch based on those feelings. Bob's food choice is the Markov process; Alice knows the parameters, but she does not know the state of Bob's feelings, which is an HMM. When they get together, Alice can determine the probability of Bob being either happy or upset based on which of the three foods he chose to eat at a given instance.
Hybrid arbitration	Techniques used to select the best ASR results (outputs) from either a local device or in cloud support system. The arbitration process generally entails two steps: (i) deciding whether to wait for the slower cloud system and (ii) selecting the best result. The first step may involve calculating a confidence measure – e.g. using a Minimum Bayes Risk (MBR) decoder– that assesses the level of confidence that the local device is decoding appropriately versus the cloud decoding appropriately.
In-context learning	A recently emerging behavior in large language models (LMs) where the LM performs a task just by conditioning on input-output examples, without optimizing any parameters. It is used in the "few-shot task learning" environment. In-context learning was part of the original GPT-3 design as a process to use LMs to learn tasks given only a few examples. During in-context learning, one gives the LM a prompt that consists of a list of input-output pairs that demonstrate a task; at the end of the prompt, one appends a test input and allows the LM to make a prediction just by conditioning on the prompt and predicting the next tokens. On many natural language processing (NLP) benchmarks, in-context learning performs well against models trained with much larger sets of labeled data. In many cases in-context learning has enabled developers to set up a system in just a few hours of labor [1].
i-Vector model	In an i-vector model, utterance statistics are extracted from features using a universal background model. The utterance is mapped to a vector in the total variability space, which is called an i-vector [4].

(Continued)

(Continued)

Term	Definition/explanation/concepts
k-Means	A commonly used clustering algorithm for a set of data points that uses an iterative approach to update the parameters (the centroids or means) of each cluster. It computes the means of each cluster and then calculate the distance of the centroids to each of the data points. The datapoint are thereafter labeled as belonging the cluster that corresponds to the closest centroid. This process is iteratively repeated until a specified convergence criterion is met, e.g. when there are no changes in the cluster assignments. The algorithm associates each point to only one cluster. It is not a probabilistic method, as would be the case in a GMM.
Language model (LM)	A model of the written letters, words and sentences or phrases. Can be trained using statistical or machine learning methods. Also see Glossary of Chapters 1 and 2.
Lattice (ASR context)	A lattice, in the ASR context, is a representation of the alternative word-sequences that are "sufficiently likely" for a given speech utterance. A lattice represents the most probable word sequence(s). A lattice can be defined as a labeled, weighted, directed acyclic graph (for example, a WFSA, with word labels).
Linear regression (LR)	Method where a linear equation (a line) is fitted to previously-collected data, specifically to pair sets of (i) an independent variable and (ii) a dependent variable, with the goal of minimizing the error between the observed values and the computed values using a given loss function, for example least squares. The linear equation has a value (a parameter) defining the slope and a "bias" the y-intercept. $$y = a_0 + a_1 * x \text{ (linear regression)}$$ LR is a type of regression analysis where the number of independent variables is one and there is a linear relationship between the independent (x) and dependent (y) variable. The goal of the LR algorithm is to find the values for a_0 and a_1 to best fit a line that approximates the distribution of the sample points under some cost (convex) function (typically to minimize the error between the predicted value and the actual value, e.g. the average squared error over all the data points). $$\text{minimize} \frac{1}{n} \sum_{i=1}^{n} \left(\text{pred}_i - y_i \right)^2$$

Term	Definition/explanation/concepts
	Gradient descent is a method that can be used for updating the LR parameters a_0 and a_1 over time to reduce the cost as measured by the cost function. The approach starts with some values for a_0 and a_1 and then varying these values iteratively to reduce the cost. In the gradient descent algorithm, the number of steps is the learning rate, which determines how fast the algorithm converges to the minimum point – a smaller learning rate could get one closer to the minimum but takes more time to reach the said minima, while a larger learning rate converges sooner but there is a chance that one could overshoot the minimum.
Logistic regression	(aka logit regression) A methodology used in regression analysis to estimate the parameters of a logistic model (the coefficients in the linear combination). The logistic model is a statistical model that establishes the probability of an event; the log-odds for the event are taken to be a linear combination of one or more independent variables.
Long short-term memory (LSTM) node	An NN node serving as a neuron that includes several gates to handle input vectors (e.g. phonemes from an utterance), a memory cell, and an output vector (e.g. contextual representation). The "input gate" and "output gate" control the information flowing into and out of the memory cell, respectively. The "forget gates" remove information from the memory cell if/as needed based on the inputs from linked cells earlier in the NN structure (cells associated with the immediate previous layer). Weights for the various gates are adjusted over the course of a training phase. The LSTM with the purpose-built memory cells that store information is well-suited for finding and utilizing long range dependencies in the data. Also see Glossary of Chapter 1 and Glossary of Chapter 2.
Mel frequency cepstral coefficients (MFCCs)	Speech features. The MFCCs of an acoustic clip define a small set of features (usually about 10–20) that concisely describe the shape of a spectral envelope. An acoustic signal $x(k)$ is processed through a Hamming window, a fast/discrete Fourier transform (FFT) operation, a power spectrum operation, a mel filtering operation, a log operation, and a discrete cosine transform (DCT) operation. Typically, the MFCCs are extracted from 25 ms time frames with a step size of 10 ms.
Minimum Variance Distortionless Response (MVDR) beamformer	A beamformer that minimizes the noise power at the output while maintaining the desired signal undistorted.
Multichannel Wiener filter (MWF)	(aka minimum mean square error [MMSE] beamformer) A filter that minimizes the variance of the error between the output and the desired signal.

(Continued)

(Continued)

Term	Definition/explanation/concepts
Multi-lingual TTS	Synthesizing speech in multiple languages with a single model.
Multi-speaker TTS	Synthesizing speech with different voices with a single model.
Natural language model	A model to processes natural language commands. Can be trained using statistical or machine learning methods.
Neural beamforming	An NN, a group of multilayer perceptrons (MLPs), coupled with a beamforming filter block; here each MLP uses information from a single frequency and a single frame of the microphone channels as input and outputs a beamforming vector element for that frequency. Hence, the aggregate group of MLPs generates a beamforming vector that is specific to each frame of the input information; the beamforming filter block outputs the target signal approximation. This enables training not with the unknown beamforming weight vectors for each frame as a target, but with the desired signal as the output target (this training resembles a form of autoencoder, but in which the decoded data is the desired signal instead of the input signal, and rather than compressing the signal as in the original use of autoencoders, these layers are utilized to analyze the mixed signal information) [16].
Noise cancellation	Reducing or canceling a magnitude of the volume of the signal of the noise comingled in the audio data.
Non-negative Matrix Factorization (NMF)	NMF is a feature extraction algorithm that is ideal when there are many features that are somewhat ambiguous. The method is used to produce patterns and themes. The method is often used in image processing and text mining. In an imaging example, the columns of B can be seen as images (the basis images, e.g. eyes, lips, nose), and C tells suggests how to sum up the basis images to reconstruct an approximation to the face in question (i.e. which features are present in the image). NMF is used to factorize a non-negative matrix, X, into the product of two lower rank matrices, W and H, such that WH approximates X. it is a methodology that allows one to automatically extract sparse and intuitive factors. NMF approximates a matrix X ($n \times p$, with n data points each with p dimensions) with a low-rank matrix approximation, namely $X \sim BC$, where B is a $p \times r$ matrix and C is an $r \times n$ matrix. The goal is to reduce p to r (a process called "rank r approximation"). In B, each column is a basis element (a component that is frequently present in all of the n data points); the columns of C provide the coordinates of a data point in the basis B. In effect, C enables one to reconstruct an approximation to the original datapoint n from a linear combination of the "building blocks" in B. NMF is an unsupervised learning algorithm utilized to reduce the dimensionality of data into lower-dimensional spaces.

Term	Definition/explanation/concepts
Out of distribution issue	Many ML solutions are not robust to "out of distribution" inputs. In the ASR context, words that are outside the vocabulary of the system might be interpreted as completely unrelated or unrecognized words.
Phone (in phonetics and linguistics context)	Any distinct speech sound or gesture (e.g. ~40 for English), regardless of whether the exact sound is critical to the meanings of words.
Phoneme	A speech sound in a given language that, if replaced with another phoneme, could change one word to another.
Pooling	A method for reducing the dimensionality of hidden layers by combining (pooling) some subsets of hidden unit activations. The combination of a set of hidden unit outputs into a summary statistic. Fixed poolings are typically used; these include average pooling and max pooling. Max pooling has been widely studied for both fully connected and convolutional DNN-based acoustic models. Pooling enables decision boundaries to be altered through the selection of relevant hidden features while keeping the parameters of the feature extractors (the hidden units) fixed. The pooling operators allow for a geometrical interpretation of the decision boundaries and how they will be affected by a constrained adaptation [52]. Also see definition in the Glossary for Chapter 2.
Regression/regression analysis	Mathematical methodology for stablishing a relationship between input variables and an output variable, which in turn is used to predict the value of a response to a new set of inputs. A method of forecasting a value of some process of interest, based on one or more independent predictors. Regression techniques may use a number of independent variables and use different types of functional relationship between the independent and dependent variables. The goal of regression is to minimize the sum of squared residuals. $y = a_0 + a_1 * x$ (linear regression) $y = a_0 + a_1 * x + a_2 * z$ (multivariate regression) $y = a_0 + a_1 * x^2 + a_2 * z^2$ (nonlinear regression, example)
Speech dereverberation	Removing reverberation from audio signals.
Speech distortion weighted multichannel Wiener filter (SDW-MWF)	SDW-MWF is a filter that generalizes the MWF and the MVDR. By identifying the two sources of error as distortion and residual noise, and weighting the residual noise component in the minimum mean square error minimization by a factor μ, it is possible to control the tradeoff between the two error sources. By setting $\mu = 1$ or $\mu = 0$, the MWF and the MVDR are obtained as special cases of the SDW-MWF, respectively [71].

(Continued)

(Continued)

Term	Definition/explanation/concepts
Speech enhancement (SE)	Approaches and algorithms aiming at improving the perceptual quality and speech intelligibility in a noisy environment. Seeks to recover clean speech from a noisy signal. It has many applications including ASR, hearing aids, and hand-free mobile communication. Single-channel SE, which utilizes a single microphone signal, can improve the speech quality but the noise reduction results in an increase in the speech's distortion. Multichannel SE is able to overcome this limitation utilizing the spatial information provided by multiple microphones.
Stochastic gradient descent (SGD) method	An iterative (heuristic) process for optimizing an objective function. The iterative process can be regarded as a stochastic approximation of gradient descent optimization, being that it replaces the actual gradient that would normally be calculated from the entire dataset, with an estimate calculated from a randomly-selected subset of the data (hence, the stochastic characterization). Also see description above for linear regression.
Sub-phonetic units	Information smaller than a phonetic unit (e.g. a recognizable speech sound) that is not directly audible or recognizable by the human ear. Sub-phonetic units can be represented using HMM. A decoding graph can then be used to interpret these sub-phonetic units to produce a text transcript.
Support vector machines (SVMs)	A linear model for classification and regression assessment. An algorithm to establish a hyperplane in an n-dimensional space (n being the number of features) that distinctly classifies data points such that the hyperplane has the maximum margin, namely the maximum distance (under a given metric) between data points of both classes – maximizing the margin provides a level of confidence that future data points can be classified with more accuracy. SVM can be used for both regression and classification tasks, but it is widely used in classification procedures. (Also see description in Chapter 1.)
Text to Speech (TTS)	Mechanisms to synthetize a text into natural sounding speech.
Transformer architecture	A highly modular DL architecture that can process long sequences of data without the need for recurrent connections; it uses self-attention mechanisms to allow the model to consider the entire input sequence when processing each element, thus capturing long-range dependencies in the input data. Also See Chapter 2 Glossary.
Voiceprint	Mechanisms to identify and/or authenticate a user based on audio input. For example, a user at a client system, such as a smartphone, may establish a voiceprint by speaking several words or phrases into a microphone of the smartphone, which may record the user's speech as audio input. Can also be used (in conjunction with ASR) to perform actions based on voice commands in the audio input.

Term	Definition/explanation/concepts
Weighted finite-state acceptor (WFSA)	A weighted finite-state acceptor (WFSA) $A = (\Sigma, Q, E, i, F, \lambda, \rho)$ over the semiring K is given by (i) an alphabet or label set Σ, (ii) a finite set of states Q, (iii) a finite set of transitions $E \subseteq Q \times (\Sigma \cup \{\varepsilon\}) \times K \times Q$, (iv) an initial state $i \in Q$, (v) a set of final states $F \subseteq Q$, (vi) an initial weight λ, and (vii) a final weight function ρ. WFSTs (see below) generalize WFSAs by replacing the single transition label by a pair (i, o) of an input label i and an output label o. While a weighted acceptor associates symbol sequences and weights, a WFST associates pairs of symbol sequences and weights, that is, it represents a weighted binary relation between symbol sequences. Models such as HMMs used in speech recognition are special cases of WFSAs.
	WFSAs accept or recognize each string that can be read along a path from the start state to a final state. Each accepted string is assigned a weight, namely the accumulated weights along accepting paths for that string, including final weights. An acceptor as a whole represents a set of strings, namely those that it accepts. As a weighted acceptor, it also associates to each accepted string the accumulated weights of their accepting paths [70].
Weighted finite-state transducer (WFST)	An FST that places weights on transitions in addition to the input and output symbols. Weights may encode probabilities, durations, penalties, or any other quantity that accumulates along paths to compute the overall weight of mapping an input sequence to an output sequence. Weighted transducers are a good choice to represent the probabilistic finite-state models prevalent in speech processing. The WFTS can be used to "decode" an utterance of T frames, that is to find the most likely word sequence and its corresponding state-level alignment.
	A WFST can be perceived to be a decoding graph $HCLG = \min(\det(H \circ C \circ L \circ G))$, where H, C, L and G represent the HMM structure, phonetic context-dependency, lexicon, and grammar respectively, and $\circ$ is WFST composition; more specifically [53], • H contains the HMM definitions; its output symbols represent context-dependent phones and its input symbols are transition-ids, which encode the pdf-id and other information • C represents the context-dependency: its output symbols are phones and its input symbols represent context-dependent phones, that is, windows of N phones; • L is the lexicon; its output symbols are words and its input symbols are phones • G is an acceptor (that is, its input and output symbols are the same) that encodes the grammar or language model In HCLG, the input labels are the identifiers of context-dependent HMM states, and the output labels represent words. The WFTS can be used to "decode" an utterance of T frames, that is to find the most likely word sequence and its corresponding state-level alignment.

(Continued)

Term	Definition/explanation/concepts
Window of audio data	Information extracted from or based on input audio data. Examples include (i) frames of audio data representing a fixed length of time of received audio (e.g. 10 ms of received audio); (ii) a time-domain representation of the magnitude and phase of the received audio; and (iii) a representation of processed received audio such as a frequency-domain representation of the received audio and/or a mel-cestrum spectrogram of the received audio. A window may include multiple frames, for example 80 frames and/or 800 ms of audio.
Zero-shot TTS	TTS systems that aim at synthetizing speech into the voice of a target speaker using just a few seconds of speech; uses zero-shot learning which is adapting the TTS model to synthesize the speech of a novel speaker without re-training the model.
Zero-shot voice conversion	Changing the voice of a given speech clip.

References

1 S. M. Xie, S. Min, "How Does In-Context Learning Work? A Framework for Understanding the Differences from Traditional Supervised Learning", The Stanford AI Lab Blog, Aug. 1, 2022, http://ai.stanford.edu/blog/understanding-incontext/. Accessed Mar. 1, 2023.

2 C. Wang, S. Chen, *et al*, "Neural Codec Language Models are Zero-Shot Text to Speech Synthesizers", Microsoft, Jan. 5, 2023, arXiv:2301.02111v1, https://github.com/microsoft/unilm.

3 M. Mohri, F. Pereira, M. Riley, "Weighted Finite-State Transducers in Speech Recognition", *Comput. Speech Lang.* 2002; 16(1): 69–88, https://doi.org/10.1006/csla.2001.0184.

4 A. Salarian, M. Cernak, *et al*, Hybrid Voice Command Processing, U.S. Patent 2022/0406305, Dec. 22, 2022. Uncopyrighted material.

5 H. Wang, T. Wang, *et al*, Complex Natural Language Processing, U.S. Patent 11,398,226, July 26, 2022. Uncopyrighted material.

6 C. Jose, Y. Mishchenko, *et al*, Wakeword Detection Using a Neural Network, U.S. Patent 1,521,599, Dec. 6, 2022. Uncopyrighted material.

7 N. I. Guarneri, V. D'Alto, Vocal Command Recognition, U.S. Patent 2022/0406298, Dec. 22, 2022. Uncopyrighted material.

8 H. Hermansky, "Perceptual Linear Predictive (PLP) Analysis of Speech", *J. Acoust. Soc. Am.* 1990; 87: 1738, https://doi.org/10.1121/1.399423.

9 P. Ghahremani, *Learning Feature Representation for Automatic Speech Recognition*, Johns Hopkins University Dissertation, Baltimore, MA, 2019.

10 S. B. Davis, P. Mermelstein, "Comparison of Parametric Representations for Monosyllabic Word Recognition in Continuously Spoken Sentences", *IEEE Trans. Acoust. Speech Signal Process.* 1980; ASSP-28(4): 357–366.

11 L. Drude, C. Boeddeker, *et al*, "Integrating Neural Network Based Beamforming And Weighted Prediction Error Dereverberation", Interspeech 2018, Hyderabad, Sept. 2–6, 2018, pp. 3043 ff.

12 N. Furnon, R. Serizel, *et al.*, "DNN-Based Mask Estimation for Distributed Speech Enhancement in Spatially Unconstrained Microphone Arrays", *IEEE/ACM Trans. Audio Speech Lang. Process.* 2021; 29: 2310–2323.

13 E. Warsitz, R. Haeb-Umbach, "Blind Acoustic Beamforming Based on Generalized Eigenvalue Decomposition", *IEEE/ACM Trans. Audio Speech Lang. Process.* 2007: 1529–1539.

14 M. Souden, J. Benesty, S. Affes, "A Study of the LCMV and MVDR Noise Reduction Filters", *IEEE Trans. Signal Process.* 2010; 58(9): 4925–4935, https://doi.org/10.1109/TSP.2010.2051803.

15 X. Feng, Y. Zhang, J. Glass, "Speech Feature Denoising And Dereverberation Via Deep Autoencoders For Noisy Reverberant Speech Recognition", in 2014 IEEE International Conference on Acoustic, Speech and Signal Processing (ICASSP), Florence, Italy, May 4–9, 2014.

16 Y. M. Kuno, B. Masiero, N. Madhu, "A Neural Network Approach to Broadband Beamforming", in Proceedings of the 23rd Congress on Acoustics, Aachen, Germany, Sept. 9–13, 2019.

17 M. Brandstein, D. Ward, (Editors), *Microphone Arrays: Signal Processing Techniques and Applications*, Springer Science & Business Media, 2001. ISBN: 978-3-662-04619-7, https://doi.org/10.1007/978-3-662-04619-7.

18 D. V. Compernolle, W. Ma, *et al.*, "Speech Recognition in Noisy Environments with the Aid of Microphone Arrays", *Speech Comm.* 1990; 9(5–6): 433–442.

19 J. Heymann, L. Drude, R. Haeb-Umbach, "A Generic Neural Acoustic Beamforming Architecture for Robust Multi-Channel Speech Processing", *Comput. Speech Lang.* 2017; 46: 374–385, ISSN: 0885-2308, https://doi.org/10.1016/j.csl.2016.11.007.

20 E. Warsitz, R. Haeb-Umbach, "Acoustic Filter-And-Sum Beam-Forming By Adaptive Principal Component Analysis", in Proceedings. (ICASSP '05). IEEE International Conference on Acoustics, Speech, and Signal Processing, 2005, https://doi.org/10.1109/ICASSP.2005.1416129.

21 S. Doclo, M. Moonen, "Design of Far-Field and Near-Field Broadband Beamformers Using Eigenfilters", *Signal Process.* 2003; 83(12): 2641–2673.

22 A. Pezeshki, B. D. Van Veen, *et al.*, "Eigenvalue Beamforming Using a Multirank MVDR Beamformer and Subspace Selection", *IEEE Trans. Signal Process.* 2008; 56(5): 1954–1967.

23 Minimum Variance Distortionless Response (MVDR) Beamformer, Material at vocal.com. Accessed Jan. 4, 2023.

24 H. L. Van Trees, *Optimum Array Processing. Detection, Estimation, And Modulation Theory*, Wiley, New York, 2002. ISBN: 0-471-09390-4.

25 M. Kajala, "A Multi-Microphone Beamforming Algorithm with Adjustable Filter Characteristics", Thesis, Tampere University, Finland, 24 Sept. 2021.

26 T. Nakatani, K. Kinoshita, "A Unified Convolutional Beamformer for Simultaneous Denoising and Dereverberation", *IEEE Signal Process. Lett.* 2019; *26*: 903–907.

27 T. Dietzen, S. Doclo, *et al.*, "Integrated Sidelobe Cancellation and Linear Prediction Kalman Filter for Joint Multi-Microphone Speech Dereverberation, Interfering Speech Cancellation, and Noise Reduction", *IEEE/ACM Trans. Audio Speech Lang. Process.* 2020; *28*: 740–754.

28 T. Nakatani, T. Yoshioka, *et al.*, "Speech Dereverberation Based on Variance-Normalized Delayed Linear Prediction", *IEEE/ACM Trans. Audio Speech Lang. Process.* 2010; *18*(7): 1717–1731.

29 S. Hashemgeloogerdi, S. Braun, "Joint Beamforming and Reverberation Cancellation Using a Constrained Kalman Filter with Multichannel Linear Prediction", in Proceedings of the IEEE International Conference on Acoustic, Speech and Signal Processing (ICASSP), Barcelona, Spain, May 4–8, 2020, pp. 481–485.

30 S. Braun, E. A. P. Habets, "Linear Prediction-Based Online Dereverberation and Noise Reduction Using Alternating Kalman Filters", *IEEE/ACM Trans. Audio Speech Lang. Process.* 2018; *26*: 1119–1129.

31 A. Cohen, G. Stemmer, "Combined Weighted Prediction Error and Minimum Variance Distortionless Response for Dereverberation", in Proceedings of the IEEE International Conference on Acoustic, Speech and Signal Processing (ICASSP), New Orleans, LA, USA, Mar. 5–9, 2017, pp. 446–450.

32 F. Tan, C. Bao, J. Zhou, "Effective Dereverberation with a Lower Complexity at Presence of the Noise", *Appl. Sci.* 2022; 12: 11819, https://doi.org/10.3390/app122211819, https://www.mdpi.com/journal/applsci.

33 F. Weninger, M. Gaudesi, *et al*, Multi-Encoder End-to-End Automatic Speech Recognition (ASR) for Joint Modeling of Multiple Input Devices, U.S. Patent 2022/0406295, 2022. Uncopyrighted material.

34 N. Shah, H. A. Patil, M. H. Soni, "Time-Frequency Mask-based Speech Enhancement Using Convolutional Generative Adversarial Network," in 2018 Asia-Pacific Signal and Information Processing Association Annual Summit and Conference (APSIPA ASC), Nov. 12–15, 2018, Taipei, Taiwan, pp. 1246–1251, https://doi.org/10.23919/APSIPA.2018.8659692.

35 S. Araki, M. Okada, *et al*, "Spatial Correlation Model Based Observation Vector Clustering and MVDR Beamforming for Meeting Recognition", in Proceedings of the IEEE International Conference on Acoustics, Speech and Signal Processing (ICASSP), Shanghai, China, Mar. 20–25, 2016, https://doi.org/10.1109/ICASSP.2016.7471702.

36 N. Ito. S. Araki, *et al*, "Relaxed Disjointness Based Clustering for Joint Blind Source Separation and Dereverberation", in Proceedings of the 14th International Workshop on Acoustic Signal Enhancement (IWAENC), Juan-les-Pins, France, Sept. 8–11, 2014, https://doi.org/10.1109/IWAENC.2014.6954300.

37 J. Heymann, L. Drude, R. Haeb-Umbach, "Neural Network Based Spectral Mask Estimation for Acoustic Beamforming," In IEEE International Conference on Acoustics, Speech and Signal Processing (ICASSP), vol. 2016, Shanghai, China, May 2016, pp. 196–200.

38 Z.-Q. Wang, D. Wang, "Mask Weighted STFT Ratios for Relative Transfer Function Estimation and Its Application to Robust ASR," in IEEE International Conference on Acoustics, Speech and Signal Processing (ICASSP). IEEE, Calgary, AB, Canada, Apr. 15–20, 2018, pp. 5619–5623.

39 H. Erdogan, J. R. Hershey, *et al*, "Improved MVDR Beamforming Using Single-Channel Mask Prediction Networks," in Proc. Interspeech 2016, San Francisco, USA, Sept. 8–12, 2016, pp. 1981–1985.

40 Bo Li, Tara N. Sainath, *et al*, "Neural Network Adaptive Beamforming for Robust Multichannel Speech Recognition", in 2016 Proc. Interspeech, International Speech Communication Association, San Francisco, USA, Sept. 8–12, 2016.

41 B. Min, R. Zbib, Z. Huang, Cross-Lingual Information Retrieval and Information Extraction, U.S. Patent 11,531,824B2, Dec. 20, 2022. Uncopyrighted material.

42 T. Nakatani, T. Yoshioka, *et al*, "Blind Speech Dereverberation With Multi-Channel Linear Prediction Based On Short Time Fourier Transform Representation," in IEEE International Conference on Acoustics, Speech and Signal Processing (ICASSP), Las Vegas, NV, USA, Mar. 31 to Apr. 4 2008.

43 T. Yoshioka, T. Nakatani, "Generalization of Multichannel Linear Prediction Methods for Blind MIMO Impulse Response Shortening", *IEEE Trans. Audio Speech Lang. Process.* 2012; 20(10): 2707–2720.

44 J. Caroselli, I. Shafran, *et al*, "Adaptive Multichannel Dereverberation For Automatic Speech Recognition", in INTERSPEECH, Stockholm, Sweden, Aug. 20–24, 2017, pp. 3877–3881.

45 T. Taniguchi, A. Shanmugam, *et al*, "Generalized Weighted-Prediction-Error Dereverberation with Varying Source Priors for Reverberant Speech Recognition", in 2019 IEEE Workshop on Applications of Signal Processing to Audio and Acoustics, New Paltz, NY, Oct. 20–23, 2019.

46 A. S. Chhetri, N. Chatlani, Systems and Methods for Noise Cancellation, U.S. Patent 11,521,635, Dec. 6, 2022. Uncopyrighted material.

47 B. C. Mateus, M. Mateus, *et al.*, "Comparing LSTM and GRU Models to Predict the Condition of a Pulp Paper Press", *Energies*. 2021; 14: 6958–6979.

48 A. Bhuvaneswari, T. J. Thomas, P. Kesavan, "Embedded Bi-directional GRU and LSTM Learning Model to Predict Disasterson Twitter Data", *Procedia Comput. Sci.* 2019; 165: 511–516, https://doi.org/10.1016/j.procs.2020.01.020.

49 A. J. Almalki, P. Wocjan, "Forecasting Method based upon GRU-based Deep Learning Model," in 2020 International Conference on Computational Science and Computational Intelligence (CSCI), Las Vegas, NV, USA, Dec. 16–18, 2020.

50 S. Hochreiter, J. Schmidhuber, "Long Short-Term Memory", *Neural Comput.* 1997; 9(8): 1735–1780.

51 S. Ioffe, C. Szegedy, "Batch Normalization: Accelerating Deep Network Training by Reducing Internal Covariate Shift". in ICML, 2015, arXiv:1502.03167v3, https://doi.org/10.48550/arXiv.1502.03167.

52 P. Swietojanski, S. Renals, "Differentiable Pooling for Unsupervised Acoustic Model Adaptation", *IEEE/ACM Trans. Audio Speech Lang. Process.* 2016; 24(10), https://doi.org/10.1109/TASLP.2016.2584700.

53 Staff, "Lattices in Kaldi", http://kaldi-asr.org/doc/graph.html. Accessed Mar. 2, 2023.

54 D. Povey, M. Hannemann, *et al*, "Generating Exact Lattices in the WFST Framework", ICASSP, 2012, 978-1-4673-0046-9/12/.

55 V. Corlat, J. J. Boutros, *et al*, "On the Decoding of Barnes-Wall Lattices", in 2020 IEEE International Symposium on Information Theory (ISIT), https://doi.org/10.1109/ISIT44484.2020.9173976.

56 J. Chorowski, D. Bahdanau, *et al*, "Attention-based Models for Speech Recognition", June 24, 2015, arXiv preprint, arXiv: 1506.07503.

57 T. H. Bui, S. Dey, *et al*, Customizable Speech Recognition System, U.S. Patent 11,538,463, Dec. 27, 2022. Uncopyrighted material.

58 D. EndalieI, G. Haile, W. Taye, "Bi-directional Long Short Term Memory-Gated Recurrent Unit Model for Amharic Next Word Prediction", *PLoS ONE*; 17(8): e0273156, https://doi.org/10.1371/journal.pone.0273156.

59 S. Patel, S. Gite, "Bi-directional Long Short-Term Memory with Convolutional Neural Network Approach for Image Captioning", *Int. J. Curr. Eng. Technol.* 2017; 7(6): 1968–1972.

60 R. Maddu, A. R. Vanga, *et al.*, "Prediction of Land Surface Temperature of Major Coastal Cities of India Using Bidirectional LSTM Neural Networks", *J. Water Clim. Change.* 2021; 12(8): 3801–3819.

61 D. K. Bone, V. Rozgic, C. Wang, System To Determine Sentiment From Audio Data, U.S. Patent 11,532,300, Dec. 20, 2022. Uncopyrighted material.

62 F. Roza, "End-to-End Learning, the (Almost) Every Purpose ML Method", Towards Data Science, May 30, 2019, https://towardsdatascience.com. Accessed Mar. 3, 2023.

63 B. Mont-Reynaud, M. Almudafar-Depeyrot, Text-to-Speech Adapted by Machine Learning, U.S. Patent 11,531,819, Dec. 20, 2022. Uncopyrighted material.

64 E. Casanova, E. Gölge, J. Weber, "YourTTS: Zero-Shot Multi-Speaker Text Synthesis and Voice Conversion", Jan. 3, 2022, https://coqui.ai/blog/tts/yourtts-zero-shot-text-synthesis-low-resource-languages. Accessed Mar. 4, 2023.

65 J. Chae, S. Han, Speech Synthesizer Using Artificial Intelligence, Method of Operating Speech Synthesizer and Computer-Readable Recording Medium, U.S. Patent 11,417,313, Aug. 16, 2022. Uncopyrighted material.

66 H. S. Heo, B.-J. Lee, *et al*, "Clova Baseline System for the VoxCeleb Speaker Recognition Challenge 2020", 29 Sept. 2020, arXiv:2009.14153, https://doi.org/10.48550/arXiv.2009.14153.

67 H. S. Heo, B.-J. Lee, *et al*, "Clova Baseline System for the VoxCeleb Speaker Recognition Challenge 2020", Sept. 29, 2020, https://doi.org/10.48550/arXiv.2009.14153.

68 G. Marks, "It Sounds Like Science Fiction But It's Not: AI can Financially Destroy Your Business". The Guardian, Apr. 9, 2023, https://www.theguardian.com/business/2023/apr/09/it-sounds-like-science-fiction-but-its-not-ai-can-financially-destroy-your-business.

69 L. Blain, "Microsoft's New VALL-E AI can Capture Your Voice in 3 Seconds", Jan. 10, 2023, New Atlas, https://newatlas.com/technology/microsoft-vall-e-speech-synthesis/. Accessed Mar. 7, 2023.

70 M. Mohri, F. Pereira, M. Riley, "Speech Recognition with Weighted Finite-State Transducers", in *Springer Handbook on Speech Processing and Speech Communication*, Springer, 2008th edition, ISBN-10: 3540491252. ISBN-13: 978-3540491255.

71 S. Markovich-Golan, S. Gannot, I. Cohen, "A Weighted Multichannel Wiener Filter for Multiple Sources Scenarios", in 2012 IEEE 27-th Convention of Electrical and Electronics Engineers in Israel, Eilat, Israel, Nov. 14–17, 2012.

72 Y. Kim, J. Bridle, *et al*, Detecting a Trigger of a Digital Assistant, U.S. Patent 11,532,306, Dec. 20, 2022. Uncopyrighted material.

4

Current and Evolving Applications to Video and Imaging

4.1 Overview and Background

This chapter explores machine learning (ML) techniques for imaging and computer vision (CV). There is an extensive and well-established body of science related to signal processing; the focus of this chapter is *not* on signal processing *per se*, which would require a voluminous book just on that topic, but on the *use of AI* to interpret or enhance images in support of applications such as classification, detection, CV, face recognition, surveillance, situational awareness, and medical imaging. In recent years, CV has been one of the most active research areas for ML applications.[1]

As discussed in previous chapters, neural networks (NNs), deep neural networks (DNNs), and ultra deep NNs (UDNNs) play important roles in automatic processing of large bodies of data, especially in the video/CV arena (including mobile vision). As previously noted, there are various types of NNs including *feed-forward networks* and *convolutional neural networks* (CNNs).

A CNN (aka ConvNet) is a specialized feed-forward NN for processing data that has a grid-like structure, such as data corresponding to images or compressed/uncompressed video. A CNN is a DNN with a convolutional structure. Many well-known CNNs are indeed UDNNs and have very deep and wide network architectures. A typical CNN includes a feature extractor (a filter[2]) comprised of a

---

1 Autonomous driving in an important application of CV; the topic deserves an entire text on all the related matters; therefore, this topic is not covered per se in this book. Additionally, medical imaging is not addressed as a special case of imaging; neither are the specific aspects of feature extraction for usage in face recognition.

2 Filters can also be modeled in an interspace, a linear space spanned by a filter basis (FB).

AI Applications to Communications and Information Technologies: The Role of Ultra Deep Neural Networks, First Edition. Daniel Minoli and Benedict Occhiogrosso.
© 2024 The Institute of Electrical and Electronics Engineers, Inc.
Published 2024 by John Wiley & Sons, Inc.

convolutional (CONV) layer and a sub-sampling layer. The nodes in the CNN input layer are organized into a set of feature detectors "filters" (loosely modeled after the receptive fields of an eye's retina). For example, the input can be a multi-dimensional array of data that describes the color components of an input image. The output of each set of filters is propagated to nodes in successive layers of the network. The computations supporting a CNN include applying the mathematical convolution operation to each filter to produce the output of that filter.

CNNs have become prevalent in the CV field in recent years: they are now often used for vision and image recognition applications (CNNs are also used for other types of pattern recognition, including speech and language processing). Image recognition enables the identification of people, objects, places, and other types of elements within an image, and enables the generation of actionable conclusions based on the analysis of the image. The typical use for CNNs is on classification tasks, where the output to an image is a single class label; however, in many visual tasks, especially in medical image processing, the desired output seeks to also include localization, that is, where a class label is supposed to be assigned to each pixel [1]. Of late, CNN models have made significant advancements in image processing and recognition, not just for entire-image classification, but also in the context of "local" tasks with structured output, e.g. bounding box object detection. Specifically, CNN models have achieved good performance in CV applications such as autonomous car vision systems, drone navigation, robotics, and medical image recognition. As indicated, CNN applications include image recognition, object detection, image classification, image segmentation, resolution-enhancement, medical imaging, mapping and localization, autonomous navigation, speech synthesis, and language translation, among others. CNNs are able to extract features from images, even when no classification layer is provided. CNNs are often used to make a prediction for the pixel labels in the context of edge detection. For resolution enhancement, CNNs can provide very high-quality results, and thus they are used in commercial application scenarios; however, these ML techniques are very computation-intensive.

Significant progress has been made from earlier coarse inference to fine inference; fine inference seeks to make a prediction at every pixel [2]. DL/NN algorithms[3] have shown significant improvement over traditional ML algorithms that are based on manual extraction of relevant features (e.g. handcrafted features); DL models perform a hierarchical feature extraction [3]. Increased model size (but with ensuing computational complexity) typically achieves

---

3 Such as the ones mentioned in Chapter 1; e.g. multi-layer perceptron (MLP); autoencoder (AE); deep belief network (DBN); CNN; recurrent neural network (RNN) including long short-term memory (LSTM) and gated recurrent units (GRUs); generative adversarial network (GAN); and deep reinforcement learning (DRL).

quality (accuracy) improvements for most tasks compared to smaller models (assuming, however, that an appropriate amount of labeled data is provided for training).

While CNNs already existed for a considerable amount of time, their practical success was initially limited because of (i) the size of the considered networks and (ii) the size of the available training sets. A key advancement was achieved in the 2010s as described in Ref. [4] by utilizing supervised training of a large NN with eight layers and millions of parameters on the ImageNet dataset, with one million training images. Since then, larger and deeper networks have been trained. Some well-known DDNs/CNNs that have been advanced in recent years include but are not limited to: AlexNet, GoogLeNet, VGG-16 and -19, ResNet-18, -34, -50, -101 and -152, Inception-v3 and Inception-v4 [5] (discussed in more detail in Section 4.4.4).

An ultra-dense CNN is typically implemented in a pipeline of multiple NN elements, that is, a stack of convolutional layers. The depth of the configuration increases as more layers are added. Increasing the width and the depth of the CNN results in higher quality NNs. However, increasing the number of layers to a large number may ultimately prevent the model from converging, either because of vanishing gradient or gradient explosion. In the former, after too many iterations the learning slows down to the point where model's weights cannot be changed. In the latter, absent a mechanism for weights rescaling, overflow or underflow occurs which can result in gradient explosion. Thus, there is a design goal to balance the width and depth of the network: in a CNN, improved performance can be reached by balancing the number of filters per stage and the depth of the network [6]. For example, to recognize a cat in the image classification task only requires describing the cat's characteristics and broad features, such as texture, color, and shape; assessing what exists in the proximity of the cat may be unnecessary. Adding more layers may not always result in improved performance; sometimes, a deeper model achieves similar training error rates compared with shallower counterparts, thus pointing to the practical possibility of using smaller models (in some cases). Optimal improvement for a constant amount of available computation power (say, on a smartphone) can be reached (i) by allocating the power in a balanced way between the width and the depth of the NN and (ii) by increasing width and depth in a related manner. Test error rates of less than 4% have been achieved of late, and improvements occur on an ongoing basis.

In the context of video and imaging data, CNNs outperform pure multi-layer perceptron (MLP) approaches because the use of the latter would require one to completely flatten the image, implying, in turn, that spatial information is lost – images are represented as arrays of pixel-related values (e.g. luminance, color depth, and so on). CNNs can retain spatial information since they ingest the images in the original format, thereby enhancing the accuracy of the analysis at hand (e.g. scene understanding); furthermore, CNNs can, in some applications, reduce the number of parameters needed by the NN, thus improving efficiency and bounding the amount of central processing unit (CPU)/ graphics processing unit (GPU) processing needed. In general, however, CNNs still require large amounts of processing capacity and memory; for example, NN-based perception applications such as autonomous driving require a large number of convolution operations and, thus, a large number of multiply-accumulate (MAC) units. Comparison metrics for NNs typically include accuracy, inference time, power consumption, memory consumption, operations count. For example, there are tradeoff between operations and power and between accuracy and throughput; small increment in accuracy typically implies a significant increase in computational time and inference time is correlated to the number of operations in the NN. Hardware accelerators have been proposed for CNNs; however, computational efficiency and low parameter count are still desirable goals, particularly for edge/mobile CV (including robotics and drone applications).

Deploying UDNNs/DNNs on embedded devices, for example, on smartphones or Internet of Things (IoT) devices, however, is still a challenge given the extensive computation and storage requirement; the number of operations and parameters increases with the complexity of the model architecture. The performance depends on the capacity of processors on which the DNN runs with varying number of layers, neurons per layer, multiple filters, filter sizes and channels, while dealing with a large dataset [3]. For example, the well-known CNN network ResNet-50 requires up to 7.7 billion floating point operations per second (FLOPS) and 25.6 million model parameters to classify a $224 \times 224 \times 3$ image; the more complex model VGG-19 with 144 million parameters model size, requires up to 39 Giga FLOPS (GFLOPs).

The rest of the chapter discusses several of these topics in greater detail.

4.2 Convolution Process

Convolution has applications that include signal processing, image processing, CV, among others. In imaging applications, convolutions are used for extracting shapes and curves in an image.

Convolution is a mathematical operation (from the field of Functional Analysis) performed by two functions to produce a third function that is a modified version of one of the two original functions. A convolution effectively "blends" one function with another. In general, it is an integral that expresses the amount of overlap of one function g as it is shifted over another function f. The goal is typically used to "clean up" a signal; in imaging, an acquired dirty map is a convolution of the "true" clean map with a noisy, dirty beam.

A convolution is defined as a product of functions f and g: the operation on f and g produces a third function denoted as $|f * g|(t)$ or $f \otimes g$ that captures how the shape of one function is modified by the other function. The nomenclature "convolution" refers to both the result function $|f * g|$ and to the process of computing it. The convolution of two functions f and g over a finite range $[0,t]$ is given by (see [7])

$$[f * g](t) \equiv \int_0^t f(\tau) g(t - \tau) d\tau$$

It is the integral of the product of the two functions after one function is reflected about the y-axis and is shifted. Convolution over an infinite range is given by

$$f * g \equiv \int_{-\infty}^{\infty} f(\tau) g(t - \tau) d\tau$$
$$= \int_{-\infty}^{\infty} g(\tau) f(t - \tau) d\tau$$

As an example, the convolution of two Gaussians functions is a Gaussians function itself:

$$f = e^{-(t - \mu_1)^2 / (2\sigma_1^2)} \Big/ \left(\sigma_1 \sqrt{2\pi} \right)$$
$$g = e^{-(t - \mu_2)^2 / (2\sigma_2^2)} \Big/ \left(\sigma_2 \sqrt{2\pi} \right)$$
$$f * g = \frac{1}{\sqrt{2\pi \left(\sigma_1^2 + \sigma_2^2 \right)}} e^{-[t - (\mu_1 + \mu_2)]^2 / [2(\sigma_1^2 + \sigma_2^2)]}$$

The concept is similar for discrete functions (data). The first function of the convolution is referred to as the *input* and the second function is referred to as the convolution *kernel*; the output is referred to as the *feature map*. Image files can be comprised of bitmap image file, JPEG (Joint Photographic Experts Group) file, SVG (Scalable Vector Graphics) file, and/or variations. A grayscale image is often represented as a 2D array (tensor); a color image captures three color channels – any color can be obtained from an independent basis of colors and the red, green and blue (RGB) are commonly used – other schemes may be used, e.g. cyan, magenta, yellow, and black (CMYK); thus, RGB image is represented as a 3D array (tensor).

Conceptually, for a grayscale image the input vector x^y could be $x^y = (x_1, x_2, \ldots x_N)$ where $x_j = (l_j, i_j)$ and l_j is the location of pixel j on a linearized coordinate and i_j is the intensity value for that pixel; conceptually, for a color image the input vector x^y could be $x^y = (x_1, x_2, \ldots x_N)$ where $x_j = (l_j, ir_j, ig_j, ib_j)$ and l_j is the location of pixel j on a linearized coordinate and ir_j is the intensity value for the red hue for that pixel, and so on (in practice other representation schemes may be utilized).

Video is a sequence of frames, and typically the YCbCr color basis is used where Y is the luma (luminance) component and Cb and Cr are the blue-difference and red-difference chroma components.

A CNN is an NN composed of multiple layers. *CNN stacks multiple convolutional (CONV) layers supporting convolutions, pooling layers, and fully connected (FC) layers.* A convolution is the application of a filter to an input that in turn results in an activation. Complex CNNs stack a large number of layers and generate very deep data representations of hundreds of feature channels per datum. The convolution kernel is typically a multidimensional array of parameters; the parameters are established and/or refined during ML training. The *features* are the *kernels* used for convolutions. The CNN is able to learn to apply various filters for various objects: the filters generate positive energy readings for those specified objects if present– each object that responds with a high energy when the applied convolution to it is an object that fits well into a defined mask; thereafter, the fully connected part of CNN learns to map the identified objects to classes. Filters are also called *kernels* or *feature detectors*. In practical terms, a convolution process can be seen as the process of performing a convolution (i) on a 2D input image or (ii) the process of performing a convolution on a feature map using a trainable filter.

For an example of a convolution process with *discrete data inputs*, refer to Figures 4.1 and 4.2, which depict an input matrix I^4 and a kernel K. The convolution result after the operator Conv(I, K), $I * K$, or $I \otimes K$ is shown for a sliding window of

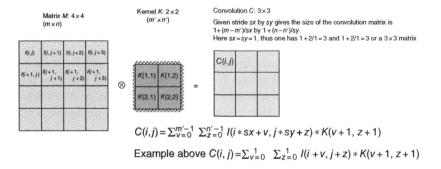

$$C(i,j) = \sum_{v=0}^{m'-1} \sum_{z=0}^{n'-1} I(i * sx + v, j * sy + z) * K(v+1, z+1)$$

Example above $C(i, j) = \sum_{v=0}^{1} \sum_{z=0}^{1} I(i + v, j + z) * K(v+1, z+1)$

Figure 4.1 Calculating the Feature map via convolutions.

---

4 The input is often represented as vector "X" but we prefer to use 'I' as we do here.

1 (in this example one slides by one pixel through each axis, such parameter being called the *stride*). Filter *K* slides over the rows then the columns of *M* from the top left side to the bottom right side. The process (in this example) entails multiplying each element of the highlighted sub-matrix *I* and the filter *K*, and then one sums all the elements together – the sub-matrix of *I* and *K* can be interpreted as vectors and the convolution operator as the dot product of these two vectors. The output matrix (the feature map) in these two figures is a 3×3 matrix because one can slide the filter through rows up to three times and through columns also up to three times, when the slide/ stride is 1 pixel. The (simple) algorithm in the figures is to slide the window to the left and undertake the dot product calculation until the entire matrix is exhausted. The size of the feature map *C* is $\{((m-m')/sx)+1)\times((n-n')/sy)+1)\}$. The individual values of the output feature map (OFM) matrix *C*, for a $n \times m$ input matrix *I*, a (single) $m' \times n'$ filter *K*, and a stride of sx ×sy, are [8],

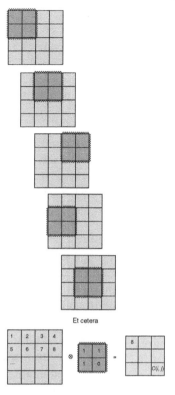

Figure 4.2 Sliding window sx×sy.

$$C(i,j) = \sum_{v=0}^{m'-1}\sum_{z=0}^{n'-1} I\left(i*sx+v, j*sy+z\right)*K\left(v+1,z+1\right).$$

For example, for an input image which is 4×4 ($n=4$, $m=4$), a 2×2 filter K ($m'=2$, n' = 2), and a 1×1 stride (sx = sy = 1), one has a 3×3 feature map

$$C(i,j) = \sum_{v=0}^{1}\sum_{z=0}^{1} I\left(i+v, j+z\right)*K\left(v+1,z+1\right)$$

For $i=1, j=1$
$i=1, j=1, v=0, z=0$: $I(1,1)*K(1,1)$
$i=1, j=1, v=0, z=1$: $I(1,2)*K(1,2)$
$i=1, j=1, v=1, z=0$: $I(2,1)*K(2,1)$
$i=1, j=1, v=1, z=1$: $I(2,2)*K(2,2)$
Or, $C(1,1) = [I(1,1)*K(1,1)+I(1,2)*K(1,2)]+[I(2,1)*K(2,1)+I(2,2)*K(2,2)]$

For $i = 1, j = 2$
$i = 1, j = 2, v = 0, z = 0$: $I(1,2) * K(1,1)$
$i = 1, j = 2, v = 0, z = 1$: $I(1,3) * K(1,2)$
$i = 1, j = 2, v = 1, z = 0$: $I(2,2) * K(2,1)$
$i = 1, j = 2, v = 1, z = 1$: $I(2,3) * K(2,2)$
Or, $C(1,2) = [I(1,2) * K(1,1) + I(1,3) * K(1,2)] + [I(2,2) * K(2,1) + I(2,3) * K(2,2)]$

For $i = 1, j = 3$
$i = 1, j = 3, v = 0, z = 0$: $I(1,3) * K(1,1)$
$i = 1, j = 3, v = 0, z = 1$: $I(1,4) * K(1,2)$
$i = 1, j = 3, v = 1, z = 0$: $I(2,3) * K(2,1)$
$i = 1, j = 3, v = 1, z = 1$: $I(2,4) * K(2,2)$
Or, $C(1,3) = I(1,3) * K(1,1) + I(1,4) * K(1,2) + I(2,3) * K(2,1) + I(2,4) * K(2,2)$

And so on.

See Appendix B at the end of this chapter for some examples of convolutions.

Inner products, or more general transformations, are also performed in an NN. An inner product is the elementary transformation undertaken by FC layers (see Figure 4.3 for a graphical view). An inner product can be perceived as a form of an aggregating transformation $\sum_{i=1}^{N} w_i x_i$, where $x^v = [x_1, x_2, ..., x_N]$ is an N-channel input vector to the neuron in question and w_i is a filter's weight for the i-th channel. This mapping can be interpreted as a combination of splitting, transforming, and aggregating. Splitting occurs when the vector x^v is sliced as a low-dimensional embedding (for the simple inner product in the above, the vector space is a one-dimensional subspace x_i); transforming occurs when the low-dimensional representation is transformed (for the simple inner product in the above it is just scaled as $w_i x_i$; aggregating occurs when the transformations are aggregated by the sum operator $\sum_{i=1}^{N}$. Advantageously, one can extend or replace the basic transformation ($w_i x_i$) with a more generic function

$$\mathcal{F}(x) \sum_{i=1}^{C} \mathcal{T}_i(x),$$

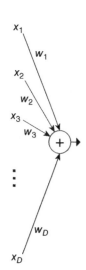

where $\mathcal{T}_i(x)$ is an arbitrary function. $\mathcal{T}_i(x)$ projects x^v into a low dimensional embedding and then transform it (C, the cardinality, is the size of the set of transformations to be aggregated).

Figure 4.3 A neuron supporting an inner product.

4.3 CNNs

As discussed earlier, CNNs are the dominant ML approach for visual object recognition (including for medical imaging) at this time. CNNs' success in large-scale image and video recognition has been facilitated by the availability of large public image repositories, such as ImageNet, and high-performance computing systems, such as GPUs or large-scale distributed clusters (e.g. but not limited to [4, 9–11]). In recent years, the imagenet large-scale visual recognition challenge (ILSVRC) has fostered the advance of deep visual recognition architectures as a testbed for large-scale image classification systems [12], and several efforts have been successful at improving the original architecture of Krizhevsky *et al* [4], for example, by utilizing smaller receptive window size and smaller stride of the first convolutional layer.

4.3.1 Nomenclature

Some basic CNN-related nomenclature is included in Table 4.1. The Glossary at the end of this chapter provides additional terminology.

4.3.2 Basic Formulation of the CNN Layers and Operation

At a broad architectural level, one needs to be cognizant of *CNN layers* and *CNN operations*.

As noted, there are three principal types of layers in a CNN: *convolutional (CONV) layers*, *pooling layers*, and *FC layers* (aka dense layers) – additionally, activation layers are added following each CONV layer and FC layer. See Figure 4.4. There are four principal types of operations in a CNN: *convolution operation, pooling operation, flattening operation, and classification operation* (or other requisite operation).

4.3.2.1 Layers
The first layer in a CNN is a CONV layer (there can be multiple convolutional layers in a CNN). It ingests the images as the input and starts to process with the goal of extracting a set of features from the image, while maintaining proximity information related to nearby pixels. There are three elements in the convolutional layer: (i) Input image, (ii) one or more filters, and (iii) feature map(s). The convolutional process was described above (with examples in Appendix B).

A CONV layer usually includes several feature maps, and each feature map may be formed by some neurons arranged in a rectangle. Neurons at the same feature map share weights; these shared weights – represented as a 2D matrix of weights – are collectively referred to as a *convolutional kernel* [13]. Thus, the convolutional

Table 4.1 Basic Nomenclature of CNNs.

Concept	Description
Classification process	A process that aims at identifying an object in an input image; the output is a vector of probabilities for different object elements, based on the training data.
Convolutional (CONV) layers	CNN layers that support/implement convolutions
Feature map	The output of the convolution via the convolution kernel is referred to as the feature map; the output matrix is the feature map. For example, for an input image which is 4×4 ($n = 4$, $m = 4$), a 2×2 filter K ($m' = 2$, $n' = 2$), and a 1×1 stride (sx = sy = 1), one has a 3×3 feature map $$C\left(i,j\right) = \sum_{v=0}^{1} \sum_{z=0}^{1} I\left(i+v, j+z\right) * K\left(v+1, z+1\right)$$ The feature map created by the immediately prior convolution is then run through a process, where a new feature map is created, but using the specified mechanism (e.g. pick the highest value, as one processes the incoming feature map with a given stride); here the filter is simply a window without elements in it and it is only used to specify a section in the incoming feature map (no parameters are learnt in the pooling layer).
Fully convolution network (FCN) (also sometimes called "completely convolutional NN")	An NN that only performs convolution (and subsampling or upsampling) operations; it is a CNN without FC layers. A typical CNN is not "fully convolutional" because it normally *also* contains FC layers (which do not perform the convolution operation). Thus, an FCN is an NN that is comprised entirely of convolutional layers, with no FC layers involved. FCN is an NN that does not contain any "dense" layers (as in traditional CNNs), but instead it contains 1×1 convolutions that perform the function of FC layers (dense layers).
Kernel size	The field of view of the convolution; for 2D imaging it is typically, but not always 3×3 cluster of pixels.
NNs with fully connected (FC) layers	NNs with layers where *neurons have full connections to all activations in the previous layer* (as is the case in a feed-forward network); traditional NN layers are fully connected in such a manner that every output unit interacts with every input unit; the output from the FC layers is used to generate an output result from the network. The FC layers apply a linear combination and an activation function to the input operand and generate an individual partial sum. By contrast, convolutional layers are often sparsely connected: the output of the convolution of a field is input to some nodes of the subsequent layer.
Object detection	Creating bounding boxes around detected objects, allowing one to see where they are in a scene.

Table 4.1 (Continued)

Concept	Description
Object instance segmentation	A process (beyond semantic segmentation) that endeavors to separate different instances from a single class: it is a cross between semantic segmentation and object detection.
Object recognition	A computer vision activity that recognizes and locates objects inside an image or video input.
Semantic segmentation process (also known as dense prediction)	The process of applying semantic labels to the pixels of an image; assigning and/or associating a label to a pixel (or small patch of pixels); marking each pixel of the input image with the class that represents a particular entity or body. Semantic segmentation is employed when the spatial information of a subject and how it interacts with it is of interest, such as for autonomous vehicle applications.
Stride	The step size of the kernel when traversing the image. Usually 1; a stride of 2 results in downsampling

kernel enables one to share weights among neurons in the same feature map. A number of convolutional kernels may be utilized in a given CONV layer to extract different image information and, in general, a larger number of convolutional kernels results in a better set of image information. In a training process of the CNN, the weights of the convolutional kernel are learned. For example, the ILSVRC training dataset may be used – e.g. ILSVRC 2014 included images of 1,000 classes and is comprised of three sets: 1.3M images for training, 50K images for validation, and 100K images with held-out class labels for testing. Another training example is ImageNet that contains more than 14 million training images across 1,000 object classes.

Each CONV layer is followed by a pooling layer. The goal of the pooling layer is to extract the most relevant features by a number of methods, such as using the maximum value where the filter is applied ("max pooling") or averaging the values in the area where the filter is applied ("average pooling"). The feature map created by the immediately-prior convolution is then run through a process where a new feature map is created, but using the specified mechanism (e.g. pick the highest value, as one processes the incoming feature map with a given stride); here, the filter is simply a window without elements in it and it is only used to specify a section in the incoming feature map (no parameters are learnt in the pooling layer). This has the effect of reducing the dimensionality (number of pixels) of the output returned from previous convolutional layers. Also, the pooling layer reduces the number of parameters in the network and may remove any noise present in the features extracted by previous convolutional layers [14].

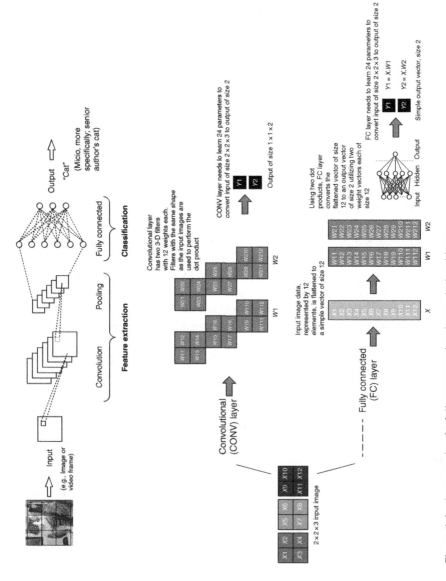

Figure 4.4 An example of a fully connected and convolutional layers.

The pooling operation occurs in each pooling layer, between a section of the feature map and the filter, and it outputs the pooled feature map.

A flatting operation takes place before processing the feature map with a fully connected MLP, using the output from the final pooling layer.

FC (dense) layers are the final layers in a CNN. The MLP is used to classify the final pooled feature map into a class label; it handles the classification (or other pertinent) task. An activation function is used in each FC layer. An FC layer flattens an input image comprising several pixels and several frames to a 1D vector and uses two 1D weight vectors to perform the dot products with the input 1D data vector to generate a 1D output vector. Each FC layer's $C_{n \times n}{}^i$ applies a convolution W^i over a $n \times n$ region of the previous layer activation a^{i-1} and offsets the result via a bias vector b^i, followed by non-linear function σ [15]:

$$a^i = C_{n \times n}{}^i \left(a^{i-1} \right) = \sigma \left(W^i * a^{i-1} + b^i \right)$$

As seen in the illustrative example of Figure 4.4, the FC layer needs to learn 24 parameters to convert input of size $2 \times 2 \times 3$ to output a new vector of size 2. As discussed, the CONV layer uses filters of the same size as input image for the dot product and generates a 1D vector output – CONV layer uses a discrete convolution operation – as seen in in the example of Figure 4.4, the CONV layer needs to learn 24 parameters to convert input of $2 \times 2 \times 3$ for output of size 2.

An FC layer can be converted to a CONV layer by redefining the parameters utilized in the FC layer: The input and output of these two layers are the same and the output $Y1$ and $Y2$ are the dot products of the input image and layer parameters (weights or filter values). This implies that an FC layer with input image of size $H \times W \times C$ and output vector of size N is equivalent to a CONV layer with the same input and kernel size of $H \times W$ with N output channels, with the proviso that the same set of parameters is utilized in convolutional filter and fully connected layer weights.

Many CNN architectures adhere to the same broad design principles of successively applying CONV layers to the input and periodically downsampling the spatial dimensions while increasing the number of feature maps. Traditional CNN architectures comprise stacked CONV layers; more modern CNN architectures aim at achieving more efficient learning [16]. Many of the new architectures acting as **rich feature extractors** utilize a repeatable unit that is employed throughout the network.

CNNs' capacity can be controlled by varying their depth and breadth; compared to standard feed-forward NNs with similarly sized layers, CNNs have much fewer connections and parameters and, thus, they are easier to train [4, 17]. The kernels associated with the CONV layers perform convolution operations, the output of which is sent to the next layer. The dimensionality reduction performed within the convolutional layers is one aspect that enables the CNN to scale to process large images. A key challenge in designing CNN models is sizing such models appropriately, since many factors need to be considered including, among others, the number of layers, feature maps, parameters, and kernel sizes. Achieving meaningful results requires an apparatus with substantial FC layers and computational resources (in addition to the CONV requirements). The recent introduction of GPU tensor cores that accelerate mixed precision matrix multiplication enables the use of DNNs for real-time processing. The CONV layers are trained using stochastic gradient descent (SGD).

Figure 4.5 provides another illustrative example of a CNN with a number of layers. In the example of Figure 4.5, the CNN is used for image processing and the input is comprised of the red, green, and blue components of the input image [9]. In CNNs, first encountered layers focus on learning low-level concepts, while later encountered layers entail more high-level (and specialized) feature mappings. In practical terms, one needs to increase the number of channels (feature maps) as one gets deeper in the NN. The input data is processed by one or more CONV layers and the output from the last convolutional is processed, in this example, by a set of FC layers where *neurons have full connections*[5] *to all activations in the previous layer* as in a feed-forward network; the output from the

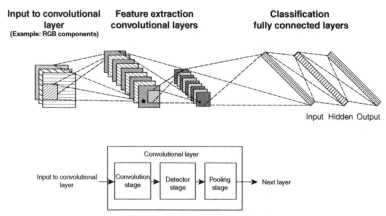

Input to a convolutional layer of a CNN can be processed in three stages of a convolutional layer: (i) a convolution stage, (ii) a detector stage, and (iii) a pooling stage. The convolution layer then outputs data to a successive convolutional layer. The final convolutional layer of the network generates an output feature map data or provide input to a fully connected layer, for example, to generate a classification value for the input to the CNN.

Figure 4.5 Example of a CNN. *Source:* Reference [9]/U.S. Patent.

---

5 The FC layers apply a linear combination and an activation function to the input operand and generate an individual partial sum.

FC layers is used to generate an output result from the network (the processing with the FC layers – where the activations are computed using matrix multiplication instead of convolution – are optional: in some implementations the second CONV layer can generate output for the CNN). Contrary to a traditional NN (where, as noted in Chapter 1, traditional NN layers are FC [fully connected] in such a manner that every output unit interacts with every input unit), the CONV layers are often sparsely connected: The output of the convolution of a field is input to the nodes of the subsequent layer (instead of the respective state value of each of the nodes in the field).

The CONV layer is a layer of neurons where convolution processing is performed on an input in the CNN. Consider a single image x_0 that is passed through a CNN. The network comprises L layers, each of which implements a non-linear transformation $H_j(\cdot)$, where j indexes the layer. Traditional convolutional feed-forward networks connect the output of the jth layer as input to the input of the $(j + 1)$th layer giving rise to the following layer transition: $x_j = H_j(x_{j-1})$, where and x_j denotes the output of the jth layer. $H_j(\cdot)$ can be a composite function of operations such as CONV, batch normalization (BN), Rectified Linear Unit (ReLU), or pooling [18]. However, for some designs (e.g. DenseNet, discussed later), in the CONV layer a neuron *may* be connected only to a subset of neurons in the preceding layer(s). Figure 4.5 depicted three stages of the convolution, the outputs of which are processed by successive stages of the convolutional layer: (i) convolution stage, (ii) detector stage, and (iii) pooling stage. The output from the CONV layer can then be processed by the next layer, this being an additional CONV layer or one of the FC layers. More specifically, in Figure 4.5 one has the following stages:

- The convolution stage computes the output of functions (at the neurons) that are connected to specific regions in the input, which can be described as the *local region* associated with the neuron. At this stage, the neurons compute a dot product between the weights of the neurons and the region in the local input to which the neurons are connected. The convolution stage typically performs multiple convolutions in parallel; these convolutions produce a set of linear activations or affine transformation (such as scaling, rotations, and so on).
- In the detector stage, each linear activation is processed de novo by a non-linear activation function to enhance the nonlinear properties of the overall network without affecting the receptive fields of the convolution layer. A number of nonlinear activation functions can be used, some of which were discussed in Chapter 1 (for example, Rectified Linear Unit – ReLU [19]).
- The pooling stage utilizes a pooling function that replaces the output of the second CONV layer with a summary statistic of the nearby outputs. Various types of pooling functions can be utilized including, for example, average

pooling or max pooling, among others [20]. The pooling function can be used to introduce translation invariance into the NN (some CNNs do not have a pooling stage but use instead an additional convolution stage having an increased stride relative to previous convolution stages).

- Next layers: FC layers or perhaps additional CONV processing.

Another way to look at the elements of a CNN is as follows: Each layer of data in a CNN is an array of size $h \times w \times d$, where h and w are spatial dimensions, and d is the feature or channel dimension (the first layer is the image, with pixel size $h \times w$, and d color channels). Locations in higher layers correspond to the locations in the image they are path-connected to, which are called their receptive fields. CNNs are built on translation invariance. Their basic components (convolution, pooling, and activation functions) operate on local input regions, and depend only on relative spatial coordinates. With x_{ij} representing the data vector at location (i, j) in a particular layer, and y_{ij} for the following layer, these functions compute outputs y_{ij} by ([2]).

$$y_{ij} = f_{ks}\left(\left\{x_{si+\delta i, sj+\delta j}\right\}_{0 \leq \delta i \leq k, 0 \leq \delta j \leq k}\right)$$

where k is the kernel size, s is the stride or subsampling factor, and f_{ks} determines the layer type: a matrix multiplication for convolution or average pooling, a spatial max for max pooling, or an elementwise nonlinearity for an activation function, and so on for other types of layers. Typical "object" recognition NNs take fixed-sized inputs and produce nonspatial outputs – the FC layers of these NNs have fixed dimensions and drop spatial coordinates.

4.3.2.2 Operations

In the convolutional computation, the input activations of a layer are structured as a set of 2-D input data, each of which is a channel. Each channel is convolved with a distinct 2-D filter from a stack of filters. The results of the convolution at each point are summed across all the channels. The result of this computation is the output activations that comprise one channel of the OFM. Repeated application of the same filter to the input data results in a map of activations called a feature map, which indicates the location and strength of a detected feature in an input image. In a convolution operation, the multiplication is performed between an array of input data and a 2-D array of weights, specifically the filter or a kernel. The filter is smaller than the input data. A 2-D convolution "convolves" along two spatial dimensions: It has a small kernel, essentially a window of pixel values that slides along those two dimensions. The specific type of multiplication applied between a filter-sized patch of the input and the filter is the dot product. As described earlier, a dot product is the element-wise multiplication between the filter-sized patch of the input and filter, which is then summed, resulting in a single numerical value. 3-D CNNs are useful for learning representations for data that can be perceived as volumetric data. 3-D CNN takes as input a sequence of 2-D frames. This model extracts features from both the spatial and temporal dimensions by performing 3-D convolutions, thereby capturing the motion information embedded in multiple adjacent frames. The model generates multiple channels of information from the input frames, and the final feature representation combines information from all channels [21].

Figure 4.6, synthesized from [22], depicts the basic operation of a convolutional operation at the functional level. Feature map FM1 is a set of data expressing various features of input data, and feature map FM2 is a data set that represents various features of output data resulting from convolution operations being performed by applying a weight map WM to the first feature map FM1 – maps FM1 and FM2 are described as having a width W (or a column), a height H (or a row), and a depth D, where the depth D is also referred to as the number of channels. In the convolution operation performed between the first feature map FM1 and the weight map WM to generate the second feature map FM2, the weight map WM (the filter or kernel) is used to filter the first feature map FM1. In the convolution, the weight map WM is shifted to slide over an entire area of the first feature map FM1, acting as a sliding window. During each shift, each of the weights included in the weight map WM is multiplied by a feature value of FM1 at a corresponding

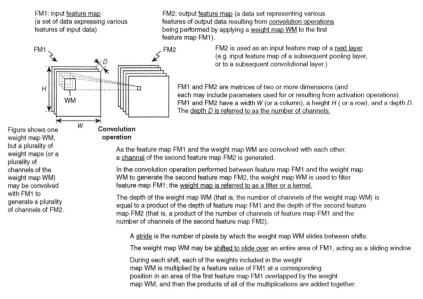

Figure 4.6 Basic operation of a convolutional operation at the functional level. *Source:* Reference [22]/U.S. Patent.

position in an area of the first feature map FM1 overlapped by the weight map WM, and then the products of all the multiplications are added together. The second feature map FM2 of the CONV layer is then used as an input feature map (IFM) of the next layer.

4.3.3 Fully Convolutional Networks (FCN)

As noted earlier in Table 4.1, FCNs are NNs that only perform convolution (and subsampling or upsampling) operations; it is a CNN without FC layers. A typical CNN is not "fully convolutional" because it normally also contains FC layers (which do not perform the convolution operation). Thus, an FCN is an NN that is comprised entirely of CONV layers, with no FC layers involved. FCNs are CNNs that take input of arbitrary size and produce correspondingly sized output with efficient inference and learning. FCNs are primarily used for semantic segmentation.

FCNs are trained end-to-end, pixels-to-pixels for the task of image segmentation. Here a traditional image classification network serves as the encoder module of the NN – producing a coarse feature map – which is complemented by a decoder module with transpose convolutional layers that upsamples the coarse feature map into a full-resolution segmentation map. Convolution, pooling, and upsampling are the only locally-linked layers employed. They have a CONV layer with a broad "receptive region" in lieu of the last FC layer. Given that dense layers are not present, there are fewer parameters, thus making the NN faster to train [23]. FCNs aim at capturing the image's overall context: what is in the image and some rudimentary indications of the locations of various things. Determining where there are more activations enables one to obtain some perspective on localization when one converts the last FC layer to a CONV layer (making the last CONV layer large enough, the localization effect can be scaled up to the size of the input image). See [24] and cited references for a lengthier discussion.

4.3.4 Convolutional Autoencoders

Autoencoder (AE) were briefly discussed in Chapter 3. AE networks are DL networks comprised of two DNNs, one called an encoder and the other a decoder. The function of an *encoder* is to generate, based on input data (say an image), a compressed feature set, this being known as the latent space representation. The encoder is responsible for deriving a correct encoding from the input data; this encoding is a compressed type of information that is a smaller version of the input data (the encoding will be used to reproduce the data). The function of the *decoder* is to recreate the input data based on the features with which it is presented. The decoder learns to reproduce the input data; the decoder uses the encoding to generate an output of the AE, similar to the input of the same AE. Thus, an AE is a type of feed-forward NN utilized to learn efficient data coding in an unsupervised setting; AEs can be trained to recreate relevant input data without the noise that may present in the input.

In addition to denoising of data (as discussed in Chapter 3), AEs can also be used for dimensionality reduction and feature extraction. Autoencoders are DNNs that apply representational learning; they impose a bottleneck on the layers such that a compressed representation of the input data is generated. Compression helps eliminate the mutually-dependent features; compression learns the dependencies such as correlation among input data, and leverages these structures during the compressed representation of data through the bottleneck of the network. Figure 4.7 top shows the architecture of an AE NN. This network can learn the representations of input data in an unsupervised way. These representations are the compressed forms of original data, as the network is trained to ignore the unnecessary part of the input. The reconstruction part of the network is responsible for the generation of output data that are as close to input data as possible by using the compressed representations [25–27]. Autoencoders are effective in generating reduced dimensional representations for the classification problems.

A "convolutional autoencoder" (CAE) is specific AE, a class of DNNs, where the *encoding and decoding structures are primarily convolutional and deconvolutional elements, respectively.* This architecture is very efficient in reconstructing image data. A CAE learn end-to-end mappings between images and is similar to a CNN: The output of this encoder is the encoding that usually represents a dense form of visual input data; the CAE decoder part has a general structure of a reversed CNN where, instead of CONV layers, there are deconvolutional layers and where one has upsampling (unpooling) layers instead of normal pooling layers. However, a CAE is a stateless feed-forward NN that does not recall (remember) information from past frames. See Figure 4.7 bottom. The first encoder block extracts a progressively compressed representation of the input x through a sequence of CONV layers followed by a pooling operator $P_{m \times m}$ that downsamples by computing the largest activation in an $m \times m$ region. The pooling operator may implement max pooling, average pooling, or other types of pooling. Starting with $e_0 = x$, successive encoder stages can be computed as [15]:

$$e_{i+1} = P_{m \times m}\left(C_{n \times n}\left(\ldots C_{n \times n}\left(e_i\right)\right)\right).$$

The last encoder stage generates a latent variables representation of the input, which is uncompressed by a succession of decoder stages:

$$d_{i+1} = U_{k \times k}\left(C_{n \times n}\left(\ldots C_{n \times n}\left(d_i\right)\right)\right).$$

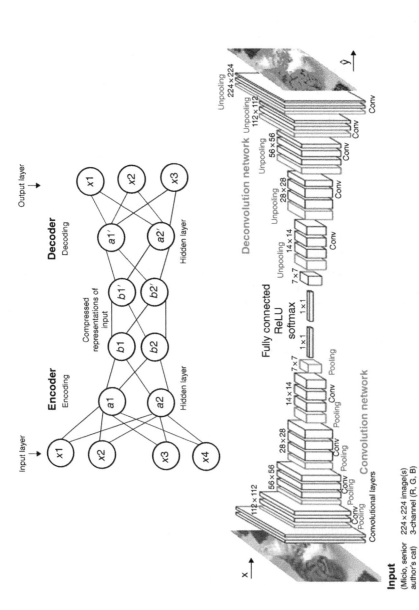

Figure 4.7 Autoencoders. Top: Regular AE. Bottom CAE. *Source:* Reference [25]. Qusay Sellat et al. 2022/with permission of Elsevier.

where $\mathcal{U}_{k \times k}$ is a $k \times k$ upsampling operator. Finally, the output image is computed as:

$$\hat{y} = \left(C_{n \times n} \left(\ldots C_{n \times n} \left(d_0 \right) \right) \right).$$

4.3.5 R-CNNs, Fast R-CNN, Faster R-CNN

As implied earlier, traditional object detection techniques generally utilize three key steps: (i) region proposal generation, (ii) feature extraction, and (iii) classification. Region proposals (usually several thousands) are candidates that could have objects within them. From each region proposal, a fixed-length feature vector is extracted utilizing various image descriptors; the feature vector aims at adequately describing an object (even if the object varies because of some transformation, such as scale or translation) [28]. The feature vector is thereafter used to assign each region proposal to either the background class or to one of the object classes. A popular model utilized for classifying the region proposals is the support vector machine (SVM).

Region-based CNN (R-CNN) (also called Regions with CNN features– see example in Figure 4.8) is a model for object detection that extracts features using a pre-trained CNN [29]. It entails a region *proposal generation section, a feature extraction section, and classification section using SVM*. Compared to the generic pipeline of the object detection techniques just described, the main value of an R-CNN is the ability to extract the features based on a CNN (all other steps are similar to the generic object detection pipeline). The original proposed R-CNN detected 80 different types of objects in images.

There are some drawbacks to the R-CNN model, including the fact that being a multi-stage model it cannot be trained end-to-end because each stage is an

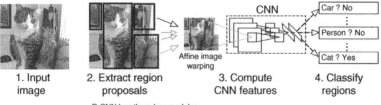

| 1. Input image | 2. Extract region proposals | 3. Compute CNN features | 4. Classify regions |

R-CNN has three key modules:
1. First module generates 2,000 region proposals using a selective search algorithm
2. Following a resize to a pre-defined size, the second module extracts a feature vector of length 4,096 from each region proposal
   Affine image warping computes a fixed-size CNN input from each region proposal, regardless of the region's shape
3. Last module utilizes a pre-trained SVM algorithm to classify the region proposal to either one the object classes or to the background

Figure 4.8 R-CNN object detection system. *Source:* Adapted from Ref. [29].

independent component; it also requires large amount of storage and significant (non-real time) computing time. Thus, an extension of the R-CNN model, the *Fast R-CNN model* (Fast R-CNN), has been proposed [30]. Fast R-CNN is a deep CNN used for object detection, that appears to the user as a single, end-to-end, network that has the ability to rapidly and accurately predict the locations of different objects; it was developed by Microsoft. Fast R-CNN builds a Region Proposal Network (RPN) that can generate region proposals that are then fed to the detection model, the Fast R-CNN, to seek out objects. The Fast R-CNN model includes an additional new layer, the Region of Interest (ROI) pooling layer, that extracts equal-length feature vectors from all proposals (i.e. all ROIs) in the same image. Fast R-CNN builds a network that has only a single stage. Using the new ROI pooling layer, the model shares the CONV layer calculations) across all proposals (i.e. ROIs) instead of doing the calculations for each proposal independently. All this makes Fast R-CNN run faster than R-CNN. The architecture of Fast R-CNN is shown in Figure 4.9. The model consists of a single stage, accepting an image as an input and outputting the class probabilities and bounding boxes of the detected objects. The architecture is trained end-to-end with a multi-task loss.

Faster R-CNN, discussed in [31], is an extension of Fast R-CNN. It uses the concept of RPN and is, by design, faster than Fast R-CNN. The key elements of this model are [28]: (i) the RPN, as a fully convolutional network that generates proposals with various scales and aspect ratios; it is an NN with attention to direct the object detection (Fast R-CNN) where to look; (ii) the use of anchor boxes, as a

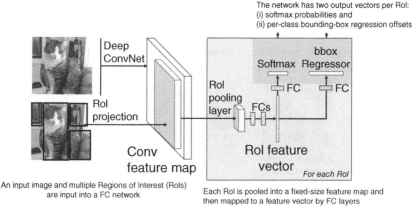

Figure 4.9 Fast R-CNN architecture. *Source:* Adapted from Ref. [30].

reference box of a specific scale and aspect ratio; with multiple reference anchor boxes, then multiple scales and aspect ratios exist for the single region; and (iii) the CONV computations are shared across the RPN and the Fast R-CNN. This arrangement reduces the computational time. The architecture of Faster R-CNN has two modules: (i) RPN: for generating region proposals; and (ii) Fast R-CNN: for detecting objects in the proposed regions. Refer to [28, 31] for additional details.

4.4 Imaging Applications

This section discusses the application of the methods and systems covered in the previous section to image analysis and processing.

Video information is represented by a set of static pictures (or "frames") arranged in a temporal sequence; frames are typically captured by a video capture device, for example, a camera. The video frames can be stored for future playback on some device (e.g. a television, a computer, a smartphone, or a tablet computer), or the video is played back in real-time, such as for surveillance, conferencing, or live broadcasting. A basic principle behind convolutional layers in a CNN, especially in the context of image processing, is that statistical information of a section of an image is the same as that of another section of the image; this implies that image information learned from one section of the image may also be applicable for another section of the image.

4.4.1 Basic Image Management

As stated, digital video is represented by a sequence of frames. Video can have a high data rate; for example, it could be comprised of 60 frames per second and each frame has $2,000 \times 2,000$ pixels or more. The final data rate (for entertainment-quality video, say the so-called 4K ultra high definition – UHD) can be 3, 6, or 12 Gbps.[6] The term "high-resolution" ("high-pixel") refers to digital images (e.g. photographs or video frames) having a large number of pixels (e.g. $1,024 \times 1,024$ or $2,000 \times 2,000$ pixels); the term "low-resolution" ("low-pixel") refers to digital images (e.g. photographs or video frames) having a small number of pixels (e.g. 512×512 pixels, or fewer) [32–43]. Given the high data rates, except for very high-end editing systems or for "contribution networks" (which move raw video between various provider elements typically within a video studio), video is invariably compressed (in a "lossy manner").

------

6 Emerging video applications – e.g. panoramic (360°), stereoscopic, and video for virtual reality (VR) – require even higher frame rates and resolutions, as well as increased dynamic range.

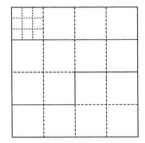

Figure 4.10 Macroblocks (simplified example).

A video source such as a video camera generates raw, unencoded video data as a sequential series of pictures (also referred to as "frames"). Frames typically include a two-dimensional matrix of samples of luminance and/or chrominance data (with Y, Cb, Cr values from 4 : 2 : 0, 4 : 2 : 2, or 4 : 4 : 4 pixel sampling), where the chrominance components may include both red hue and blue hue chrominance components (static images such as photographs may typically encode RGB data for samples of a picture).

Various video applications seek to balance the (i) coding efficiency and (ii) the level of details to be captured. To undertake desired compression, each frame picture is thus subdivided into "subregions." The standards define various video-related blocks that may be arranged in rows and columns. Processing a video frame entails subdividing the frame into independent subregions known by various terms such as *macroblocks, tiles, Basic Processing Units (BPUs), Coding Tree Unit* (CTU), or other comparable terms in various video coding standards.[7] The macroblocks can have various shapes and sizes of pixel groupings, such as 128×128, 64×64, 32×32, 16×16, 4×8, 16×32, 8×8, or other shapes and sizes; see illustrative Figure 4.10, where an exemplary subdivision into 4×4 BPUs (refer to the dash lines); also see Figure 4.11. Thus, the macroblocks of a color picture include a luminance (luma) component (Y) representing brightness information, and two chroma components (e.g. Cb and Cr) representing color information.

A $n \times n$ CTU has n samples in a vertical direction and n samples in a horizontal direction. A tile is a rectangular (or square) region of CTUs within a particular column and a particular row in a picture. Image tiles constituting a mosaic of an image. Tiles can be encoded as separate decodable streams: Encoders can code *tiles* at separate qualities or bitrates, and decoders can decode tiles in parallel. Tiles are simple to express using standard encoding libraries, such as FFmpeg and are supported by many video codecs [44–46]. Some additional nomenclature follows. A *tile column* refers to a rectangular region of CTUs having a height equal to the height of the picture and a width specified by syntax elements (for example, such as in a picture parameter set). A *tile row* refers to a rectangular region of CTUs having a height specified by syntax elements (e.g. such as in a picture parameter

---

7 Related terms with slightly different nuanced meaning include (i) *Coding Tree Blocks* (CTBs)/*Coding Tree Units* (CTUs) in some video coding (or compression) standards (e.g. MPEG family, H.261, H.263, or H.264/AVC), and (ii) *Coding Units* (CUs) in other video coding standards (e.g. H.265/HEVC or H.266/VVC).

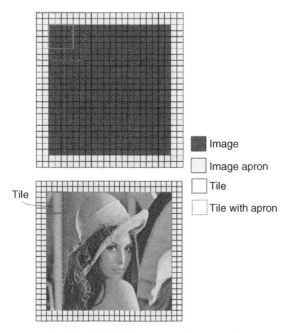

Image

Image apron

Tile

Tile with apron

Tile

Figure 4.11 Tiles. The ROI pooling layer segments each region proposal into a grid of cells. The max pooling operation is applied to each cell to return a single value; the combined set of values from all cells is the feature vector. *Source:* Conor Lawless/Flickr/ CC BY 2.0.

set) and a width equal to the width of the picture. A tile may be partitioned into multiple *bricks*, each of which may include one or more CTU rows within the tile. The bricks in a picture may also be arranged in a *slice*. A slice is an integer number of bricks of a picture that may be exclusively contained in a single Network Abstraction Layer (NAL) unit (the NAL formats the Video Coding Layer representation of the video and supplies header information in a manner appropriate for transmission using a variety of transport layers, or provides other capabilities for video storage).

Macroblocks are processed by the coding algorithm in an effort to greatly reduce the raw data rate; the reduction is typically 100-1, so that a 3 Gbps UHD signal is reduced to 30 Mbps or less via a "lossy process." High definition video has a lower compressed data rate of 4–12 Mbps. Video coding entails a number of stages.[8] As noted, video devices may encode, decode, transmit, receive, and/ or store digital video information more efficiently by implementing video

---

8 For coding each stage, the size of the basic processing units may still be too large and thus can be further divided into segments referred to as "basic processing *sub*-units" or "blocks."

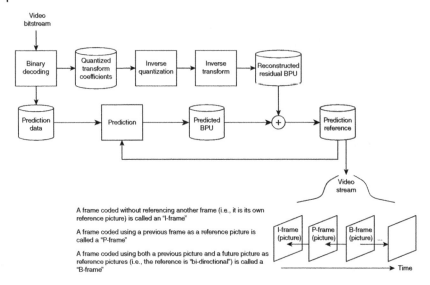

Figure 4.12 Encoding process example. *Source:* Reference [47]/U.S. Patent.

coding techniques such as MPEG-2/4, ITU-T H.264/MPEG-4, Advanced Video Coding (AVC), and ITU-T H.265/High Efficiency Video Coding (HEVC). To reduce the storage space and/or the network transmission bandwidth, the video is typically compressed before storage and transmission by an encoder and then decompressed by a decoder before the display. The encoder and decoder are collectively referred to as a "codec." Figure 4.12 illustrates an example of an encoding process.

To reduce or remove redundancy inherent in video sequences, video coding techniques utilize spatial (intra-picture) prediction and/or temporal (inter-picture) prediction. Typically, as noted, video codecs subdivide the picture into basic segments, and encode or decode the picture segment by segment. To achieve encoding efficiency, the information of a picture being encoded (a "current picture") includes changes with respect to a reference picture (e.g. a picture previously encoded and reconstructed). Changes typically include position changes, luminosity changes, or color changes of the pixels. Changes in position are of critical importance: position changes of a group of pixels that represent an object reflect the motion of the object between the reference picture and the current picture. A frame coded without referencing another frame (i.e. it is its own reference picture) is called an intra-coded picture, an "I-frame"; a frame coded using a previous frame as a reference picture is called a "P-frame" (predicted frame); and a frame coded using both a previous picture and a future picture as reference pictures (i.e. the reference is "bi-directional") is called a "B-frame." In particular, P-frames enable macroblocks to be compressed using temporal prediction and spatial prediction; for motion estimation, P-frames use frames that have been previously encoded. See Figure 4.13.

4.4.2 Image Segmentation and Image Classification

A basic image processing need is to determine (i) what elements are in a given image, and (ii) where are the elements in the image located. These problems deal with image segmentation and with image classification.

Image segmentation is a CV task intended to label given regions of an image according to what is being shown. More specifically, semantic image segmentation seeks to label *each pixel* of an image with a corresponding class of what is being represented.[9] Segmentation models are used, for example, in autonomous vehicles and medical image diagnostics. Semantic segmentation problems are classification problems where each pixel is assigned to one of several object classes (e.g. for traffic management: roads, trees, buildings; for facial segmentation skin, hair, eyes, nose, mouth elements). Semantic segmentation, however, embodies an inherent tension between semantics and location: global information resolves "what" while local information resolves "where" [2]. In image segmentation, one is not separating *instances* of the same class; the process is only concerned with the category of each pixel.[10] One can create an output channel for each of the possible classes; when one overlays a *single channel* of the target (or prediction), one refers to this as

Video compression modes:

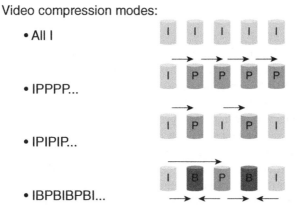

- • All I

- • IPPPP...

- • IPIPIP...

- • IBPBIBPBI...

Figure 4.13 *I, P, B* frame encoding.

---

9 Since one is predicting for every pixel in the image this task is also known as dense prediction.

10 Note that for two objects of the same category in the input image, the segmentation map does not intrinsically distinguish these as being separate objects. Other models, such as *instance segmentation models* do distinguish between separate objects of the same class.

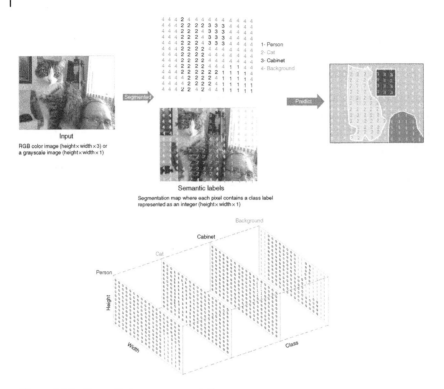

Figure 4.14 Semantic image segmentation.

it "being a *mask*," which highlights the regions of an image where a specific class is present [24]. See Figure 4.14 for some key concepts. Classical image processing approaches make use of CNNs for semantic segmentation (as seen in the figure) in which each pixel is labeled with the class of its enclosing object or region. Image segmentation and edge detection are important perception issues; these two problems are related because a segmentation contour defines a closed boundary of a region; edges, however, they do not always form closed contours.

In *image classification* problems, one is only concerned with *what* the image contains and its position in the field of view – therefore, in image classification problems, one can alleviate computational burden of traditional CNNs by downsampling the feature maps utilizing pooling techniques.

Both image classification problems and image segmentation problems can be supported by an NN with a number of CONV layers which enable the output of a final segmentation map that can be used by some downstream application; the NN learns a mapping from the input image to its corresponding segmentation via the successive transformation of feature mappings. All pixels are determined simultaneously. See Figure 4.15.

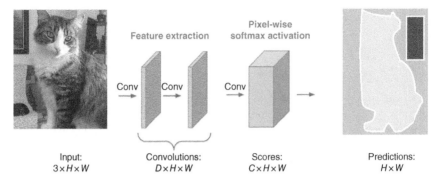

Figure 4.15 Basic CNN architecture for image classification and image segmentation problems.

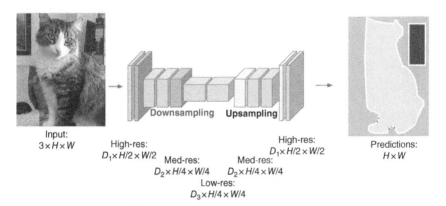

Figure 4.16 Encoder/decoder structure.

For image segmentation, a desideratum is to produce a *full-resolution* semantic prediction. The challenge is that this approach is computationally overtaxing when one wants to preserve the full resolution throughout the network. One approach for image segmentation models is to utilize an encoder/decoder structure. The approach here is to *downsample* the *spatial resolution of the input* image(s); this results in lower-resolution feature mappings while remaining very efficient at discriminating between classes; a later step is to *upsample* feature representations into a full-resolution segmentation map. See Figure 4.16 for an example.

Figure 4.17 depicts some basic "static" upsampling methods. Typically, one uses a dot product of the values currently in the filter's view to generate a value for the corresponding output position. There also are "*learned upsampling*" methods, utilizing transpose convolutions; the transpose convolution method takes a single value from the low-resolution feature map and multiplies all the weights in the filter by that value, thus projecting those weighted values into the OFM.

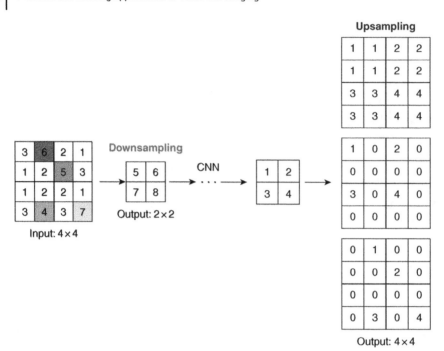

Figure 4.17 Examples of static upsampling.

4.4.3 Illustrative Examples of a Classification DNN/CNN

Figure 4.18 illustrates *an* example of a DNN/CNN that is *trained to receive images and output classifications of objects in the images* [48]. As discussed, the DNN/CNN includes a sequence of (i) convolutional layers, (ii) pooling layers, and (iii) FC layers. These layers, as a group, execute tensor computation that includes a plethora of tensor operations, such as convolution (e.g. MAC operations, etc.), pooling operations, elementwise operations (e.g. elementwise addition, elementwise multiplication, etc.), and/or other types of tensor operations. The following components as typical, as already discussed [48]:

- The CONV layers summarize the presence of features in the input image. The convolution may be (i) a standard convolution or (ii) a depthwise convolution. In the *standard convolution*, the entire filter slides across the IFM and all the input channels are combined to produce an output tensor (also referred to as output feature map [OFM]). In the *depthwise convolution*, the input channels are not combined; instead, MAC operations are performed on an individual input channel and an individual kernel and produce an output channel. Depthwise convolution produces a depthwise output tensor.

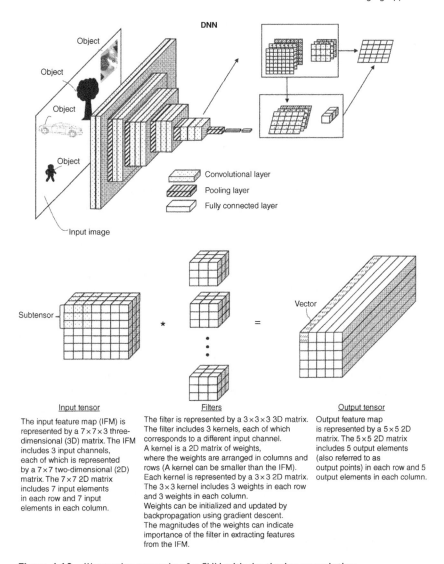

Figure 4.18 Illustrative example of a CNN with depthwise convolution. *Source:* Reference [48]/U.S. Patent.

- In the illustrative figure, the depthwise output tensor is represented by a $5 \times 5 \times 3$ 3D matrix. The depthwise output tensor includes three output channels, each of which is represented by a 5×5 2D matrix. The 5×5 2D matrix includes five output elements in each row and five output elements in each column. Each output channel is a result of MAC operations of an input channel of the IFM and a kernel of the filter – e.g. the *first output channel* shown with diagram dots is a result of MAC operations of the first input channel, shown with diagram dots and the first kernel shown with diagram dots; the second output channel shown with diagram horizontal strips is the result of MAC operations of the *second input channel* shown with diagram horizontal strips and the second kernel shown with diagram horizontal strips; and so on. The number of input channels equals the number of output channels, and each output channel corresponds to a different input channel. After the depthwise convolution, a pointwise convolution is performed on the depthwise output tensor and a $1 \times 1 \times 3$ tensor to produce the OFM. The OFM is then passed to the next layer in the sequence, for example, an activation function (e.g. ReLU) – recall that ReLU returns the value provided as input directly, or the value zero if the input is zero or less. In this example, the DNN includes 16 convolutional layers.

- As noted earlier, the pooling layers down-sample feature maps generated by the convolutional layers, e.g. by summarizing the presence of features in the patches of the feature maps. The pooling operation reduces the size of the feature maps while preserving their important characteristics, thus improving the efficiency of the DNN. Recall from Chapter 3 that the pooling layer(s) may perform the pooling operation through average pooling (calculating the average value for each patch on the feature map), max pooling (calculating the maximum value for each patch of the feature map), or a combination of both. The size of the pooling operation is smaller than the size of the feature maps (e.g. 2×2 pixels applied with a stride of 2 pixels, so that the pooling operation reduces the size of a feature map by a factor of 2, e.g. the number of pixels or values in the feature map is reduced to one-quarter the size).

- As noted earlier, the FC layers are the last layers of the CNN/DNN. The FC layers apply a linear combination and an activation function to the input operand and generate an individual partial sum. The individual partial sum may contain as many elements as there are classes: Element i represents the probability that the image belongs to class I, and the aggregate sum is 1. These probabilities are calculated by the last FC layer using a logistic function (binary classification) or a softmax function (multi-class classification) as an activation function.

4.4.4 Well-Known Image Classification Networks

Table 4.2[11] depicts some of the more well-known press time CNN architectures (partially based on the concepts cited Ref. [16]); this table is a summary only and interested readers should consult the cited papers.

---

11 Regarding ResNet note that in traditional deep networks, too-high learning rates may result in gradients that explode or vanish and getting stuck in poor local minima. Batch normalization (BN) helps address these issues. By normalizing activations throughout the network, it prevents small changes to the parameters from amplifying into larger and suboptimal changes in activations in gradients; for instance, it prevents the training from getting stuck in the saturated regimes of nonlinearities. Batch normalization also makes training more resilient to the parameter scale [56].

Table 4.2 Well-known press time CNN architectures.

Model	Date	General approach	Parameters	References
LeNet-5	1998	Convolutional layers utilize a subset of the previous layer's channels for each filter to reduce computation and institute a break of symmetry in the network. Subsampling layers use average pooling methods.	60,000	[20]
AlexNet	2012	*Image classification* NN: an image-focused model where convolutional layers utilize a subset of the previous layer's channels for each filter to reduce computation. (See Figure 4.19). *Quoted from Ref. [4]:* "The net contains eight layers with weights; the first five are convolutional and the remaining three are FC. The output of the last FC layer is fed to a 1,000-way softmax which produces a distribution over the 1,000 class labels. Our network maximizes the multinomial logistic regression objective, which is equivalent to maximizing the average across training cases of the log-probability of the correct label under the prediction distribution. The kernels of the second, fourth, and fifth convolutional layers are connected only to those kernel maps in the previous layer which reside on the same graphics processing unit (GPU). The kernels of the third CONV layer are connected to all kernel maps in the second layer. The neurons in the FC layers are connected to all neurons in the previous layer. Response-normalization layers follow the first and second convolutional layers. Max-pooling layers follow both response-normalization layers as well as the fifth convolutional layer. The ReLU non-linearity is applied to the output of every convolutional and FC layer. The first CONV layer and filters it with 256 kernels of size 5×5×48. The third, fourth, and fifth convolutional layers are connected to one another without any intervening pooling or normalization layers. The third CONV layer has 384 kernels of size 3×3×256 connected to the (normalized, pooled) outputs of the second convolutional layer. The fourth CONV layer has 384 kernels of size 3×3×192, and the fifth CONV layer has 256 kernels of size 3×3×192. The FC layers have 4,096 neurons each." (The model utilizes relatively large receptive fields in the first convolution layers of 11×11 with stride 4.)	~60,000,000 and 650,000,000 neurons	[4]

(*Continued*)

Table 4.2 (Continued)

Model	Date	General approach	Parameters	References
Visual Geometry Group's VGG 16, VGG-19 (aka ConvNet) (aka VGGNet)	2014	Well-accepted CNN network family for extracting image features, although both models are expensive in terms of the computational requirements and the number of model parameters. They have significant depth (16–19 weight layers) using an architecture with small (3×3) convolution filters. For example VGG-19 has 16 convolutional (CONV) layers and 3 FC layers: (in the siglum the first number is the receptive field size and the second number is the number of layers) CONV-3-64, CONV-3-64, CONV-3-128, CONV-3-128, CONV-3-256, CONV-3-256, CONV-3-256, CONV-3-256, CONV-3-512, CONV-3-512, CONV-3-512, CONV-3-512, CONV-3-512, CONV-3-512, CONV-3-512, CONV-3-512, maxpool, FC-4,096, FC-4,096, FC-1,000, softmax. (See Figure 4.20.) Quoted from Ref. [50]: "A fixed-size RGB image is used as input. The preprocessing entails subtracting the mean RGB value from each pixel. The image is passed through a stack of convolutional layers with filters with a 3×3 receptive field (which is the smallest size to capture the notion of left/right, up/down, center – in one version 1×1 convolution filters are used, representing a linear transformation of the input channels, followed by non-linearity). The convolution stride is fixed to 1 pixel; the spatial padding of the convolution layer input is such that the spatial resolution is preserved after convolution. Spatial pooling is carried out by five max-pooling layers, which follow some of the convolution layers. Max-pooling is performed over a 2×2 pixel window with stride 2. A stack of convolutional [CONV] layers (which has a different depth in different architectures) is followed by three FC layers: the first two have 4,096 channels each, the third performs 1,000-way ILSVRC classification, and thus contains 1,000 channels (one for each class). The final layer is the softmax layer. The configuration of the FC layers is the same in all networks. All hidden layers are equipped with the ReLU non-linearity. However, architectural simplicity results in a high cost because evaluating the network requires a large amount of computation."	~144 million	[50]

| Inception architecture/ GoogLeNet | 2014 | Network learns to extract features at different scales from the input. To save computation, smaller convolutions are used in a series of "Inception cell" units which perform a set of convolutions at different scales and then aggregate the results (e.g. each 5×5 convolution is replaced by two 3×3 convolution, and so on). Model makes extensive use of dimension reduction, this being a special case of factorizing convolutions in a computationally efficient manner. Specifically, the network uses various techniques for (i) factorization into smaller convolutions; (ii) spatial factorization into asymmetric convolutions; (iii) efficient grid size reduction; and (iv) model regularization via label smoothing. (See Figure 4.21.) | 5–23 million | [6, 51] |

Quoted from Ref. [6]: "Inception networks are fully convolutional; each weight corresponds to one multiplication per activation … reduction in computational cost results in reduced number of parameters. This means that with suitable factorization, [one] can end up with more disentangled parameters and therefore with faster training. Also, [one] can use the computational and memory savings to increase the filter-bank sizes of our network while maintaining [the] ability to train each model replica on a single computer… . factorized the traditional 7×7 convolution into three 3×3 convolutions … the Inception part of the network [has] 3 traditional inception modules at the 35×35 with 288 filters each. This is reduced to a 17×17 grid with 768 filters using the grid reduction technique. This is followed by five instances of the factorized inception modules. At the coarsest 8×8 level, [one has] two Inception modules, with a concatenated output filter bank size of 2,048 for each tile. The output size of each module is the input size of the next one."

(Continued)

Table 4.2 (Continued)

Model	Date	General approach	Parameters	References
ResNet		A modularized architecture that stacks building blocks of the same connecting shape. The system uses many sequentially stacked "residual units" (aka *residual blocks*) with which intermediate layers of a block can learn a residual function with reference to the block input. Each residual unit can have two or three layers – see Figure 4.22. More specifically, each unit residual unit can be expressed in a general form $y_\ell = h(x_\ell) + F(x_\ell, W_\ell), x_{\ell+1} = f(y_\ell)$, where x_ℓ and $x_{\ell+1}$ are input and output of the ℓ-th unit, and F is a residual function (e.g. a stack of two 3×3 convolutional layers). For example, $h(x_\ell) = x_\ell$ is an identity mapping and f is a ReLU function. The central concept of ResNets is to learn the additive residual function F with respect to $h(x_\ell)$. The residual function is, thus, a refinement element where the NN learns how to adjust the input feature map (rather than learning new and distinct feature maps) – if no refinement is required the residual block effectively becomes just an identity function. ResNets can have 200–1,000 layers; ResNets that are over 100-layer deep have shown state-of-the-art accuracy for several challenging recognition tasks on ImageNet datasets. *Note:* Although deep residual networks are able to scale up to thousands of layers and still demonstrate improving performance, each fraction of a percent of improved accuracy requires doubling the number of layers; consequently, training very deep residual networks has a challenge of diminishing feature reuse, which makes the training of these networks slow and onerous. Wide residual networks, [53], can address these problems by decreasing depth and increasing width of residual networks. There are three simple ways to increase power of residual blocks: (i) add more convolutional layers per block, (ii) widen the convolutional layers by adding more feature planes, and (iii) increase filter sizes in convolutional layers. Wide residual networks have shown to be superior over their commonly used thin and very deep counterparts: even a simple 16-layer-deep wide residual network outperforms in accuracy and efficiency all previous deep residual networks, including thousand-layer deep networks.	~25–57 million (ResNet50/ ResNet152)	[52, 54] (also see [53])

ResNeXt		The ResNeXt architecture is an extension of the deep residual network by replacing the residual block with a new residual block that makes use of a "split-transform-merge" approach that is employed in the Inception architecture of GoogLeNet. Here, instead of performing convolutions over the full input feature map, the block's input is projected into a series of lower (channel) dimensional representations on which then one separately applies smaller convolutional filters before merging the results. See Figure 4.23. Quoted for Ref. [55] "Carefully designed topologies are able to achieve compelling accuracy with low theoretical complexity. The Inception models have evolved over time, but an important common property is a split-transform-merge strategy. In an Inception module, the input is split into a few lower-dimensional embeddings (by 1×1 convolutions), transformed by a set of specialized filters (3×3, 5×5, etc.), and merged by concatenation. It can be shown that the solution space of this architecture is a strict subspace of the solution space of a single large layer (e.g. 5×5) operating on a high-dimensional embedding. The split-transform-merge behavior of Inception modules is expected to approach the representational power of large and dense layers, but at a considerably lower computational complexity."	25 million	[55]
DenseNet	2016	A dense convolutional network that references feature maps from earlier stages in the network: each layer's feature map is concatenated to the input of *every downstream layer* within a given dense block. This enables layers further along within the network, in the given block to *directly* utilize, as needed, the features from earlier layers, fostering feature reuse within the network and improved efficiency. Thus, DenseNets combine features by concatenating them, whereby the jth layer has j inputs, consisting of the feature-maps of all preceding convolutional blocks; its own feature-maps are passed on to all $L - j$ subsequent layers. This introduces $L(L + 1)/2$ connections in an L-layer network, instead of just L, as in traditional architectures. Owing to the fact that the network is capable of directly utilizing any previous feature map, the system can operate with small output channel depths, say 12 filters per layer, significantly *reducing* the total number of parameters needed. DenseNets are able to achieve good performance with reduced complexity. See Figure 4.24.	1–40 million	[18]

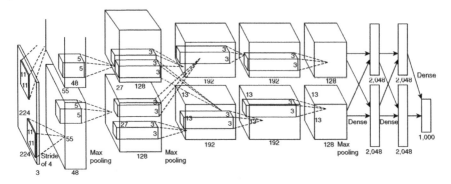

Figure 4.19 AlexNet CNN. *Source:* With permission from ACM, Ref. [4].

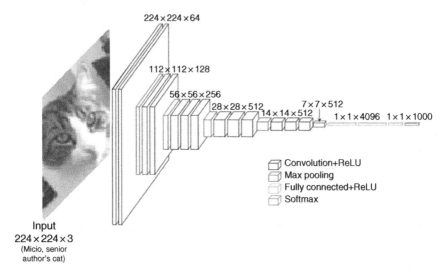

Figure 4.20 VGG16 CNN (13 CONV and 3 FC layers). *Source:* Adapted from Ref. [49].

Conv	3×3/2		299×299×3
Conv	3×3/1		149×149×32
Conv padded	3×3/1		147×147×32
Pool	3×3/2		147×147×64
Conv	3×3/1		73×73×64
Conv	3×3/2		71×71×80
Conv	3×3/1		35×35×192
3×Inception	Inception modules where each 5×5 convolution is replaced by two 3×3 convolution (*)		35×35×288
5×Inception	Inception modules after the factorization of the n×n convolutions: n=7 for the 17×17 grid (*)		17×17×768
2×Inception	Inception modules with expanded filter bank outputs. Used on the coarsest (8×8) (*)		8×8×1280
Pool	8×8		8×8×2048
Linear	Logits		1×1×2048
Softmax	Classifier		1×1×1000

(*) see reference for details

Figure 4.21 Inception (GoogLeNet) architecture summary. *Source:* Adapted from Ref. [6].

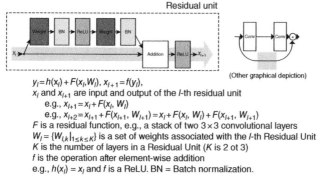

$y_l = h(x_l) + F(x_l, W_l), \ x_{l+1} = f(y_l),$
x_l and x_{l+1} are input and output of the l-th residual unit
e.g., $x_{l+1} = x_l + F(x_l, W_l)$
e.g., $x_{l+2} = x_{l+1} + F(x_{l+1}, W_{l+1}) = x_l + F(x_l, W_l) + F(x_{l+1}, W_{l+1})$
F is a residual function, e.g., a stack of two 3×3 convolutional layers
$W_l = \{W_{l,k}|_{1 \le k \le K}\}$ is a set of weights associated with the l-th Residual Unit
K is the number of layers in a Residual Unit (K is 2 ot 3)
f is the operation after element-wise addition
e.g., $h(x_l) = x_l$ and f is a ReLU. BN = Batch normalization.

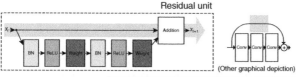

Two consecutive 3×3 convolutions with batch normalization and with ReLU preceding convolution: conv3×3 – conv3×3

Figure 4.22 ResNet: Top Original Residual Unit; Bottom: Improved Residual Unit. *Source:* Adapted from Ref. [52].

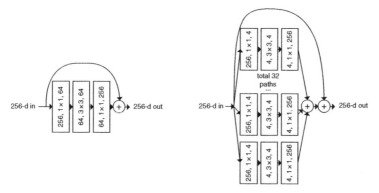

Stage	Output	ResNet-50			Stage	Output	ResNeXt-50 (32 × 4d)	
Conv1	112 × 112	7 × 7, 64, stride 2			Conv1	112 × 112	7 × 7, 64, stride 2	
		3 × 3 max pool, stride 2					3 × 3 max pool, stride 2	
Conv2	56 × 56	1 × 1, 64 3 × 3, 64 1 × 1, 256	× 3		Conv2	56 × 56	1 × 1, 128 3 × 3, 128, C = 32 1 × 1, 256	× 3
Conv3	28 × 28	1 × 1, 128 3 × 3, 128 1 × 1, 512	× 4		Conv3	28 × 28	1 × 1, 256 3 × 3, 256, C = 32 1 × 1, 512	× 4
Conv4	14 × 14	1 × 1, 256 3 × 3, 256 1 × 1, 1,024	× 6		Conv4	14 × 14	1 × 1, 512 3 × 3, 512, C = 32 1 × 1, 1,024	× 6
Conv5	7 × 7	1 × 1, 512 3 × 3, 512 1 × 1, 2,048	× 3		Conv5	7 × 7	1 × 1, 1,024 3 × 3, 1,024, C = 32 1 × 1, 2,048	× 3
	1 × 1	Global average pool 1000-d fc, softmax				1 × 1	Global average pool 1000-d fc, softmax	

Figure 4.23 ResNeXt. Left: A block of ResNet. Right: A block of ResNeXt with approximately the same complexity. A layer is shown in the graphic as (# in channels, filter size, # out channels). *Source:* Adapted from Ref. [55].

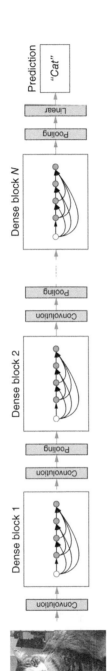

Input
(Micio, senior author's cat)

Dense block 1 → Pooling → Convolution → Dense block 2 → Convolution → Pooling → Dense block N → Pooling → Linear

Prediction "Cat"

DenseNet architectures for ImageNet. The growth rate for all the networks is $k = 32$.

Layers	Output size	DenseNet-121	DenseNet-169	DenseNet-201	DenseNet-264
Convolution	112×112	7×7 conv, stride 2			
Pooling	56×56	3×3 max pool, stride 2			
Dense block (1)	56×56	[1×1 conv / 3×3 conv] ×6	[1×1 conv / 3×3 conv] ×6	[1×1 conv / 3×3 conv] ×6	[1×1 conv / 3×3 conv] ×6
Transition layer (1)	56×56	1×1 conv			
	28×28	2×2 average pool, stride 2			
Dense block (2)	28×28	[1×1 conv / 3×3 conv] ×12	[1×1 conv / 3×3 conv] ×12	[1×1 conv / 3×3 conv] ×12	[1×1 conv / 3×3 conv] ×12
Transition layer (2)	28×28	1×1 conv			
	14×14	2×2 average pool, stride 2			
Dense block (3)	14×14	[1×1 conv / 3×3 conv] ×24	[1×1 conv / 3×3 conv] ×32	[1×1 conv / 3×3 conv] ×48	[1×1 conv / 3×3 conv] ×64
Transition layer (3)	14×14	1×1 conv			
	7×7	2×2 average pool, stride 2			
Dense block (4)	7×7	[1×1 conv / 3×3 conv] ×16	[1×1 conv / 3×3 conv] ×32	[1×1 conv / 3×3 conv] ×32	[1×1 conv / 3×3 conv] ×48
Classification layer	1×1	7×7 global average pool			
		1000D fully-connected, softmax			

Note that each "conv" layer shown in the table corresponds the sequence BN-ReLU-Conv.

Figure 4.24 Example of DenseNet and elements. *Source:* Adapted from Ref. [18].

4.5   Specific Application Examples

There are a large number of applications documented in the vast research and patent literature. A handful of such applications are described below, in no particular order.

4.5.1   Semantic Segmentation and Semantic Edge Detection

There is an interest in end-to-end learning frameworks that unify both *semantic segmentation* and *semantic edge detection*. The learning framework enables multi-tasking to produce high quality edge and segmentation using a single backbone network and with mutual improvement. A Coupled Segmentation and Edge Learning (CSEL) system is described in [57]. CSEL processes images and utilizes computers to understand the images and automate visual tasks that are typically performed by humans. The image processing includes extracting data from the images and using the extracted data for the automated visual tasks. The CSEL processor extracts segmentation and edge information from an input image and provides a refined semantic feature map. The method includes: (i) receiving an input image, (ii) generating from the input image a semantic feature map, an affinity map, and a semantic edge map from a single backbone network of a CNN, and (iii) producing a refined semantic feature map by smoothing pixels of the semantic feature map using spatial propagation and by controlling the smoothing using both affinity values from the affinity map and edge values from the semantic edge map. Figure 4.25 illustrates a block diagram of an example of a CV system, say, for autonomous or semiautonomous driving systems or a smart home.

The CSEL processor described in [57] includes a backbone network and a dynamic gap propagation (DGP) layer. The CSEL processor is typically a CNN that provides the semantic edge map, the affinity map, and the semantic edge map to the DGP layer utilizing convolution operations. The backbone network includes three processing streams: a feature stream, an affinity stream, and an edge stream. The feature stream produces encoded semantic features from an input image. The encoded semantic features, provided as a semantic feature map, encodes segmentation information from the input image and is represented by "F" in Figure 4.25. The affinity stream produces an affinity map that encodes the affinity of pairwise neighboring pixels of the input image. The edge stream produces a semantic edge map from the shared concatenation of dense side features and edge classification from the input image. The affinity map and the semantic edge map are represented by "A" and "E." The backbone network provides the semantic feature map, the affinity map, and the semantic edge map to the DGP layer. The DGP receives the outputs from the three processing

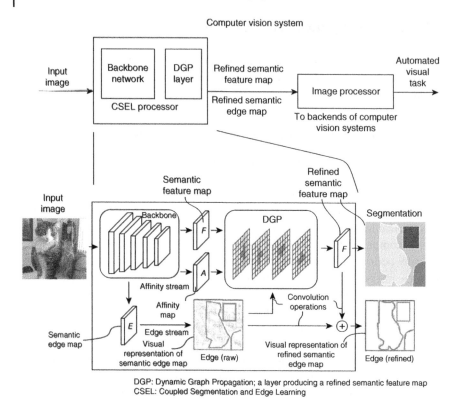

DGP: Dynamic Graph Propagation; a layer producing a refined semantic feature map
CSEL: Coupled Segmentation and Edge Learning

Figure 4.25 Example of CV system. *Source:* Adapted from Reference [57]/U.S. Patent.

streams and uses learnable message passing to produce a refined semantic feature map for both segmentation and edge refinement. The DGP layer transforms sparse edge signals into dense region-level ones to interact with segmentation. A visual representation of the semantic edge map is provided to illustrate the raw edges of the semantic edge map before refinement. In addition to being provided to the DGP layer, the semantic edge map is also summed with an output of the DGP layer, a refined semantic feature map, to provide the refined semantic edge map. The refined semantic feature map can also be provided for segmentation. A visual representation of the refined semantic feature map and the refined semantic edge map are also illustrated in Figure 4.25. The refined semantic feature map and the refined semantic edge map can be provided to a backend of a CV system.

4.5.2   CNN Filtering Process for Video Coding

For block-based video coding, a video slice (e.g. a video picture or a portion of a video picture) may be partitioned into video blocks, for example, CTUs) and/ or coding nodes. As discussed earlier, video blocks in an intra-coded (I) slice of a picture (frames) are encoded using spatial prediction with respect to reference samples in neighboring blocks in the same picture. Video blocks in an inter-coded (P or B) slice of a picture may use spatial prediction with respect to reference samples in neighboring blocks in the same picture or temporal prediction with respect to reference samples in other reference pictures. Video encoders and video decoders may typically implement in-loop filters that improve the quality of reconstructed pictures deblocking filters and Adaptive Loop Filters (ALFs). Recently, CNN based filters have been proposed; these filters include an NN that takes blocks of a reconstructed picture as input and outputs filtered blocks.

The CNN-based filter approach proposed in [46] uses a Leaky Rectified Linear Unit (LeakyReLU) activation function; see Figure 4.26 (this filter is not a very deep NN). Note that during training the LeakyReLU activation provides better performance than the ReLU activation function and it provides greater stability than the Parametric Rectified Linear Unit (PReLU) activation function – also shown in Figure 4.26.

Figure 4.27 is a diagram illustrating an example CNN-based filter with multiple hidden layers and LeakyReLU as an activation function (the input layer and each

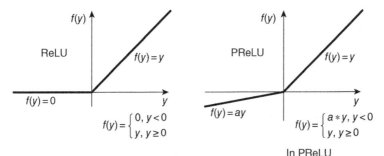

Figure 4.26   LeakyReLU.

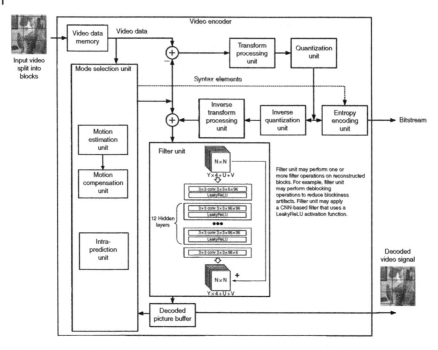

Figure 4.27 Use of CNNs on reconstruct a block of video to reduce artifacts. *Source:* Adapted from Ref. [46].

of the hidden layers is associated with an activation function that is applied to outputs of the neurons of the layer – the activation function has a significant impact on the performance and trainability of the NN).

4.5.3 Virtual Clothing

Virtual dressing entails performing fusion on a body image of a user and a clothes image, including target clothes, by using an image fusion technology. The goal is to obtain an image in which the user wears the target clothes, such that the user can get a sense of wearing them without actually trying them on.

In a virtual dressing process, feature extraction is usually performed on the body image and on the clothes image separately by using an image fusion model, and a new image; the image in which the user wears the target clothes is generated based on two extracted image features. Rough image features are extracted by using the image fusion model; however, the newly-generated image is likely to lack detailed information, which leads to distortion and results in a poor virtual dressing effect.

An improved exemplary process is described in [58] entails,

i) Obtaining a first body image including a target body and a first clothes image including target clothes;

ii) Transforming the first clothes image based on a posture of the target body in the first body image to obtain a second clothes image, the second clothes image including the target clothes, and a posture of the target clothes matching the posture of the target body;

iii) Performing feature extraction on the second clothes image, an image of a bare area in the first body image, and the first body image to obtain a clothes feature, a skin feature, and a body feature respectively, the bare area being an area of the target body in the first body image that is not covered by clothes; and

iv) Generating a second body image based on the clothes feature, the skin feature, and the body feature, the target body in the second body image wearing the target clothes.

As shown in Figure 4.28, the image fusion model includes (i) a clothes area prediction network, (ii) a spatial transformer network, (iii) a clothes coding network, (iv) a skin coding network, (v) a portrait coding network, and (vi) a decoding network. The clothes area prediction network is configured to determine an area covered by target clothes after a target body wears the target clothes. The spatial transformer network is configured to perform affine transformation on a clothes image based on a posture of the target body. The clothes coding network,

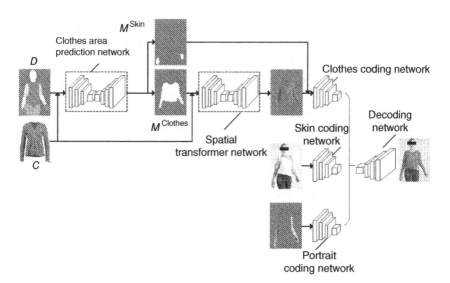

Figure 4.28 Image fusion model. *Source:* Reference [58]/U.S. Patent.

the skin coding network, and the portrait coding network are configured to extract features of a body image and the clothes image in different dimensions. The decoding network is configured to decode the extracted features in a plurality of dimensions to generate a new body image. The body area recognition network is constructed based on a CNN/FCN, and it includes at least one operation layer for extracting body semantic features.

4.5.4 Example of Unmanned Underwater Vehicles/Unmanned Aerial Vehicles

Classical techniques of object recognition from imagery from unmanned underwater vehicle (UUN) or unmanned aerial vehicle (UAV) are based on traditional CV algorithms and rely on standard image enhancing methods such as, but not limited to, contrast adjustment, brightness adjustment, and fast Fourier Transform methods. However, these methods are not typically accurate enough to replace a human operator, thus there is a need for detecting and classifying specific objects as targets of interest in environments including an underwater environment. Figure 4.29 [59] illustrates a possible ML approach to the matter.

4.5.5 Object Detection Applications

Object detection is important for many applications such as, but not limited to, identifying faces (e.g. for facial recognition or tracking) and identifying vehicles (e.g. for autonomous driving). An object detector can be implemented using CV-based detectors, or using an ML-based model that is configured to identify/classify specific classes of image features. For example, there is extensive ML support for Intelligent Transportation Systems (ITSs). ITS entails three category of tasks [60]: (i) perception tasks that try to detect, identify, and recognize data patterns to extract, understand, and present relevant information – this task deals a lot with CV; (ii) prediction tasks endeavoring to try to predict future states given historical and real-time data; and (iii) management tasks responsible for dictating the behavior in ITS. In large measure, perception tasks are typically widely supported by CNNs, and less so by recurrent neural networks (RNNs), long short-term memory (LSTM) models, R-CNNs, You Only Look Once (YOLO), and graph neural networks (GNNs). Prediction tasks are supported by the same type of DNNs just listed, but (currently) more so using LSTMs.

A method of processing image data for object detection can include the following steps (as in the system described in [61][12]): (i) obtaining, from an image

---

12 This entire section is summarized from reference [61].

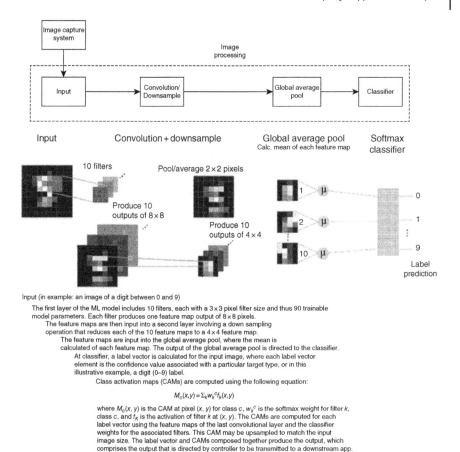

Input (in example: an image of a digit between 0 and 9)

The first layer of the ML model includes 10 filters, each with a 3×3 pixel filter size and thus 90 trainable model parameters. Each filter produces one feature map output of 8×8 pixels.

The feature maps are then input into a second layer involving a down sampling operation that reduces each of the 10 feature maps to a 4×4 feature map.

The feature maps are input into the global average pool, where the mean is calculated of each feature map. The output of the global average pool is directed to the classifier.

At classifier, a label vector is calculated for the input image, where each label vector element is the confidence value associated with a particular target type, or in this illustrative example, a digit (0–9).

Class activation maps (CAMs) are computed using the following equation:

$$M_c(x,y) = \Sigma_k w_k{}^c f_k(x,y)$$

where $M_c(x, y)$ is the CAM at pixel (x, y) for class c, $w_k{}^c$ is the softmax weight for filter k, class c, and f_K is the activation of filter k at (x, y). The CAMs are computed for each label vector using the feature maps of the last convolutional layer and the classifier weights for the associated filters. This CAM may be upsampled to match the input image size. The label vector and CAMs composed together produce the output, which comprises the output that is directed by controller to be transmitted to a downstream app.

Figure 4.29 ML system for object classification from unmanned underwater vehicle/unmanned aerial vehicle inputs. *Source:* Reference [59]/U.S. Patent.

capture device, a first image frame comprising an object; (ii) determining, using an object detector, an object validation score associated with detection of the object in the first image frame; (iii) determining the object validation score is less than a validation threshold; and (iv) based on the object validation score being less than the validation threshold, tracking the object for one or more image frames received subsequent to the first image frame. The object detector can detect (e.g. identify and/or classify) objects of interest in one or more image frames. Based on the detection objects of interest in an image frame, the object detector can output a detection or classification output. In some examples, the detection or classification output can include (i) information indicating a ROI (e.g. a bounding region, such as a bounding box) associated with a detected object or portion of the object and (ii) a confidence level or score corresponding to the detected object or portion of the object. In some implementations, a confidence level or score can include a value between 0 and 1 – a confidence level/score closer to 0 indicating a lower confidence that an object is accurately detected and a confidence level/score closer to 1 indicating a higher confidence that an object is accurately detected. In some cases, the detection or classification output can include a size of the ROI (e.g. bounding box or other bounding region) associated with an object detected in the image frame, a location of the ROI within the image frame in which the corresponding object is detected, and/or motion vector information associated with the object associated with the ROI. Additionally, in some cases, the detection or classification output can include a class associated with a detected object (e.g. a face, a vehicle, or other classification).

One example on an NN-based detector is CIFAR-10 (Canadian Institute For Advanced Research), as illustrated in Figure 4.30; the NN includes various convolutional layers (Conv1 layer, Conv2/ReLU2 layer, and Conv3/ReLU3 layer), several pooling layers (Pool1/ReLU1 layer, Pool2 layer, and Pool3 layer), and ReLU layers mixed in. Normalization layers Norm1 and Norm2 are also provided. A final layer is the interspace pruning/IP1 layer. The CIFAR-10 per se is dataset is a collection of images (60,000 32 × 32 color images of 10 classes – airplane, automobile, bird, cat, deer, dog, frog, horse, ship, and truck), that are commonly used to train ML and CV algorithms.

Another DL-based object detector to detect or classify objects in image frames is the single-shot detector (SSD), which is a fast single-shot object detector that can be applied for multiple object categories or classes. The SSD model utilizes multi-scale convolutional bounding box outputs attached to multiple feature maps at the top of the NN. Such a representation allows the SSD to efficiently model diverse box shapes.

Figure 4.31a from [61] includes an image frame and Figure 4.31b, c include diagrams illustrating how an SSD detector (with the VGG deep network-based model) operates; for example, SSD matches objects with default boxes of different aspect ratios (shown as dashed rectangles in Figure 4.31b, c). Each element of the feature map has a number of default boxes associated with it. Any default box with an intersection-over-union with a ground truth box over a threshold (e.g. 0.4, 0.5, 0.6) is considered a match for the object. For example, two of the

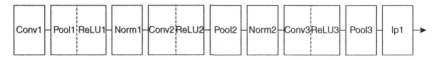

Figure 4.30 CIFAR-10 Object recognition NN.

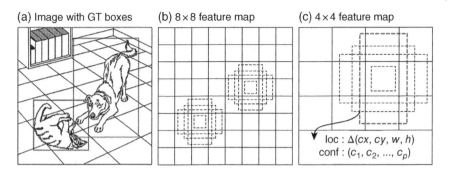

Figure 4.31 Single-shot detector. (a) Image with GT boxes. (b) 8×8 feature map. (c) 4×4 feature map. *Source:* Reference [61]/U.S. Patent.

8×8 boxes in Figure 4.31b are matched with the cat, and one of the 4×4 boxes in Figure 4.31c is matched with the dog. SSD has multiple features maps, with each feature map being responsible for a different scale of objects, allowing it to identify objects across a large range of scales. For example, the boxes in the 8×8 feature map of Figure 4.31b are smaller than the boxes in the 4×4 feature map of Figure 4.31c. For each default box in each cell, the SSD NN outputs a probability vector of length c, where c is the number of classes, representing the probabilities of the box containing an object of each class. In some cases, a background class is included that indicates that there is no object in the box. The SSD network also outputs (for each default box in each cell) an offset vector with four entries containing the predicted offsets required to make the default box match the underlying object's bounding box. The vectors are given in the format (ex, cy, w, h), with ex indicating the center x, cy indicating the center y, w indicating the width offsets, and h indicating height offsets. The vectors are only meaningful if there actually is an object contained in the default box. For the image frame shown in Figure 4.31a, all probability labels would indicate the background class with the exception of the three matched boxes (two for the cat, one for the dog).

Another DL-based detector that can be used to detect or classify objects in image frames includes the YOLO detector. Figure 4.32a from [61] includes an image frame and Figure 4.32b, c include diagrams illustrating how the YOLO detector operates. The YOLO detector can apply a single NN to a full image frame. The YOLO network divides the image frame into regions and predicts bounding boxes and probabilities for each region. These bounding boxes are weighted by the predicted probabilities. For example, as shown in Figure 4.32a, the YOLO detector divides up the image frame into a grid of 13×13 cells. Each of the cells is responsible for predicting five bounding boxes.

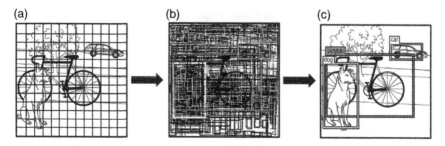

Figure 4.32 YOLO. (a) Object(s). (b) Predicted bounding boxes. (c) Final predicted bounding boxes and classes. *Source:* Reference [61]/U.S. Patent.

A confidence score is provided that indicates how certain it is that the predicted bounding box actually encloses an object. This score does not include a classification of the object that might be in the box, but indicates if the shape of the box is suitable. The predicted bounding boxes are shown in Figure 4.32b. The boxes with higher confidence scores have thicker borders. Each cell also predicts a class for each bounding box. For example, a probability distribution over all the possible classes is provided. Any number of classes can be detected, such as a bicycle, a dog, a cat, a person, a car, or other suitable object class. The confidence score for a bounding box and the class prediction are combined into a final score that indicates the probability that that bounding box contains a specific type of object. For example, the box with thick borders on the left side of the image frame in Figure 4.32b is 85% sure it contains the object class "dog." There are 169 grid cells (13×13) and each cell predicts five bounding boxes, resulting in 945 bounding boxes in total. Many of the bounding boxes will have very low scores, in which case only the boxes with a final score above a threshold (e.g. above a 30% probability, 40% probability, 50% probability) are kept. Figure 4.32c shows an image frame with the final predicted bounding boxes and classes, including a dog, a bicycle, and a car.

4.5.6 Classifying Video Data

There are many methods for classifying video data. A number of these methods depend on (i) extracting video tokens, which are a representation of spatiotemporal information in the video data and (ii) providing the video tokens as input to a video understanding model which then provides a classification output.

Reference [62] discusses a transformer-based ML model (MLM) architectures for video understanding; the model can operate on extracted spatiotemporal tokens from input video, which are then encoded by a series of transformer layers. The transformer-based model includes a self-attention mechanism that computes self-attention on a sequence of extracted spatiotemporal tokens (transformers and attention were discussed in Chapter 2). The video data can include a plurality of image frames and the image frames can depict one or more objects – the video data can be stored in a suitable format, such as, but not limited to an .mp4 file format or a.wav file format. The video data can include a number of image frames (e.g. T), a height (e.g. H), a width (e.g. W), and/or a number of channels (e.g. C). As an example, the video data can include three channels, such as a red channel, a green channel, a blue channel, and/or other color channels. The video is a point vector in a $T \times H \times W \times C$ space. In some implementations, positional embeddings can be added to the sequence of tokens. This sequence of spatiotemporal tokens is then passed through the video understanding model.

Figure 4.33 (synthetized from [62]) depicts a block diagram of an example video understanding model. The video understanding model receives input data (e.g. video data) and produces a classification output). As a first step, the model extracts a plurality of video tokens from the video data. The video tokens are a representation (e.g. an embedding representation) of spatiotemporal information of the video data. Processing a video can involve a large number of extracted tokens (each frame can be broken into one or more "patches"; for instance, each patch

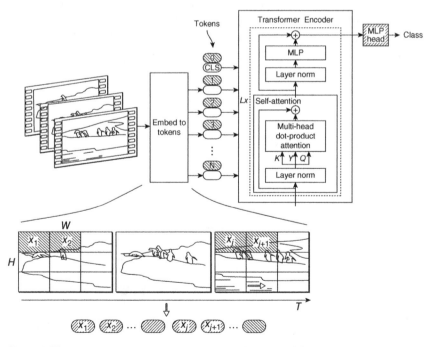

Figure 4.33 Video understanding model. *Source:* Reference [62]/U.S. Patent.

can span a subset of the length and/or width of a single frame). The video understanding model can include a video transformer encoder model. The transformer encoder model can include an attention mechanism (e.g. a self-attention mechanism), at least one normalization layer and/or at least one MLP layer; the output from the transformer encoder is provided to a classification model (e.g. an MLP head) that can classify the output and provide the video classification output.

4.5.7 Example of Training

As noted in Chapter 1, ML gives computers the ability to "learn" a specific task without expressly programming the computer for that particular task. An NN may be used to convert images that are natively generated (e.g. by a rendering engine) into images of a higher resolution. In the imaging context, learning takes place by using thousands or millions of photos to "train" the network to recognize when an object is in a photograph. The training process typically includes determining weights for the model that achieves the indicated goal (e.g. identifying cats [Micio] within a photo).

Training an NN may involve using one or more loss functions. The training process may include backpropagation using a loss function in a way that seeks to train the model that will minimize the loss represented by the function. Different loss functions can be used, including $L1$ (least absolute deviations) and $L2$ (least square errors) loss functions. See Figure 4.34, partially based on [63] for a quick description of $L1$ ($L2$ is the Euclidean distance). In some applications, an NN is

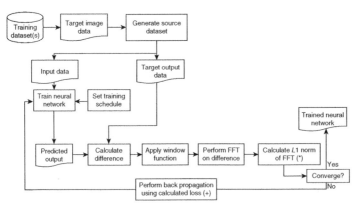

(+) Back propagation on the weights and other values of the neural network; e.g. the weights of the neural network are updated by using, for example, gradient decedent that uses the derivative of the loss function that is measured at the data points of the training data set. The training process is repeated until this convergence is reached.

(*) Algorithm takes the $L1$ Norm of the absolute value of each coefficient in the two dimensional array of complex numbers. $L1$ Norm is the sum of the magnitudes of the vectors in a space. It is a basic way to measure distance between vectors. L^1-norm of a vector $\mathbf{x} = (x_1, x_2, ..., x_n)$ is L^1-norm $= |\mathbf{x}|_1 = \sum_{r=1}^{n} |x_r|$.

Figure 4.34 Training and $L1$ loss function. *Source:* Reference [63]/U.S. Patent.

trained to upscale images from one resolution to another resolution, where the system can be configured to generate, from the plurality of images, input image data and then apply the input image data to an NN to generate predicted output image data. A difference between the predicted output image data and target image data is calculated and that difference may then be transformed in frequency domain data. The $L1$-loss is then used on the frequency domain data calculated from the difference, which is then used to train the NN using backpropagation (e.g. stochastic gradient descent). Using the $L1$ loss may encourage sparsity of the frequency domain data, which may also be referred to, or part of, Compressed Sensing.

4.5.8   Example: Image Reconstruction is Used to Remove Artifacts

Image reconstruction can be used to remove artifacts (e.g. missing pixels, aliasing, noise, and low resolution) from input images. Input data of interest can include images data, data acquired by depth sensors, data acquired by temperature sensors, density data (e.g. medical imaging and geological), and so on. Advances have been made in recent years in adaptive sampling for rendering images, reducing the number of samples needed to produce each image; however, the sample reduction often results in the loss of high-frequency details and, when applied to sequences of images (e.g. video), temporal artifacts may be introduced. Real-time rendering is limited by machine cycles/time to compute additional samples per pixel and, so, it relies on approximate solutions based on spatio-temporal image reconstruction filters applied to one sample per-pixel images.

Although there is a current dearth DNNs use in real-time rendering because of computational demands, NN-based rendering techniques have been proposed, such as in [15]: antialiasing may be modeled as an image reconstruction problem and a CAE may be used as a starting point to develop the encoder/decoder NN model.

A sample predictor NN learns spatio-temporal sampling strategies, such as placing more samples in dis-occluded regions and tracking specular highlights. Temporal feedback enables a denoiser NN to boost the effective input sample count and increases temporal stability. The sample predictor and denoiser are trained and run at interactive rates to achieve significantly improved image quality and temporal stability compared with conventional adaptive sampling techniques.

In this proposal, a warped external RNN is used for reconstructing data [15]. The warped external RNN is not recurrent at each layer and has a feed-forward flow-only warping external state output by the final layer (in contrast, in a conventional RNN, hidden state generated at each layer is provided as a feedback input to the generating layer); the warped external RNN is trained end-to-end to minimize the errors, between pairs of aliased and antialiased images. During supervised training, the warped external RNN learns to identify aliased image features and to adaptively remove (i.e. filter out) the undesirable artifacts (e.g. aliased image features) and/or modify areas with missing and incorrect information. After being trained, the warped external RNN may be deployed to reconstruct data. The warped external RNN includes an encoder/decoder NN model, a temporal warp function, and a combiner function. The encoder/decoder NN receives input data at time t and warped external state from the previous iteration, i.e. at time $t-1$. The input data includes artifacts that are removed during the reconstruction process to produce output data that approximates the input data without the artifacts. Output data that approximates the input data without the artifacts has fewer artifacts compared with the input data. The warped external state from time $t-1$ includes warped reconstructed data from time $t-1$. The encoder/decoder NN model processes the input data and the warped external state using multiple layers to produce at least one filter kernel. The combiner function receives at least one filter kernel, the input data from time t, and warped reconstructed data from time $t-1$. The combiner function applies at least a first portion of at least one filter kernel to the reconstructed data to produce filtered first input data. The combiner function applies at least a second portion of at least one filter kernel to the input data to produce filtered second input data.

Figure 4.35 (synthetized form [15]) illustrates a block diagram of the *encoder/decoder* NN model; each stage of the encoder portion of the encoder/decoder NN model uses one CONV layer and a pooling layer. In a specific implementation, 3×3 convolutions are used in the encoder/decoder NN model. In a specific implementation the convolutional layers of the *encoder* portion are $N \times N$, each followed by a 2×2 max pooling layer, where $N = 32, 64, 96, 128, 160$ in a feed-forward sequence with the output of each max pooling layer input to each convolutional layer. Each stage of the *decoder* portion of the encoder/decoder NN model uses the nearest upsampling layer followed by two convolutional layers. In some implementations, the output of each encoder stage is propagated to the corresponding decoder stage via residual skip connections and accumulated with the output of an upsampling layer (the residual skip connections enable faster training of deep convolutional networks by improving the backpropagation of gradients). The figure also illustrates an external NN layer without hidden state recursion. Instead of receiving hidden state generated by the first CONV layer during processing of a previous frame, the first CONV layer receives the warped external state generated by the encoder/decoder NN model during processing of a previous frame. The encoder/decoder NN model functions as a single recurrent layer. The warped external state generated by the encoder/decoder NN model, combiner function, and temporal warp function acts as the hidden state for the encoder/decoder NN model and is used as input with the next data frame. See refence [15] for a more inclusive discussion of this system.

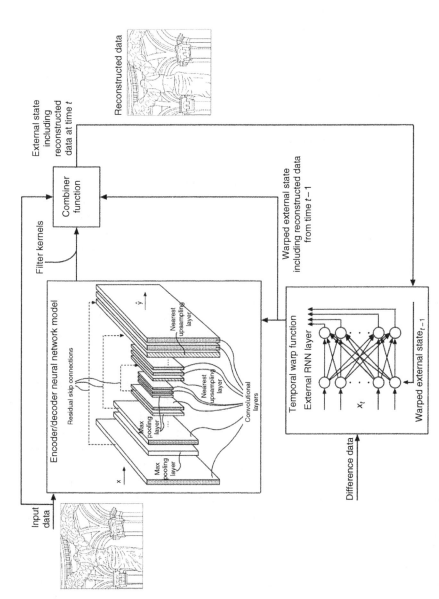

Figure 4.35 Neural Network System with Temporal Feedback for Denoising of Rendered Sequences. *Source:* Reference [15]/U.S. Patent.

4.5.9 Example: Video Transcoding/Resolution-enhancement

When different user devices (e.g. a smartphone, a tablet computer, a desktop computer, or the like) connect to the server for viewing the video, the service provider can transcode the video into different resolutions to adapt to the different transmission bandwidths of the user devices or different quality demands of users. A number of resolution-enhancement techniques can be used to enhance video [47]: (i) interpolation-based techniques, (ii) reconstruction based techniques, and (iii) learning-based techniques such as DNN/RNN/CNN for taking input data and outputting a hierarchical representation of the input data. DNN-based techniques generally provide high-quality results, making the technique attractive for commercial application scenarios.

4.5.10 Facial Expression Recognition

Facial *expression recognition* has become increasingly prevalent. However, many facial expressions are asymmetric; for example, for many facial expressions, the left side of the face may be more expressive than the right side of the face (or vice versa). As noted in Ref. [64] basic expression recognition techniques may have difficulty analyzing such asymmetric facial expressions; to improve expression recognition the training set is augmented to increase the number of asymmetric samples in the training dataset. Image data augmentation is the most well-known type of data augmentation and involves creating transformed versions of images in the training dataset that belong to the same class as the original image. Transforms may include a range of operations from the field of image manipulation, such as shifts, flips, zooms, and much more. The intent is to expand the training dataset with new, plausible examples. Modem DL algorithms, such as CNN, are capable of learning features that are invariant to their location in the image. Augmentation can further aid in this transform invariant approach to learning and can aid the model in learning features that are also invariant to transforms such as left-to-right and top-to-bottom ordering, light levels in photographs, and more.

An asymmetry loss may be introduced during model training to help refine asymmetrical expressions [64]. The asymmetry loss may act as an indirect augmentation of the original input. A first loss function for the original expression parameters and ground truth labels associated with the original expression parameters may be determined. A second loss function for asymmetrical expression parameters and ground truth labels associated with the asymmetrical expression parameters may be determined.

Figure 4.36 (synthetized from [64]) illustrates an example system for controlling an animation based on predictions of facial expression output from a trained facial expression recognition model [64]. The system applies an NN (e.g. a trained

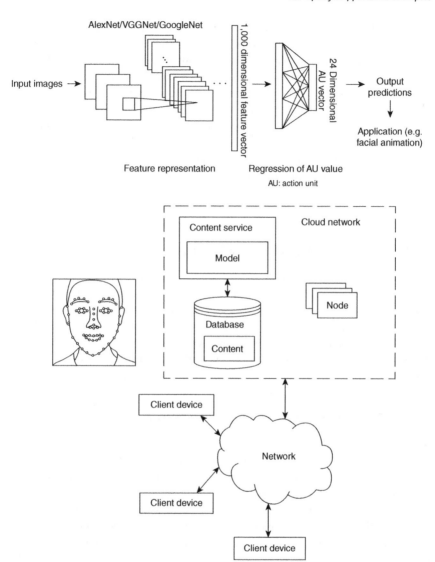

Figure 4.36 Facial expression recognition. *Source:* Adapted from Ref. [64].

facial expression model) to generate a feature representation (e.g. feature vector) associated with each image. The NN includes for example VGGnet (as noted earlier, VGGnet has three FC layers: the first two have 4,076 channels each and the third has 1,000 channels, 1 for each class), or AlexNet (as noted, AlexNet is a CNN that contains eight layers with weights; the first five are CONV and the remaining three are FC– the output of the last fully connected layer is fed to a 1,000-way softmax which produces a distribution over the 1,000 class labels), or GoogLeNet (GoogLeNet is a 22-layer deep CNN), and/or any other suitable type of NN. The initiation output of the system could be a high dimensional (e.g. 1,000 dimensional) feature vector, for each input image. Regression (for example, linear regression) may be performed on the high dimensional feature vector(s). Linear regression is a supervised learning algorithm used to predict a real-valued output. The linear regression model is a linear combination of the features of the input examples. For example, the real-valued output may be a 24-dimensional feature vector. The 24-dimensional feature vector may be, for example, a facial action unit (AU) vector. Each 24-dimensional feature vector may indicate a predicted expression associated with an image among the input images.

4.5.11 Transformer Architecture for Image Processing

As discussed in previous chapters, transformers have the ability to convert data from one-dimensional strings (such as sentences) to two-dimensional arrays (such as images). Researchers are applying transformers to more difficult tasks, such as the creation of new images. Another potential use of transformers is for multimodal processing – models that can process multiple types of data simultaneously, such as raw images, video, and languages. Transformers could soon rival generative adversarial networks (GANs) as a technique for generating images and video [65].

4.5.12 Example: A GAN Approach/Synthetic Photo

GAN-based approaches for image processing have been suggested for some applications, for example [13]. In this example, the generator is a network that is learning to perform the task of producing a synthetic photo. The generator receives a random noise z as input and generates an output $G(z)$. The discriminator is an NN that is learning to discriminate whether a photo is a real-world photo. The discriminator receives the input x, where x represents a possible photo. An output $D(x)$ generated by the discriminator represents the probability that x is a real-world photo. If $D(x)$ is 1, it indicates that x is a real-world photo. If $D(x)$ is 0, it indicates that x is not a real-world photo. In training the GAN, an objective of the generator is to generate a photo as real as possible (to avoid detection by discriminator), and an objective of the discriminator is to try to discriminate between a real-world photo and the photo generated by the generator. This training constitutes a dynamic adversarial process between the generator and the discriminator. The aim of the training is for the generator to learn to generate a photo that the discriminator cannot discriminate from a real-world photo (ideally, $D(G(z)) = 0.5$). The trained generator is then used for model application, which is generation of a synthetic photo.

4.5.13   Situational Awareness

Situational awareness[13,14] (SA) refers to one's capability to efficiently comprehend their physical environment. SA can be conceived as knowing what is going on around us [69]. A useful definition from [70] is "The perception of the elements in the environment within a volume of time and space, the comprehension of their meaning, and the projection of their status in the near future." More specifically, SA involves the timely acquisition of knowledge about real-world events, distillation of those events into higher-level conceptual constructs, and their synthesis into a coherent context-sensitive view [71]. In practical terms, SA can be defined as being in a state of awareness. For public safety and law enforcement teams, this means having full, real-time visibility on the status of prospective threats – and possible targets. SA entails the ability to develop and deploy traditional or AI-based mechanisms to assess, recognize, anticipate, and intercept events specific to the use case of interest.

SA is defined in a number of ways, including the following "up-to-the-minute cognizance or awareness required to move about, operate equipment, or maintain a system" [72]. More formally, SA is described as knowing what is "going on" around a reference entity and within that knowledge of the entity's surroundings, knowing what is important [73]; the process of perceiving the elements in the environment, understanding the elements in the environment, and the projection of their status into the near future [74]. SA originates from human factor and cognitive studies [75]. It has been extensively studied and applied to several disciplines, including psychology, aviation and military operations, starting with the important work of M. R. Endsley, e.g. [74] and numerous other publications. Endsley's SA reference model defines three levels of SA: perception, comprehension and projection. McGuinness and Foy [76] provided extension of Endsley's SA model that includes resolution of the perceived state as the level four SA component. SA spans many domains and applications. A short list of illustrative examples includes military applications, behavioral science, emergency medical call-outs, vehicle driving, search-and-rescue, law enforcement, and cybersecurity threat operations/security incident response. ML and fuzzy logic are two of the many underlying technologies used in SA.

In practical terms SA deals with capturing incoming data, analyzing the data and generating actionable intelligence. Recent technology trends in video sensing, edge computing, big data storage and ML can be aligned to create a shared real-time information system for SA in a number of environments including physical security, law enforcement, and vehicular systems. Ultimately, many smart city

---

13 This section is a summarization of papers [66–68] published by these authors.
14 Some researchers have used the phrase "situation awareness."

applications including livability, infrastructure management, traffic management, utilities management, policing, and physical security, which are all key aspects of a city's operation, entail some aspect of SA, which in turn entail big data analytics, image processing, and cloud computing, all of which can benefit from ML/DL mechanisms. Some underlying trends of relevance that impact SA include the following, among others [71]:

i) The growing deployment and acceptance of always-on video cameras in public spaces. Some have predicted that "It will soon be possible to find a camera on every human body, in every room, on every street, and in every vehicle" [77].

ii) Another important trend is real-time video analytics using edge computing [78, 79]: to facilitate scalability, such analytics need be performed close to the point of capture because transferring video to the cloud from a large set of cameras typically results in excessive bandwidth requirements on the ingress networks; thus, the solution is to perform the video analytics on dispersed edge gateways that have wired or wireless LAN connectivity to associated cameras.

SA helps decision makers spanning a large set of environments have the distilled information and focused understanding to make effective decisions in the course of their work; clearly, this is desirable in the law enforcement context. Figures 4.37–4.39 depict the broad SA ecosystem. When properly designed,

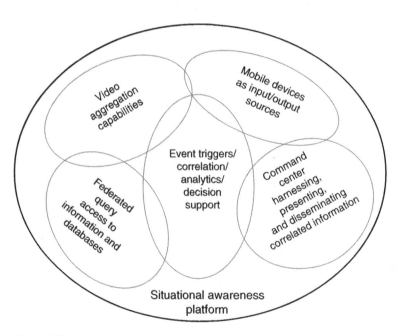

Figure 4.37 A situational awareness ecosystem.

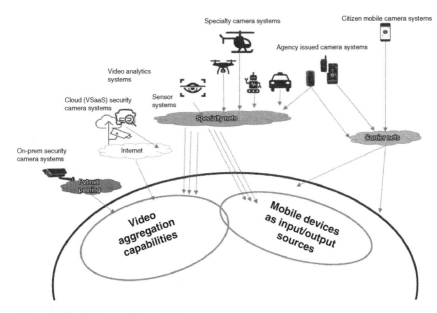

Figure 4.38 Plethora of input devices.

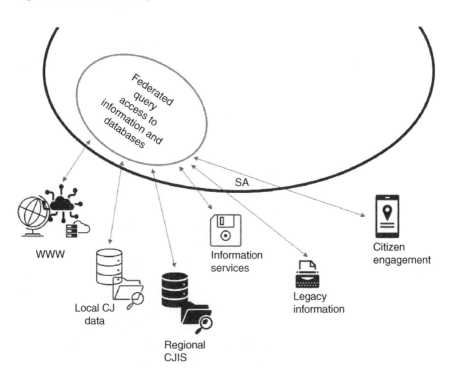

Figure 4.39 Federated queries.

configured, and deployed, these systems are harnessed law enforcement and public safety environment to provide incident commanders as well as analysts the ability to assess threats with the stated goal of anticipating event in advance of their actual occurrence (i.e. providing timely and accurate intelligence). SA provides ongoing value during and after events' occurrence, but admittedly depending on the situation this may offer less value than advance warning.

As covered in other chapters in this book, ML predicts and classifies data using various algorithms optimized to the dataset in question. ML techniques are increasingly utilized to analyze, cluster, associate, classify, and apply regression methods to SA data (as well as to many other IoT environments). ML techniques endeavor to examine and establish the internal relationships of a set of data collected from the plethora of input devices collecting visual, audio and signal data from the field. A short list of references includes, but is certainly not limited to, [80–89].

In the SA context, ML techniques are applicable to image processing and analysis, CV, speech recognition, and natural language understanding. There are six major types of technology platform categories that are used to present real time information to law enforcement/public safety. This information incorporates visual images acquired from numerous cameras (both real time and historical); data contained in countless databases/systems; and event triggers provided by a range of sources including emergency 911 calls, sensors, and video analytics. The technologies used to deliver these capabilities are specific to their mission, but also overlap in some capabilities, which complicates what could otherwise be a straightforward correlation of needs to capabilities. These six technologies are:

1) Video Management System (VMS);
2) Physical Security Information Management (PSIM);
3) Computer-Aided Dispatch/Records Management Systems (CAD/RMS);
4) Correlation/Data Mining/Analytics;
5) Visualization/Presentation; and,
6) Hybrid SA platform.

There are three parts to a modern VMS, which are (i) cameras, (ii) recorders/access-managers, and (iii) viewers. The modern VMS has evolved from its Closed-Circuit Television (CCTV) physical security roots, although many still see the VMS as a simplistic video recording platform for security related video. One need only look at the solutions from the leading vendors to understand that these VMS platforms offer a wealth of functions that go far toward meeting the needs for real-time information and the deep analysis to support decision making. However, the single overwhelming limitation of these VMS offerings is that they must manage the connection to the camera, either directly or through some federation process. Unfortunately, the capability to federate VMS platforms is limited to federation within a specific vendor's platform. Thus, Vendor A can only federate with Vendor A, Vendor B strictly with Vendor B, Vendor C exclusively with Vendor C, and so on, but none can federate with others. Because all public/private partnerships entail accessing video content from a multitude of different VMS platforms, the advantages presented by VMS federation cannot be exercised.

Despite the technological advancements of the technology, with ML methods in particular, a number of ethics questions related to SA exist. Recently, a number of scientists and human rights activists have raised ethical concerns about a number of AI-based technologies, including face recognition technology in particular. A relatively large research investment of the CV community has focused in recent years on advancing and improving facial recognition methods. An increasing number of scientists are now urging researchers to avoid working with firms or universities linked to unethical projects, to re-evaluate how they collect and distribute facial recognition data sets and to rethink the ethics of their own studies. Some institutions are already taking steps in this direction and several journals and an academic conference have announced that they would undertake additional ethics checks on studies on these topics [90]. There are acceptable applications of face and biometric recognition: technology that recognizes or analyses human faces has a number of useful applications, such as to assist in the process of finding lost children; of tracking criminals and terrorists by law enforcement agencies; of accessing smartphones and cash machines; and in telemedicine to help diagnose or remotely track consenting participants. However, there must be an ethical recognition that technologies that can remotely identify or classify people without their knowledge is fundamentally perilous.[15]

For facial recognition algorithms to work effectively, they must be trained and tested on large data sets of images, ideally captured many times under different lighting conditions and at different angles. The issue is that it is difficult to train accurate facial recognition algorithms without vast data sets of photos. Earlier in time, researchers used volunteers to pose for these photos, but now most

---

15 AI-powered chatbots such as ChatGPT — and their visual image-creating counterparts like DALL-E – have been in the news for fear that they could replace human jobs. Such AI tools work by scraping the data from millions of texts and pictures, refashioning new works by remixing existing ones in intelligent ways that make them seem almost human; a lesser-known AI-driven database is scraping images from social media for millions of people. For example, Clearview AI is a tech company that specializes in facial recognition services and markets its facial recognition database to law enforcement "to investigate crimes, enhance public safety, and provide justice to victims". The database has been used by US law enforcement nearly a million times since 2017. Clearview states that "Clearview AI's database is used for after-the-crime investigations by law enforcement, and is not available to the general public ... Every photo in the dataset is a potential clue that could save a life, provide justice to an innocent victim, prevent a wrongful identification, or exonerate an innocent person." However, revelations as to how the company obtains images for their database of nearly 30 billion photos have caused concerns [91].

researchers collect facial images without obtaining any consent, e.g. taken at university campuses or taken from the image-sharing site Flickr. Some of these data sets have then been used by companies that worked on military projects in China, and large online image collections such as MSCeleb are still distributed among researchers who continue to use them [90, 92]. In many states in the US (but not all), it is illegal for commercial firms to use a person's biometric data without their consent, and some have argued that the European Union's General Data Protection Regulation (GDPR) also precludes researchers from doing so (although additional clarifications of these regulations are still outstanding as of press time). Recently, Facebook agreed to settle a class-action lawsuit related to the use of non-public photographic data for facial recognition; other companies have also been sued including IBM, Google, Microsoft, and Amazon.

4.6 Other Models: Diffusion and Consistency Models

Diffusion models (also known as score-based generative models) have experienced advances in recent years, surpassing the performance of other generative models such as GANs and variational autoencoders (VAEs), enjoying more stable training and less susceptibility to the problem of mode collapse [93–99]. Diffusion models have achieved good success in image generation, video generation, and audio synthesis. Unlike GANs, these models do not rely on adversarial training and are thus less prone to issues such as unstable training and mode collapse. Furthermore, diffusion models do not impose the same strict constraints on model architectures as in autoregressive models, VAEs, or normalizing flows. Fundamental to success is their iterative sampling process, which progressively removes noise from a random noise vector. This iterative refinement procedure repetitively evaluates the diffusion model, allowing for the trade-off of computation for sample quality: by using extra compute for more iterations, a small-sized model can unroll into a larger computational graph and generate higher quality samples. Iterative generation is also crucial for the zero-shot data editing capabilities of diffusion models, enabling them to solve complex inverse problems including image inpainting, colorization, and medical imaging.

However, the generation of content relies on very deep generative models. Compared to single-step generative models like GANs, VAEs, and normalizing flows, the iterative generation procedure of diffusion models typically requires 10–2,000 times more computation causing slow inference and limiting their real time applications. Indeed, in a diffusion model, to generate a realistic sample, it is necessary to solve an ordinary (for score-based models) or stochastic differential equation that entails the score function of the data, which is estimated via an NN. Reference [93] proposes an improvement by creating generative models that facilitate efficient, single-step generation without sacrificing important advantages of iterative refinement. This is achieved with *consistency models*, a new type of generative model that support single-step generation at the core of its design, while still allowing iterative generation for zero-shot data editing and trade-offs between sample quality and compute. Consistency models can be trained in either the distillation mode or the isolation mode. In the former case, consistency models distill the knowledge of pre-trained diffusion models into a single-step sampler, significantly improving other distillation approaches in sample quality, while allowing zero-shot image editing applications. In the latter case, consistency models are trained in isolation, with no dependence on pre-trained diffusion models. This makes them an independent new class of generative models [93].

A. Basic Glossary of Key AI Terms and Concepts

(Based on a variety of references, but specifically including [3, 8, 9, 13, 23, 48, 56, 58, 66, 68, 93, 100, 101]).

Term	Definition/explanation/concept
Affine transformation	Any transformation that can be defined as a linear transformation with an additional translation. Affine transformations include rotations, translations, scaling, and combinations of these transformations.
Atrous convolution	(also known as à trous) A convolution with dilation.
Batch normalization (BN)	A process that aims at reducing internal covariate shift, thus accelerating the training of a DNN. It accomplishes this via a normalization step that fixes the means and variances of layer inputs. BN also has a beneficial effect on the gradient flow through the network by reducing the dependence of gradients on the scale of the parameters or of their initial values. This enables much higher learning rates without the risk of divergence.
Classification process	A process that aims at identifying an object in an input image; the output is a vector of probabilities for different object elements, based on the training data.
Computer vision (CV) applications	CV applications supported by CNN techniques range from reproducing human visual abilities (e.g. face recognition, object classification, object detection), to enabling new categories of visual abilities (e.g. CV applications can be configured to recognize sound waves from the vibrations induced in objects visible in a video).
Consistency models	Newly defined type of generative models that support single-step generation at the core of its design, while still allowing iterative generation for zero-shot data editing and trade-offs between sample quality and compute. Consistency models can be trained in either the distillation mode or the isolation mode. In the former case, consistency models distill the knowledge of pre-trained diffusion models into a single-step sampler [93].
Convolution	A mathematical operation (from the field of functional analysis) performed by two functions to produce a third function that is a modified version of one of the two original functions.
Convolution dilation rate	The spacing between the values in a kernel. Dilation factor controls the spacing between the kernel points. For example, a 3×3 kernel (using only nine parameters) and a dilation rate of 2 has the same field of view as a 5×5 kernel (it can be perceived as a 5×5 kernel but where the second and fourth column and second and fourth row is deleted or ignored.

Term	Definition/explanation/concept
Convolution value of location (i, j) of the feature map	The size of the feature map C is $((m - m')/sx) + 1 \times ((n - n')/sy) + 1)$. The individual values of the output feature map matrix C, for an $n \times m$ input matrix I, a (single) $m' \times n'$ filter K, and a stride of $sx \times sy$, are, $$C(i,j) = \sum_{v=0}^{m'-1} \sum_{z=0}^{n'-1} I\left(i * sx + v, j * sy + z\right) * K\left(v+1, z+1\right)$$
Convolutional (CONV) layers	CNN layers that support/implement convolutions.
Convolutional neural network (CNN)	A specialized feed-forward NN for processing data that has a grid-like structure, such as data corresponding to images or compressed/uncompressed video.
DALL·E 2	An ML system that can create realistic images and art from a description in natural language.
Diffusion models	DL models that use an iterative sampling process, which progressively removes noise from a random noise vector. This iterative refinement procedure repetitively evaluates the model, allowing for the trade-off of computation for sample quality. By using extra compute for more iterations, a small-sized model can unroll into a larger computational graph and generate higher quality samples.
Dot product	It is a scalar, a numerical value. Given vectors, multiply corresponding components and then add: $a^v(a_1, a_2, a_3)$ and $b^v(b_1, b_2, b_3)$, $a^v \mathbf{o}\ b^v = a_1 b_1 + a_2 b_2 + a_3 b_3$. More generally, $a^v \mathbf{o}\ b^v = |a^v||b^v| \cos(a)$ where $|a^v|$ is the magnitude of vector a and a is the angle between them. When $a = 0$, the two vectors point in the same direction and $\cos(a) = 1$, thus the dot product is the highest. When the two vectors are perpendicular, $a = p/2$ and $\cos(a) = 0$. It can be thought of as a "directional multiplication."
	A dot product is the elementwise multiplication between the kernel-sized patch of the input feature map (IFM) and the corresponding kernel, which is then summed resulting in a single value – being a single value, the operation is often referred to as the "scalar product."
Edge detection	Process for finding the boundaries for the objects of the image. CNN are often used to make a prediction for the pixel labels.
Fast R-CNN model	An extension of R-CNN; a deep CNN used for object detection, that appears to the user as a single, end-to-end network. It builds a Region Proposal Network that can generate region proposals that are fed to the detection model (Fast R-CNN) to seek out objects. The Fast R-CNN model includes an additional new layer, the Region of Interest (ROI) pooling layer, that extracts equal-length feature vectors from all proposals in the same image.

(Continued)

(Continued)

Term	Definition/explanation/concept
Feature map	The output of the convolution via the convolution kernel is referred to as the feature map; the output matrix is the feature map. For example, for an input image which is 4×4 ($n = 4$, $m = 4$), a 2×2 filter K ($m' = 2$, $n' = 2$), and a 1×1 stride ($sx = sy = 1$), one has a 3×3 feature map $$C(i,j) = \sum_{v=0}^{1} \sum_{z=0}^{1} I(i+v, j+z) * K(v+1, z+1)$$ The feature map created by the immediately prior convolution is then run through a process where a new feature map is created, but using the specified mechanism (e.g. pick the highest value, as one processes the incoming feature map with a given stride); here, the filter is simply a window without elements in it and it is only used to specify a section in the incoming feature map (no parameters are learnt in the pooling layer).
Fully convolution network (FCN)	(also sometimes called "completely convolutional NN") is an NN that only performs convolution (and subsampling or upsampling) operations; it is a CNN without fully connected layers. A typical CNN is not "fully convolutional" because it normally also contains FC layers (which do not perform the convolution operation). Thus, an FCN is an NN that is comprised entirely of convolutional layers with no FC layers involved.
Fully connected NNs (FNN)	NNs with FC layers. NNs with layers where neurons have full connections to all activations in the previous layer (as is the case in a feed-forward network); traditional NN layers are fully connected in such a manner that every output unit interacts with every input unit; the output from the FC layers is used to generate an output result from the network. The FC layers apply a linear combination and an activation function to the input operand and generate an individual partial sum. By contrast, CONV layers are often sparsely connected; the output of the convolution of a field is input to some nodes of the subsequent layer.
Generalization gap	The difference between training and test accuracy. Generalization can be improved by regularizations, enabling the model to use geometrical prior knowledge about the scene, shifting the model back to an area where it generalizes well, and by pruning the network.
Graph neural networks (GNNs)	An NN based on a graph structure that models a set of nodes (entity) and edges (relationship). They utilize non-Euclidean data structures for DL.

Term	Definition/explanation/concept
Graphics processing unit (GPU)	High end processors that support parallel computing. High-end GPUs have thousands of cores (e.g. GeForce GTX TITAN Z included 5760 cores). GPUs became valuable DL tools, given that the operations required to run a DNN/CNN algorithm can be accomplished with parallel computing/ computations.
Hyperparameters	Parameters associated with a convolutional layer, such as (i) the number of kernels, (ii) the size F kernels (e.g. a kernel is of dimensions FxFxD pixels), (iii) the S step with which the window corresponding to the kernel is dragged on the image (e.g. a step of one means moving the window one pixel at a time), and (iv) the zero-padding P (e.g. adding a black contour of P pixels thickness to the input image of the convolutional layer). Also see glossary of Chapter 1.
Image classification	Imaging problems where one is only concerned about *what* the image contains, and not where it is located in the frame under consideration.
Image manipulation	The process of changing a digitized image to transform it into a desired image, including image editing/enhancement, movie post-production, animation (where [manipulated] images are shown in sequence to appear as moving images), and human-computer interaction. Methods typically require parametric 3D face/object modeling and then perform animation in the 3D domain.
Image processing and analysis	Processing such as, but not limited to, image filtering (removal of noise and artifacts); image enhancement (highlight features of interest); image segmentation (object recognition); and quantitative image feature description (object measurements). At a more macro level: image enhancement, image restoration, image transformation, image analysis, image recognition, or image compression.
Image segmentation	A CV task intended to label specific regions of an image according to what is being shown.
Image semantic information	Information obtained by representing each image part of a scene by using different pixel values.
Inner product	The elementary transformation $\sum_{i=1}^{N} w_i x_i$, where $x^v = [x_1, x_2, ..., x_N]$ is an N-channel input vector to the neuron in question and w_i is a filter's weight for the i-th channel.
Input and output channels of a convolutional layer	A convolutional layer takes a number of input channels (I) and calculates a specific number of output channels (O). The number of parameters for a convolutional layer is $I \times O \times k$, where k is the kernel size.

(Continued)

(Continued)

Term	Definition/explanation/concept
Interspace pruning (IP)	A general tool to improve existing pruning methods; it uses filters represented in a dynamic interspace by linear combinations of an underlying adaptive filter basis (FB). For IP, FB coefficients are set to zero while un-pruned coefficients and FBs are trained jointly. IP keeps the CNN flexible and leads to an improved information flow and better trainable models.
Kernel size	The field of view of the convolution; for 2D imaging it is typically, but not always 3 × 3 pixels.
Mask	A mechanism that highlights the regions of an image where a specific class is present; a filter.
Midjourney	An independent research lab that produces an ML program that creates images from (short) textual descriptions; it similar to OpenAI's DALL-E and Stable Diffusion.
NNs with fully connected (FC) layers	NNs with layers where *neurons have full connections to all activations in the previous layer (as is the case in a feed-forward network)*; traditional NN layers are fully connected in such a manner that every output unit interacts with every input unit; the output from the FC layers is used to generate an output result from the network. The FC layers apply a linear combination and an activation function to the input operand and generate an individual partial sum. By contrast, convolutional layers are often sparsely connected: the output of the convolution of a field is input to some nodes of the subsequent layer.
Object detection	Creating bounding boxes around detected objects, enabling one to see where they are in a scene.
Object detection techniques	Generally, a three-step process: (i) region proposal generation, (ii) feature extraction, and (iii) classification. The feature vector is used to assign each region proposal to either the background class or to one of the object classes.
Object instance segmentation	A process (beyond semantic segmentation) that endeavors to separate different instances from a single class. It is a cross between semantic segmentation and object detection.
Object recognition	A CV activity that recognizes and locates objects inside an image or video input.
Optimization techniques (DL context)	Techniques that aim at modifying DL algorithms to make them more hardware-friendly with minimal loss of accuracy, such as precision reduction, pruning, low-rank approximation.
Pooling (a feature map)	An operation that downsamples the resolution of a video or image (frame) by summarizing a local area with a single value (e.g. average or max pooling).

Term	Definition/explanation/concept
Pruning	Process that sets parts of the CNN's weights to zero. This reduces the model's complexity and memory requirements, speeds up inference and may lead to an improved generalization ability.
Region (on an image)	A portion of an image; for example, one-pixel block or a group of pixel blocks (a pixel block means a group of adjacent pixels including at least one pixel).
Region-based CNN (R-CNN)	(also called Regions with CNN features) A model for object detection that extracts features using a pre-trained CNN; it entails a region proposal generation section, a feature extraction section, and classification section using SVM.
Scene understanding	(Complex) process where one not only needs to identify objects but also understand the distribution of objects in a scene. Traditional scene recognition methods have utilized manually designed algorithm to extract the feature to construct a Bag-of-Words (BOW) model; newer methods use deep CNNs.
Semantic feature map (SFM)	Mechanism used by NNs to model the spatial object contexts to process image and video contents. Using a CNN one can extract high-level semantic object features on input image for every object proposal, and organize the features in an SFM to preserve spatial information among objects.
Semantic segmentation (process)	(also known as dense prediction) The process of applying semantic labels to the pixels of an image; assigning a label to a pixel; the process of marking each pixel of the input image with the class that represents a particular entity or body; employed when the spatial information of a subject and how it interacts with it is of interest, such as for an autonomous vehicle applications.
Semantic segmentation problems	Classification problems where each pixel is assigned to one of several object classes (e.g. for traffic management, identifying roads and buildings; for facial segmentation skin, hair, eyes, nose, and mouth elements).
Semantic segmentation process (also known as dense prediction)	The process of applying semantic labels to the pixels of an image; assigning and/or associating a label to a pixel (or small patch of pixels); marking each pixel of the input image with the class that represents a particular entity or body. Semantic segmentation is employed when the spatial information of a subject and how it interacts with it is of interest, such as for an autonomous vehicle application.
Spatial information	Information describing (or capturing) the relationships between the nearby pixels.

(Continued)

(Continued)

Term	Definition/explanation/concept
Spatial transformer network (STN)	An operation network that can spatially transform data, such as feature images, in a network without introducing additional data labels.
Stable diffusion	A recently introduced text-to-image DL model that can be used to generate detailed images based on text descriptions and to generate image-to-image translations subtended by a text prompt. It is a kind of deep GNN that has been released publicly and it can run on most consumer hardware equipped with a modest GPU – for comparison text-to-image models such as DALL-E and Midjourney are text-to-image models such that are accessible only via cloud services.
Stride	The number of pixels by which the weight map WM slides between shifts. The step size of the kernel when traversing the image. Usually 1; a stride of 2 results in downsampling.
Strided convolutions	Convolutions that entail compressing the spatial resolution.
Thin plate spline (TPS)	An interpolation algorithm that can be used in tasks, such as image warping, to drive, by using a few control points, an image to change.
Translation invariance	Mechanisms that ensure that small translations to the input do not change the pooled outputs; used in scenarios where the presence of a feature in the input data is more important than the precise location of the feature.
Translation invariance	The ability to recognize an object regardless of where it appears in an image. Can be achieved by sequentially reducing the image dimensions in both x and y dimensions (pooling) while increasing the depth (e.g. by increasing the number of convolutional kernels).
Transposed convolutional layer	A layer that carries out a convolution but reverts its spatial transformation. Similar to upscaling an image using a convolution.
Unpooling (upsampling) (a feature map)	Operation that upsamples the resolution of a video or image (frame) by distributing a single value into a higher resolution.
Unstructured pruning (UP)	Process used to reduce the memory footprint of convolutional neural networks (CNNs), both at training and inference time. CNNs contain parameters arranged in $K \times K$ filters. Standard UP reduces the memory footprint of CNNs by setting filter elements to zero, thereby specifying a fixed subspace that constrains the filter. If pruning is applied before or during training, this induces a strong bias.

Term	Definition/explanation/concept
Variational autoencoder (VAE)	An NN architecture that utilizes probabilistic graphical models and variational Bayesian methods [101]. A VAE is similar to an autoencoder model but it is different in the goal and mathematical machinery. VAEs are probabilistic generative models that require NN as only a part of their overall structure. The first NN (encoder) maps the input variable to a latent space that corresponds to the parameters of a variational distribution, allowing it to produce multiple different samples that all come from the same distribution. The second NN (decoder) maps from the latent space to the input space, to produce or generate data points. Both NNs are trained in unison, however, the variance of the noise model can be learned distinctly.
Video formats	Standard definition (SD), high definition (HD), full HD, or ultra HD image (4K, SK, 16K, 32K, etc.). Also: compressed form by MPEG (e.g. MP2, MP4, MP7, etc.), AVC, H.264, HEVC.
Visual computing systems	Computer systems used to gain a high-level of understanding from images and utilize the understanding to automate tasks typically performed by a human's visual system. A CV system includes processing and analyzing images to extract information and use the information to make decisions and perform tasks. Processing and analyzing an image often use ML systems.
Weigh	A parameter used to calculate an output activation in each NN channel; a value assigned to a connection between channels.

B. Examples of Convolutions

The convolution operation takes place between a section of the image and the filter. The convolution operation creates and outputs the feature map, which can be seen as a reduced image.

The size of the image section is typically equal to the size of the filter(s) being used. The convolution operation occurs in each convolutional layer; Figure 4.B1 depicts the elementwise multiply-sum calculation between an image section and one filter (a row-wise multiply/add is shown, but a column-wise multiply/add can be used with the same result).

During the convolution (the calculation) the filter(s) slide vertically and horizontally (by the value of the stride) on the input image to create different image sections. The number of image sections that are created depends on the stride. In Figure 4.B1, the stride was 1 and the number of sections was 9. If the stride is larger, the number of image sections and the feature map will be smaller.

The filter (also called kernel or feature detector) is a small matrix and the elements of the filter define the filter configuration. The same-sized filters are used within a convolutional layer. Each filter has a specific function and multiple filters are used to identify a different set of features in the image. The size of the filter and the number of filters is specified by the user as hyperparameters and is smaller than the size of the input image. These elements are a type of parameters in the CNN and are learned during the training [14].

The feature map captures the outputs of different convolution operations between different image sections and the filter(s). The feature map is the input for the next pooling layer. The number of elements in the feature map is equal to the number of unique image sections obtained by sliding the filter(s) on the image. In Figure 4.B1, this was a 3 × 3 matrix, with 9 entries. When there are several convolutional layers in the CNN, the size of the feature map will be further reduced along the way. To avoid having an unusable feature map, one applies padding to the input image, adding additional pixels with zero values to each side of the image; the feature map will then retain the same size as the input. See Figure 4.B2 (with a stride of 1 and a 2 × 2 filter the feature map would be 4 × 4, the same as the original image).

Figure 4.B3 depicts an example of convolution operation with two 2 × 2 kernel.

A color image has three color channels, so three filters are needed to do the calculations, one for each channel. The appropriate filter acts on the corresponding input channel between using image sections as per the grayscale example. The final feature map is obtained by adding all outputs of each channel's calculations. See Figure 4.B4.

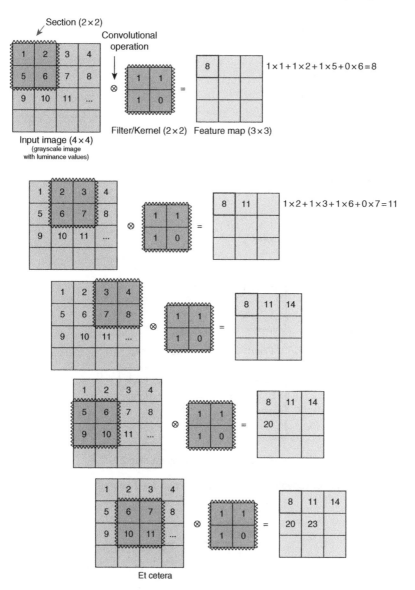

Figure 4.B1 Example of convolution operation with 2×2 kernel and stride of 1.

Convolution operations on an RGB images with multiple filter sets is common in practical scenarios. See Figure 4.B5.

Figure 4.B6 depicts an example of the max pooling operation.

0	0	0	0	0	0
0	1	2	3	4	0
0	5	6	7	8	0
0	...				0
0					0
0	0	0	0	0	0

Padding = 1

Grayscale image showing luminance

Figure 4.B2 Padding.

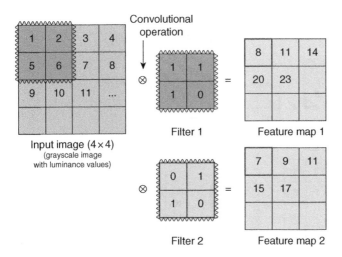

Convolutional operation

Input image (4 × 4)
(grayscale image with luminance values)

Filter 1   Feature map 1

Filter 2   Feature map 2

Figure 4.B3 Example of convolution operation with two 2 × 2 kernels and stride of 1.

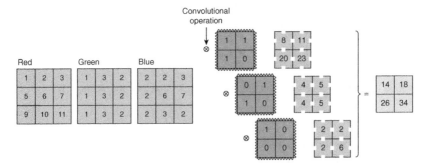

Convolutional operation

Red   Green   Blue

Figure 4.B4 Example of convolution operation with three channels.

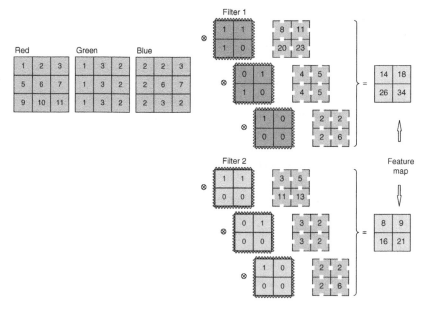

Figure 4.B5 Example of convolution operation with three channels and multiple filter sets.

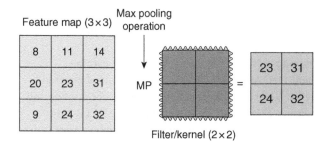

Figure 4.B6 Max pooling operation.

References

1 O. Ronneberger, P. Fischer, T. Brox, "U-Net: Convolutional Networks for Biomedical Image Segmentation", May 18, 2015, arXiv:1505.04597, https://doi.org/10.48550/arXiv.1505.04597.

2 J. Long, E. Shelhamer, T. Darrell, "Fully Convolutional Networks for Semantic Segmentation", Mar. 8, 2015, arXiv:1411.4038, https://doi.org/10.48550/arXiv.1411.4038.

3 M. Dhouibi, A. K. Ben Salem, *et al.*, "Accelerating Deep Neural Networks Implementation: A Survey", *IET Comput. Digit. Techn.* 2021, https://doi.org/10.1049/cdt2.12016.

4 A. Krizhevsky, I. Sutskever, G. E. Hinton, "ImageNet Classification with Deep Convolutional Neural Networks", *Commun. ACM.* 2017; 60(6): 84–90, https://doi.org/10.1145/3065386 (The original version of this paper was published in the Proceedings of the 25th International Conference on Neural Information Processing Systems, Lake Tahoe, NV, Dec. 2012, pp. 1097–1105).

5 A. Canziani, E. Culurciello, A. Paszke, "An Analysis of Deep Neural Network Models for Practical Applications", Apr. 14, 2017, arXiv:1605.07678v4.

6 C. Szegedy, V. Vanhoucke, *et al*, "Rethinking the Inception Architecture for Computer Vision", arXiv:1512.00567, https://doi.org/10.48550/arXiv.1512.00567.

7 Staff, Wolfram MathWorld, "Convolution", 2023, https://mathworld.wolfram.com. Accessed Jan. 21, 2023.

8 S. Chouri, Demystifying the Math and Implementation of Convolutions: Part I, Mar. 26, 2019, https://praisethemoon.org/demystifying-the-math-and-implementation-of-convolutions-part-i/. Accessed Dec. 1, 2022.

9 S. Yan, J. Li, Z. Liu, Slimming of Neural Networks in Machine Learning Environments, U.S. Patent 11,537,892, Dec. 26, 2022. Uncopyrighted material.

10 P. Sermanet, D. Eigen, *et al*, "OverFeat: Integrated Recognition, Localization and Detection Using Convolutional Networks". in Proceeding of the [ICLR 2014] International Conference on Learning Representations, April 14th to April 16th 2014, in Banff, Canada, 2014.

11 K. Simonyan, A. Zisserman, "Two-Stream Convolutional Networks for Action Recognition in Videos", CoRR, abs/1406.2199, Published in Proc. NIPS, 2014.

12 O. Russakovsky, J. Deng, *et al*, "ImageNet Large Scale Visual Recognition Challenge", CoRR, abs/1409.0575, 2014.

13 Z. Yi, Q. Tand, *et al*, Methods and Systems for High Definition Image Manipulation with Neural Networks, U.S. Patent 20230019851, Jan. 19, 2023. Uncopyrighted material.

14 R. Pramoditha, "Convolutional Neural Network (CNN) Architecture Explained in Plain English Using Simple Diagrams", 2013, https://towardsdatascience.com, https://towardsdatascience.com/convolutional-neural-network-cnn-architecture-explained-in-plain-english-using-simple-diagrams-e5de17eacc8f. Accessed Feb. 8, 2023.

15 C. J. Munkberg, J. N. T. Hasselgren, *et al*, Neural Network System with Temporal Feedback for Denoising of Rendered Sequences, U.S. Patent 2023/0014245, Jan. 19, 2023. Uncopyrighted material.

16 J. Jordan, "Common Architectures in Convolutional Neural Networks", Apr. 19, 2018, https://www.jeremyjordan.me/convnet-architectures/. Accessed Dec. 4, 2022.

17 D. Eigen, J. Rolfe, *et al*, "Understanding Deep Architectures Using a Recursive Convolutional Network", 2013, arXiv:1312.1847, https://doi.org/10.48550/arXiv.1312.1847.

18 G. Huang, Z. Liu, *et al*, "Densely Connected Convolutional Networks", Jan. 28, 2018, arXiv:1608.06993, https://doi.org/10.48550/arXiv.1608.06993.

19 X. Glorot, A. Bordes, Y. Bengio, "Deep Sparse Rectifier Neural Networks", in AISTATS, Apr. 11–13, 2011. Ft. Lauderdale, FL, USA, 2011.

20 Y. LeCun, L. Bottou, *et al.*, "Gradient-based Learning Applied to Document Recognition", *Proc. IEEE*. 1998; 86(11): 2278–2324.

21 F. Zhou, X. Jin, *et al*, Adaptive MAC Array Scheduling in a Convolutional Neural Network, U.S. Patent 2023/0013599, Jan. 19, 2023. Uncopyrighted material.

22 S. Ha, G. Kim, D. Lee, Method and Apparatus with Neural Network Parameter Quantization, U.S. Patent 2023/0017432 Al, Jan. 19, 2023. Uncopyrighted material.

23 A. Sarangam, "Fully Convolutional Network (FCN): A Basic Overview", Apr. 17, 2021, https://www.jigsawacademy.com/blogs. Accessed Dec. 4, 2022.

24 J. Jordan, "An Overview of Semantic Image Segmentation", May 21, 2018, https://www.jeremyjordan.me/semantic-segmentation/. Accessed Dec. 4, 2022.

25 Q. Sellat, R. Priyadarshini, *et al.*, "Semantic Segmentation for Self-Driving Cars Using Deep Learning", in *Cognitive Big Data Intelligence with a Metaheuristic Approach*, Copyright © 2022 Elsevier Inc., 2022, https://doi.org/10.1016/C2020-0-02004-9.

26 A. Dhiman, D. K. Sharma, *et al.*, "An Introduction to Deep Learning Applications in Biometric Recognition", in *Trends in Deep Learning Methodologies*, V. Piuri, S. Raj, *et al.*, (Editors), Academic Press, London, UK, 2021.

27 N. Koul, S. K. S. Manvi, "Computational intelligence techniques for cancer diagnosis", in *Recent Trends in Computational Intelligence Enabled Research*, S. Bhattacharyya, P. Dutta, *et al.*, (Editors), Copyright © 2021 Elsevier Inc, 2021, https://doi.org/10.1016/C2019-0-05399-X.

28 A. F. Gad, "Faster R-CNN Explained for Object Detection Tasks", https://blog.paperspace.com/faster-r-cnn-explained-object-detection/. Accessed Dec. 11, 2022.

29 R. Girshick, J. Donahue, *et al*, "Rich Feature Hierarchies for Accurate Object Detection and Semantic Segmentation", in 2014 IEEE Conference on Computer Vision and Pattern Recognition (CVPR). June 23, 2014 to June 28, 2014. Columbus, OH, USA, 2014, arXiv:1311.2524 (cs.CV), https://doi.org/10.48550/arXiv.1311.2524.

30 R. Girshick, "Fast R-CNN", in 2015 IEEE International Conference on Computer Vision (ICCV), Santiago, Chile, 2015, pp. 1440–1448, https://doi.org/10.1109/ICCV.2015.169.

31 S. Ren, K. He, et al, "Faster r-CNN: Towards Real-Time Object Detection With Region Proposal Networks", in Advances in Neural Information Processing Systems, 2015, arXiv:1506.01497, https://doi.org/10.48550/arXiv.1506.01497.

32 D. Minoli, *Imaging in Corporate Environments – Technology and Communication*, McGraw-Hill, 1994.

33 D. Minoli, *IP Multicast with Applications to IPTV and Mobile DVB-H*, Wiley, 2008.

34 D. Minoli, *Linear and Non-Linear Video and TV Applications Using IPv6 and IPv6 Multicast*, Wiley, 2012.

35 D. Minoli, *Mobile Video with Mobile IPv6*, Wiley, 2012.

36 D. Minoli, *Advances in Satellite Communication*, Wiley, 2015.

37 D. Minoli, *3D Television (3DTV) Technology, Systems, and Deployment*, Taylor and Francis, 2011.

38 D. Minoli, *3D Television (3DTV) Content Capture, Encoding, and Transmission*, Wiley, 2011.

39 D. Minoli, *Distance Learning: Technology and Applications*, Artech House, 1996.

40 D. Minoli, *Multimedia over the Broadband Network: Business Opportunities and Technology – A comprehensive Report, International Engineering Consortium*, (co-editor/co-advisor), 1996.

41 D. Minoli, *Video Dialtone Technology: Digital Video over ADSL, HFC, FTTC, and ATM*, McGraw-Hill, 1995.

42 D. Minoli, *Distributed Multimedia Through Broadband Communication Services (Co-authored)*, Artech House, 1994.

43 D. Minoli, in *The Telecommunications Handbook*, K. Terplan, P. Morreale, (Editors), IEEE Press, 2000.

44 A. Mazumdar, B. Haynes, *et al.*, *Perceptual Compression for Video Storage and Processing Systems, SoCC '19*, Association for Computing Machinery, ACM, Santa Cruz, CA, USA, 2019. ISBN: 978-1-4503-6973-2/19/11. https://doi.org/10.1145/3357223.3362725.

45 F. Bellard, FFmpeg, https://fmpeg.org. Accessed Dec. 8, 2022.

46 H. Wang, J. Chen, M. Karczewicz, Activation Function Design in Neural Network-Based Filtering Process for Video Coding, U.S. Patent 2023/0012661, Jan. 19, 2023. Uncopyrighted material.

47 T. M. Bae, S. Xie, *et al*, Method and System for Video Transcoding Based on Spatial or Temporal Importance, U.S. Patent 11,528,493, Dec. 13, 2022. Uncopyrighted material.

48 M.-T. Grymel, D. T. Bernard, *et al*, Write Combine Buffer (WCB) For Deep Neural Network (DNN) Accelerator, U.S. Patent 2023/0020929, Jan. 19, 2023. Uncopyrighted material.

49 Datagen Staff, "Understanding VGG16: Concepts, Architecture, and Performance", https://datagen.tech/guides/computer-vision/vgg16/. Accessed Dec. 10, 2022.

50 K. Simonyan, A. Zisserman, "Very Deep Convolutional Networks for Large-Scale Image Recognition", ICLR 2015 San Diego, CA, USA, May 7–9, 2015, arXiv:1409.1556, https://doi.org/10.48550/arXiv.1409.1556. Published as a conference paper at ICLR 2015.

51 C. Szegedy, W. Liu, *et al*, "Going Deeper with Convolutions", in 2015 IEEE Conference on Computer Vision and Pattern Recognition (CVPR). DOI: 10.1109/CVPR31182.2015. June 7–12, 2015, pp. 1–9, 2015.

52 K. He, X. Zhang, *et al*, "Identity Mappings in Deep Residual Networks", July 25, 2016, arXiv:1603.05027, https://doi.org/10.48550/arXiv.1603.05027.

53 S. Zagoruyko, N. Komodakis, "Wide Residual Networks", 2016, arXiv:1605.07146, https://doi.org/10.48550/arXiv.1605.07146.

54 K. He, X. Zhang, *et al*, "Deep Residual Learning for Image Recognition", in 2016 IEEE Conference on Computer Vision and Pattern Recognition (CVPR), June 27–30, 2016, Las Vegas, NV, USA, 2016. doi: 10.1109/CVPR.2016.90.

55 S. Xie, R. Girshick, *et al*, "Aggregated Residual Transformations for Deep Neural Networks", 2017, arXiv:1611.05431, https://doi.org/10.48550/arXiv.1611.05431.

56 S. Ioffe, C. Szegedy, "Batch Normalization: Accelerating Deep Network Training by Reducing Internal Covariate Shift" in ICML, 2015, arXiv:1502.03167v3, https://doi.org/10.48550/arXiv.1502.03167.

57 Z. Yu, R. Huang, *et al*, Image Processing Using Coupled Segmentation and Edge Learning, U.S. Patent 2023/0015989, Jan. 19, 2023. Uncopyrighted material.

58 Y. Song, C. Ge, Image Generation Method and Apparatus, U.S. Patent 2023/0017112, Jan. 19, 2023. Uncopyrighted material.

59 D. J. Gebhardt, K. N. Parikh, I. P. Dzieciuch, Machine Learning Based Automated Object Recognition for Unmanned Autonomous Vehicles, U.S. Patent 11,551,032. Jan. 10, 2023. Uncopyrighted material.

60 T. Yuan, W. Da Rocha Neto, *et al*, "Machine Learning for Next-Generation Intelligent Transportation Systems: A Survey", in Wiley's Transactions on Emerging Telecommunications Technologies, Dec. 14, 2021, https://doi.org/10.1002/ett.4427. Accessed Dec. 11, 2022.

61 W. C. Feng, H. S. Kao, J. J. Su, Hybrid Object Detector and Tracker, U.S. Patent 2023/0021016, Jan. 19, 2023. Uncopyrighted Material.

62 A. Arnab, M. Dehghani, *et al*, Systems and Methods for Improved Video Understanding, U.S. Patent 2023/0017072, Jan. 19, 2023. Uncopyrighted material.

63 P. Cordani, A. Delattre, System and Methods of Neural Network Training, U.S. Patent 2023/0019874, Jan. 19, 2023. Uncopyrighted material.

64 M. L. H. Tay, W. Ma, *et al*, Asymmetric Facial Expression Recognition, U.S. Patent 20230046286, Feb. 16, 2023. Uncopyrighted material.

65 Datagen Staff, "What is the Transformer Architecture and How Does It Work?", https://datagen.tech/guides/computer-vision/transformer-architecture/. Accessed Dec. 12, 2022.

66 P. Wimmer, J. Mehnert, A. Condurache, "Interspace Pruning: Using Adaptive Filter Representations to Improve Training of Sparse CNNs", Mar. 15, 2022, arXiv:2203.07808, https://doi.org/10.48550/arXiv.2203.07808.

67 D. Minoli, B. Occhiogrosso, A. Koltun, "Situational Awareness for Law Enforcement and Public Safety Agencies Operating in Smart Cities – Part 2: Platforms", in *Springer's Book IoT and WSN Based Smart Cities: A Machine Learning Perspective*, S. Rani, V. Sai, R. Maheswar, (Editors), Springer, 2022. ISBN: 978-3-030-84181-2.

68 D. Minoli, B. Occhiogrosso, A. Koltun, "Situational Awareness for Law Enforcement and Public Safety Agencies Operating in Smart Cities – Part 1: Technology", in *Springer's Book IoT and WSN Based Smart Cities: A Machine Learning Perspective*, S. Rani, V. Sai, R. Maheswar, (Editors), Springer, 2022. ISBN: 978-3-030-84181-2.

69 M. Endsley, *SAGAT: A Methodology for the Measurement of Situation Awareness*", NOR DC 87-83, Northrop Corporation, Los Angeles, CA, 1987.

70 M. R. Endsley, "Toward a Theory of Situation Awareness in Dynamic Systems", *Human Factors*. 1995; 37(1): 32–64.

71 M. Satyanarayanan, "Edge Computing for Situational Awareness", in 2017 IEEE International Symposium on Local and Metropolitan Area Networks (LANMAN), Osaka, 2017, pp. 1–6, https://doi.org/10.1109/LANMAN.2017.7972129. Available online on Dec. 1, 2020 at satya-lanman2017-camready.pdf (cmu.edu). Accessed Dec. 10, 2022.

72 R. Pew, A. Mavor, (Editors), *Modeling Human and Organizational Behavior: Application to Military Simulations*, National Academy Press, 1998.

73 A. D'Amico and M. Kocka, "Information Assurance Visualization for Specific Stages of Situational Awareness and Intended Uses: Lessons Learned", in Workshop on Visualization for Computer Security, USA, 2005.

74 M. R. Endsley, D. J. Garland, (Editors), *Situation Awareness Analysis and Measurement*, Lawrence Erlbaum Associates, Mahwah, NJ, 2000. ISBN: 0-8058-2133-3.

75 C. Onwubiko, "Designing Information Systems and Network Components for Situational Awareness", *Situational Awareness in Computer Network Defense*, pp. 104–123, 2012;. Available online on Dec. 1, 2020 at https://doi.org/ 10.4018/978-1-4666-0104-8.ch007. https://www.c-mric.com/wp-content/ uploads/2019/12/CyberSA_Design_Requirements.pdf. Accessed Dec. 10, 2022.

76 B. McGuinness, L. Foy, "A Subjective Measure of SA: The Crew Awareness Rating Scale (CARS)", in First International. Conference on Human Performance, Situation. Awareness and Automation, Savannah, GA, October 15–19, 2000, 2000.

77 S. Banerjee, D. O. Wu, *Final Report from the NSF Workshop on Future Directions in Wireless Networking*, National Science Foundation, 2013.

78 D. Minoli, B. Occhiogrosso, "Blockchain-enabled Fog and Edge Computing: Concepts, Architectures and Smart City Applications", in *Blockchain-enabled Fog and Edge Computing: Concepts, Architectures and Applications*, M. H. Rehmani, M. M. Rehan, (Editors), CRC Press, Taylor & Francis Group, Boca Raton, FL, pages 29–77, 2020.

79 P. Simoens, Y. Xiao, *et al.*, "Edge Analytics in the Internet of Things", *IEEE Pervasive Computing.* 2015; 14(2): 24–31.

80 Y. Wang, M. Narasimha, R. W. Heath, "MmWave Beam Prediction with Situational Awareness: A Machine Learning Approach", in 2018 IEEE 19th International Workshop on Signal Processing Advances in Wireless Communications (SPAWC), June 25–28, 2018, Kalamata, Greece, 2018, pp. 1–5. doi: 10.1109/SPAWC.2018.8445969.

81 Y. Wang, A. Klautau, *et al.*, "MmWave Vehicular Beam Selection with Situational Awareness Using Machine Learning", *IEEE Access.* 2019; 7: 87479–87493, https://doi.org/10.1109/ACCESS.2019.2922064.

82 A. Shahsavari, M. Farajollahi, *et al.*, "Situational Awareness in Distribution Grid Using Micro-PMU Data: A Machine Learning Approach", *IEEE Transactions on Smart Grid.* 2019; 10(6): 6167–6177, https://doi.org/10.1109/TSG.2019.2898676.

83 Y. Nikoloudakis, I. Kefaloukos, *et al.*, "Towards a Machine Learning Based Situational Awareness Framework for Cybersecurity: An SDN Implementation", *Sensors.* 2021; 21: 4939, https://doi.org/10.3390/s21144939.

84 B. D. Little, C. E. Frueh, "Space Situational Awareness Sensor Tasking: Comparison of Machine Learning with Classical Optimization Methods", ARC, Published Online: Dec. 20, 2019, https://doi.org/10.2514/1.G004279.

85 Z. Yifan, "Application of Machine Learning in Network Security Situational Awareness", in 2021 World Conference on Computing and Communication Technologies (WCCCT), Dalian, China, 2021, pp. 39–46, https://doi.org/10.1109/WCCCT52091.2021.00015.

86 L. J. Wong, W. H. Clark, *et al.*, "An RFML Ecosystem: Considerations for the Application of Deep Learning to Spectrum Situational Awareness", *IEEE Open Journal of the Communications Society.* 2021; 2: 2243–2264, https://doi.org/10.1109/OJCOMS.2021.3112939.

87 E. Maltezos, L. Karagiannidis, *et al*, "Preliminary Design of a Multipurpose UAV Situational Awareness Platform Based on Novel Computer Vision and Machine Learning Techniques", in 2020 5th South-East Europe Design Automation, Computer Engineering, Computer Networks and Social Media Conference (SEEDA-CECNSM), Corfu, Greece, 2020, pp. 1–8, https/doi.org/10.1109/SEEDA-CECNSM49515.2020.9221786.

88 P. Barford, M. Dacier, *et al.*, "Cyber SA: Situational Awareness for Cyber Defense", in *Cyber Situational Awareness. Advances in Information Security*, Vol. 46, S. Jajodia, P. Liu, *et al.*, (Editors), Springer, Boston, MA, 2010. https://doi.org/10.1007/978-1-4419-0140-8_1.

89 M. McKenzie, S. C. Wong, "Subset Selection of Training Data for Machine Learning: A Situational Awareness System Case Study", in Proceedings Volume 9494, Next-Generation Robotics II; and Machine Intelligence and Bio-inspired Computation: Theory and Applications IX, 94940U (2015), 11 May 2015, https://doi.org/10.1117/12.2176536, Event: SPIE Sensing Technology + Applications, Baltimore, MA, USA.

90 R. Van Noorden, "The Ethical Questions That Haunt Facial-Recognition Research", Nature, Nov. 18, 2020. Available online on Jan. 5, 2021 at https://www.nature.com/articles/d41586-020-03187-3. Accessed Dec. 11, 2022.

91 N. Karlis, "AI Company Harvested Billions of Facebook Photos for a Facial Recognition Database It Sold to Police", Salon, Apr. 6, 2023, https://www.salon.com/2023/04/06/ai-company-harvested-billions-of-facebook-photos-for-a-facial-recognition-database-it-sold-to-police/. Accessed Apr. 5, 2023.

92 M. Murgia, "Who's Using Your Face? The Ugly Truth About Facial Recognition", Financial Times, Sept. 18, 2019. Available online on Jan. 5, 2021 at https://www.ft.com/content/cf19b956-60a2-11e9-b285-3acd5d43599e.

93 Y. Song, P. Dhariwal, *et al*, "Consistency Models", 2023, https://arxiv.org/pdf/2303.01469.pdf.

94 J. Sohl-Dickstein, E. Weiss, *et al*, "Deep Unsupervised Learning Using Nonequilibrium Thermodynamics", in The 32nd International Conference on Machine Learning (ICML 2015) was held in Lille, France, on July 6–July 11, 2015, pp. 2256–2265, 2015.

95 Y. Song, S. Ermon, "Generative Modeling by Estimating Gradients of the Data Distribution", in 2019 Conference on Neural Information Processing Systems, Dec. 8, 2019–Dec. 14, 2019, Vancouver Convention Centre, pp. 11918–11930, 2019.

96 Y. Song, S. Ermon, "Improved techniques for training score-based generative models", in *Advances in Neural Information Processing Systems*, Vol. 33, H. Larochelle, M. Ranzato, *et al*., (Editors), Curran Associates, Inc., Red Hook, NY, pages 12 438–12 448, 2020.

97 J. Ho, W. Chan, *et al*, "Imagen Video: High Definition Video Generation with Diffusion Models", arXiv preprint, arXiv:2210.02303, 2022.

98 Y. Song, J. Sohl-Dickstein, "Score-based Generative Modeling Through Stochastic Differential Equations", in International Conference on Learning Representations. ICLR 2021. Vienna, Austria May 04, 2021, 2021. https://openreview.net/forum?id=PxTIG12RRHS.

99 S. Benaïchouche, "Open AI Proposes Consistency Models: A New Family of Generative Models That Achieve High Sample Quality Without Adversarial Training", Mar. 10, 2023, https://www.marktechpost.com/. https://www.marktechpost.com/2023/03/10/open-ai-proposes-consistency-models-a-new-family-of-generative-models-that-achieve-high-sample-quality-without-adversarial-training/. Accessed Mar. 12, 2023.

100 A. Krizhevsky, I. Sutskever, G. E. Hinton, "ImageNet Classification with Deep Convolutional Neural Networks", in Neural Information Processing Systems (NIPS 2012), Dec. 3–6, 2012, Lake Tahoe, Nevada, USA, pp. 1106–1114, 2012.

101 D. P. Kingma, M. Welling, "Auto-encoding Variational Bayes", in International Conference on Learning Representations, 2014, arXiv:1312.6114, https://doi.org/10.48550/arXiv.1312.6114.

5

Current and Evolving Applications to IoT and Applications to Smart Buildings and Energy Management

5.1 Introduction

The fundamental concept of Internet of Things (IoT),[1] now well understood, is to provide intelligent capabilities to dispersed devices – these devices being stationary or mobile, high-end or basic – to support automated (machine-to-machine [M2M] or people-to-machine) data aggregation, and also, as appropriate, to reliably transmit control commands to end-systems. The data can be end-system data (e.g. device status, patient monitoring), near-environment sampled data (e.g. site temperature, traffic patterns, site parameters), or interactive data (e.g. multimedia streams, video surveillance). The commands can be actuation information, for example, reset a device or system parameter or perform some action (for example, for a road signal, a dam door, a grid transfer switch, a drone function, or a remote robot action).

Artificial intelligence (AI), machine learning (ML), and deep learning (DL) techniques enable one to address the very large amounts of data collected by the multitude of IoT sensors deployed in the environment and to reach meaningful conclusions and suggest actionable follow-ons from the data.

5.1.1 IoT Applications

In broad terms, IoT applications span (i) smart buildings and/or smart homes; (ii) smart cities including surveillance and smart grids (SG); (iii) industrial environments; (iv) medical environments; and (v) a (very) large set of wireless,

---

1 The authors have published extensively on this topic and the introductory sections of this chapter are liberally based and/or synthetized from these publications [1–21].

AI Applications to Communications and Information Technologies: The Role of Ultra Deep Neural Networks, First Edition. Daniel Minoli and Benedict Occhiogrosso.

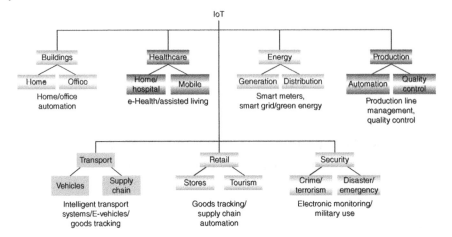

Figure 5.1 Nonexhaustive overall taxonomy of IoT applications.

distributed sensor-based systems with distances ranging from 10 m to thousands of kilometers (e.g. for satellite-based applications [1]).

Figures 5.1–5.4 (from Ref. [2]) provide a pictorial view of some key IoT environments and applications. Figure 5.1 provides a nonexhaustive overall taxonomy of IoT applications (also see Table 5.1); Figure 5.2 depicts a typical distributed environment comprising IoT-enabled devices; Figures 5.3 and 5.4 show a smart home IoT setting. Figures 5.5 through 5.10 depict other typical classes of IoT applications.

5.1.2   Smart Cities

In reference to the first two categories listed in the previous paragraph, in 2008, the world's population reached a 50–50 split in the distribution of populations between urban and non-urban environments. Seventy percent of the human population is expected to live in cities by the year 2050. The largest growth in urban landscapes is occurring in developing countries. There are now more than 400 cities with over one million inhabitants and 20 cities with over 10 million; the United States, for example, has more than 19,000 cities and towns. It is self-evident that technological solutions are needed to manage the increasingly scarce infrastructure resources under the limitations imposed by population growth, limited financial resources, and political inertia. AI/ML/DL-based systems are expected to facilitate the improved management of cities, moving them toward the goal of becoming full-fledged smart cities. The IoT (in conjunction with AI) offers the promise of improving the resource management of many assets related to city life, the flow of goods, people, and vehicles, and the greening of the environment by optimizing energy consumption and maximizing life-activity efficiency. Smart cities applications of the IoT, improving and optimizing quality of life (QoL) are a clear benefit to society (livability) and the environment (efficient use of resources, such as energy).

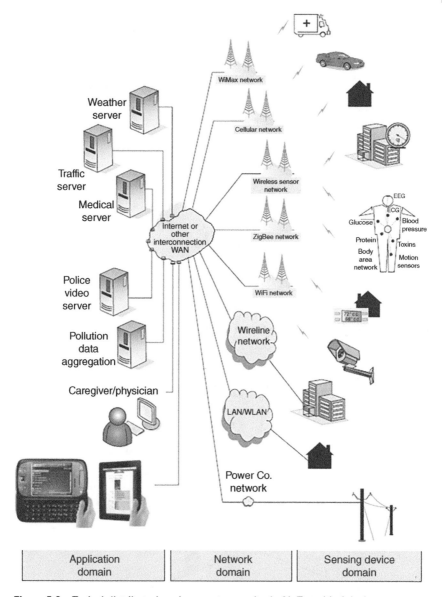

Figure 5.2 Typical distributed environment comprised of IoT-enabled devices.

At the technology level, there is interest in establishing how the evolving technology itself can be effectively used, the requisite architectures, standards, and the critical requirements for IoT cybersecurity. The applicable technology spans sensors, networking (especially wireless technologies for personal area networks, fogs, and cores, such as 5G cellular), and analytics; generally speaking,

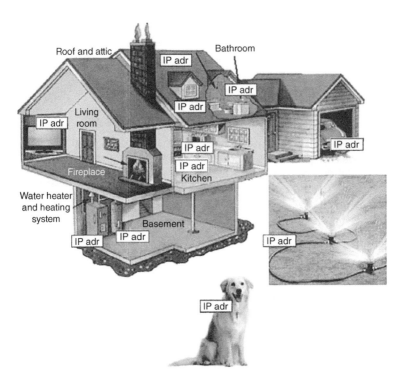

Figure 5.3 An exemplary IoT-based smart home.

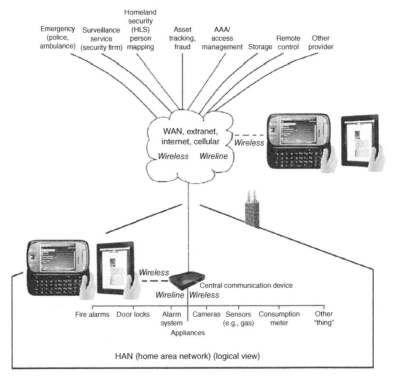

Figure 5.4 Another exemplary IoT-based smart home.

analytics is where ML is typically applied. Architecture deals with how things are assembled, including hierarchy. Standards relate to the ability to deploy the technology in a commodity fashion, assuring simple and reliable end-to-end interoperability – these also include M2M standards and approaches.[2] IoT cybersecurity, in canonical form, spans confidentiality, integrity, and availability (more on this topic in Chapter 6). Application areas include, but are not limited to, (i) Intelligent transportation systems (ITSs) (including smart mobility), (ii) vehicular automation and traffic control), (iii) SG, (iv) smart building, (v) goods and products logistics (including smart manufacturing), (vi) sensing (including crowdsensing and smart environments), (vii) surveillance/intelligence, and (viii) smart services. There is a plethora of issues, technical solutions, and technical challenges associated with a broad-based deployment of IoT in urban settings to enjoy the benefits of a smart city.

---

2 M2M is a set of technologies and an underlying architecture that incorporates standards for direct communication between intelligent between machines, devices, systems, controllers, and people. It represents a major enhancement of the older supervisory control and data acquisition (SCADA) system. The European Telecommunications Standards Institute (ETSI) has published a set of definitions and requirements of an M2M system, including the usage of gateways communication between M2M applications in the network and the M2M devices [1]. A number of specific M2M standards have emerged including: management protocols such as (i) Open Mobile Alliance Device Management (OMA DM) and OMA LightweightM2M; (ii) messaging protocols such as Message Queuing Telemetry Transport (MQTT – a Client/Server publish/subscribe light weight messaging transport protocol); (iii) application layer protocols such as TR-069 (Technical Report 069); (iv) data discovery protocols such as HyperCat; (v) communications protocols such as OneM2M and Google Thread; and (vi) open source software frameworks such as AllJoyn. The applications for M2M are practically unlimited. A short list of devices and applications include as industrial instruments, electric meters, vending machines, smart home devices, thermostats, fitness monitors, remote healthcare monitoring, security systems, data collection sensors, asset tracking, smart grid sensors, radio-frequency identification (RFID) and near field communication (NFC) systems, digital wallets including Google Wallet and Apple Pay, and any other telemetry application. Thus, M2M communication refers to communication between machines: M2M supports the communication between two devices without human intervention. M2M systems have seen broad presence throughout a large number of industries, especially in view of the ubiquitous emergence of wireless technology – such as Wi-Fi and cellular – and ML, including autonomic computing software to enable a network device to process and interpret the data and reach actionable decisions. M2M and IoT have slightly different connotations. Both technologies address the communication of connected devices, but M2M systems are typically isolated, well-defined, and well-identified standalone networked devices, while IoT systems elevate M2M to the "next level," aggregating disparate systems into one large, connected ecosystem [22]. M2M systems typically use point-to-point communications between machines, sensors, and hardware over cellular or wired networks, while IoT systems rely more typically on the Internet or other IP-based networks to transmit data collected from IoT-connected devices to edge gateways that use the cloud to send data to middleware platforms.

Table 5.1 A view to the Scope of IoT.

Service sector	Application group	Location (partial list)	Devices ("Things") of interest (partial list)
Real estate (industrial)	Commercial/institutional	Office complex, school, retail space, hospitality space, hospital, medical site, airport, stadium	UPS, generator, HVAC, environmental health and safety (EHS), lighting, security monitoring, security control/access
	Industrial	Factory, processing site, inventory room, clean room, campus	
Energy	Supply providers/consumers	Power generation, power transmission, power distribution, energy management, advanced metering infrastructure (AMI)	Turbine, windmills, uninterruptible power supplies (UPS), batteries, generators, fuel cells
	Alternative energy systems	Solar systems, wind system, co-generation systems	
	Oil/gas operations	Rigs, well heads, pumps, pipelines, refineries	
Consumer and home	Infrastructure	Home wiring/routers, home network access, home energy management	Power systems, HVAC/thermostats, sprinklers, dishwashers, refrigerators, ovens, eReaders, washer/dryers, computers, digital videocameras, meters, lights, computers, game consoles, TVs, PDRs
	Safety	Home fire safety system, home environmental safety system (for example, CO_2), home security/intrusion detection system, home power protection system, remote telemetry/video into home, oversight of home children, oversight of home-based babysitters, oversight of home-bound elderly	
	Environmentals	Home heating, ventilation, and air conditioning (HVAC), home lighting, home sprinklers, home appliance control, home pools and jacuzzis	
	Entertainment	TVs, PDRs	

Healthcare	Care	Hospitals, ERs, mobile POC, clinic, labs, doctor's office	MRIs, PDAs, implants, surgical equipment, BAN devices, power systems
	In vivo/home	Implants, home monitoring systems, body area networks (BANs)	
	Research	Diagnostic lab, pharmaceutical research site	
Industrial	Resource automation	Mining sites, irrigation sites, agricultural sites, monitored environments (wetlands, woodlands, etc.)	Pumps, valves, vets, conveyors, pipelines, tanks, motors, drives, converters, packaging systems, power systems
	Fluids management	Petrochemical sites, chemical sites, food preparation site, bottling sites, wineries, breweries	
	Converting operations	Metal processing sites, paper processing sites, rubber/plastic processing sites, metalworking site, electronics assembly site	
	Distribution	Pipelines, conveyor belts	
Transportation	Non-vehicular	Airplanes, trains, busses, ships/boats, ferries	Vehicles, ships, planes, traffic lights, dynamic signage, toll gates, tags
	Vehicles	Consumer and commercial vehicle (car, motorcycle, etc.), construction vehicle (for example, crane)	
	Transportation subsystems	Toll booths, traffic lights and traffic management, navigation signs, bridge/tunnel status sensors	
Retail	Stores	Supermarkets, shopping centers, small stores, distribution centers	POS (Point of Sale) terminals, cash registers, vending machines, ATMs, parking meters
	Hospitality	Hotel, restaurants, café, banquet halls, shopping malls	
	Specialty	Banks, gas stations, bowling, movie theaters	

(Continued)

Table 5.1 (Continued)

Service sector	Application group	Location (partial list)	Devices ("Things") of interest (partial list)
Public safety and security	Surveillance	Radars, military security, speed monitoring systems, security monitoring systems	Vehicles, ferries, subway trains, helicopters, airplanes, video cameras, ambulances, police cars, fire trucks, chemical/radiological monitors, triangulation systems, UAVs,
	Equipment	Vehicles, ferries, subway trains, helicopters, airplanes	
	Tracking	Commercial trucks, postal trucks, ambulances, police cars	
	Public infrastructure	Water treatment sites, sewer systems, bridges, tunnels	
	Emergency services	First responders	
IT systems and networks	Public networks	Network facilities, central offices, data centers, submarine cable, cable TV headends, telco hotels, cellular towers, poles, teleports, ISP centers, lights-off sites, NOCs	Network elements, switches, core routers, antenna towers, poles, servers, power systems, backup generators
	Enterprise networks	Data centers, network equipment (for example, routers)	

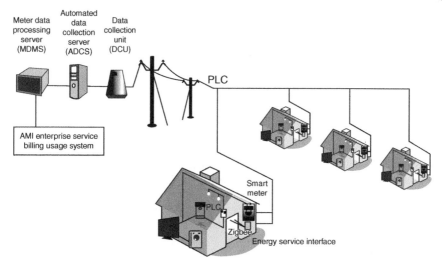

Figure 5.5 Advanced metering infrastructure.

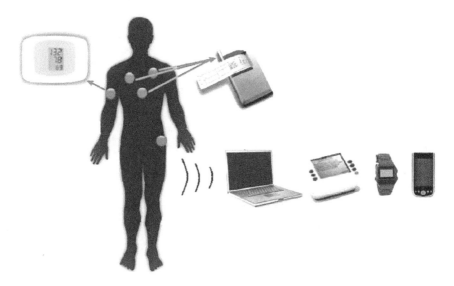

Figure 5.6 Medical wireless body area network.

Table 5.2 provides a short list of typical smart city applications, although many more such applications have already been scrutinized [3]. The table *also includes ML methods* that have been proposed in the literature smart city applications, including, but not limited to long short-term memory (LSTM), support vector machine (SVM), random forest (RF), convolutional neural networks (CNNs),

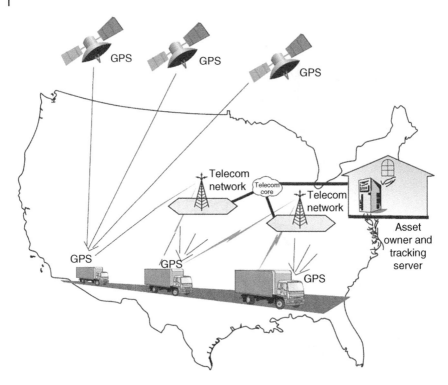

Figure 5.7 Vehicular IoT asset tracking on a national scale.

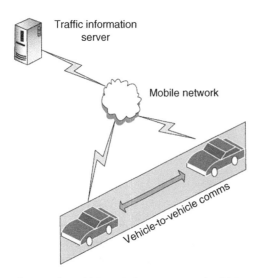

Figure 5.8 Vehicle-to-infrastructure and vehicle-to-vehicle IoT applications.

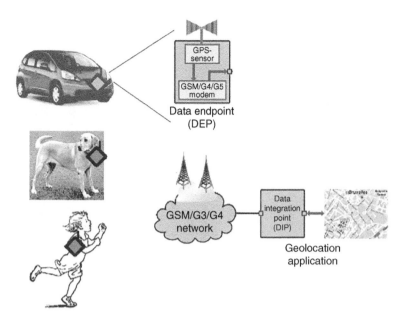

Figure 5.9 Geolocation and tracking IoT applications.

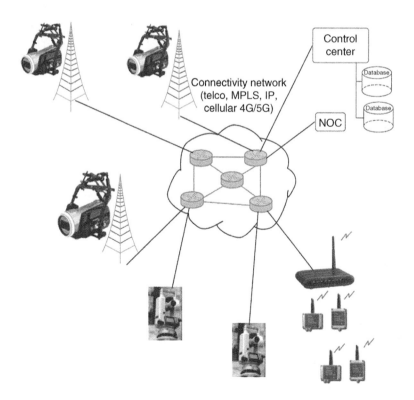

Figure 5.10 Distributed IoT environment.

Table 5.2 Partial list of Smart City Applications.

Application	Examples	Short list of ML methods
Intelligent Transportation Systems (ITSs)	For example, traffic monitoring, to assess traffic density and vehicle movement patterns, e.g. to adjust traffic lights to different hours of the day, special events and public safety (e.g. ambulances, police and fire trucks).	Extensive literature, including but not limited to Ref. [23] which provides an excellent press time survey of the extensive ML support for ITSs. Minimally, ITS covers three categories of tasks: (i) Perception tasks that try to detect, identify and recognize data patterns to extract, understand, and present relevant information; (ii) Prediction tasks endeavoring to try to predict future states given historical and real-time data; and (iii) Management tasks responsible for dictating the behavior in ITS. A long, detailed list of ML research references is included in [23], as follows: • Traffic sign and marking recognition; uses SVM, RF for classification with hand-crafted features; • Road signs recognition with RGB single image; uses CNN for classification and recognition method; • Road signs recognition with LAB color space and in moving vehicles; uses ELM, SVM for classification and recognition method; • Road detection and road scene understanding; uses CNN to distinguish different image patches; • Road lane detection; uses CNN for detection position of lane; • Obstacle detection; uses CNN, SVM to solve as regression; • Detect parking occupancy; uses CNN for a classifier of parking; • Road surface state and road crack recognition; uses SVM, CNN for classification of surface state and estimate cracks; • Vehicle detection using appearance features; uses SVM, R-CNN, Adaboost for classification method of vehicles;

		• Vehicle classification; uses SVM, RF for classification algorithm; • Vehicle identification with license plate recognition; uses SVM, CNN for character recognition of license plates; • Vehicle re-identification; uses CNN, SNN for feature extraction and classification; • Brake, vehicle steering, lane change, orientation, drive behavior detection; uses CNN, RF, SVM for classification method for driving behaviors; • Recognize driving styles of drivers; uses k-means, SVM, k-NN, RF, RNN for classification of driving styles into groups; • Pedestrian detection using handcrafted features; uses SVM, AdaBoost to distinguish pedestrians from the background of images; • Pedestrian detection using deep features; uses UL, CNN, R-CNN for feature learning and classification of pedestrians; • Cluster or rank network messages or nodes; uses k-means, SL for classifying network messages and nodes; • Network safety hazard detection; uses LSTM, DRL, RF for feature extraction and classification. Also [17].
Smart Grids	For example, advanced metering infrastructure (AMI) and demand response (DR).	Extensive literature, including but not limited to [24-28]. (Also [9, 10].)
Lighting Management	Control light intensity when area is empty or sparsely populated and/or when background light is adequate (e.g. depending on lunar phases, seasons, etc.).	Buildings are one of the significant sources of energy consumption and greenhouse gas emission in urban areas all over the world. Lighting control is an effective measure in reducing overall energy consumption and carbon emission. SVM and LSTM methods, among others have been proposed [29-32] (also [11]).

(Continued)

Table 5.2 (Continued)

Application	Examples	Short list of ML methods
Smart Building	For example, building service management, specifically for city-owned real estate to remotely monitor and manage energy utilization	In broad terms one can group solutions into two main classes: occupant-centric versus energy/devices-centric [33]. The first class of solutions uses ML for issues related to the occupants, including (i) occupancy estimation and identification, (ii) activity recognition, and (iii) estimating preferences and behavior. The second class of solutions uses ML to estimate aspects related either to energy or devices, including (i) energy profiling and demand estimation, (ii) appliances profiling and fault detection, and (iii) inference on sensors. Extensive literature, including but not limited to [33–42] (also [12, 13])
Municipal Solid Waste Management (MSWM)	For example, for disposition of public containers or city-owned properties	Reference [43] assesses more than 200 articles and proposals related to municipal solid waste management
Sensing (including Crowdsensing, Smart Environments, and Drones)	Environmental monitoring, for example, sensors on city vehicles to monitor environmental parameters. Also, crowdsensing, where the citizenry at large uses smartphones, wearable, and car-based sensors to collect and forward for aggregation a variety of visual, signal, and environmental data	Extensive literature on these topics. Crowdsensing: among others: [16, 44–46] Smart environments: among others: [47–49] Drones: among many others: [50–54]
Water Management	For example, to manage water usage, for example, for sprinklers, considering rain events	Among others: [55–57]

| Surveillance/Intelligence | For example, law enforcement applications | Extensive literature on this topic. Among others: [58–62] |
| Smart Services | As an example, the New York City Transit Department of Buses recently introduced advanced computer-aided dispatch automatic vehicle location to track the position of buses in real-time via global positioning system (GPS) using cellular overlay and provide advanced fleet management and next-bus time of arrival notification at bus stops throughout the region and on customer smartphones. | Among others: [63–65] |

extreme learning machines (ELM), region-based CNN (R-CNN), spiking neural networks (SNNs) for feature extraction, k-nearest neighbors (k-NNs) algorithm, RFs, recurrent neural networks (RNNs), unsupervised learning (UL), supervised learning (SL), and deep reinforcement learning (DRL). Many IoT applications for smart cities involve video and multimedia streams; these applications include classical video surveillance, drone-based video surveillance, drone-based multimedia sensors, crowdsensing, and social networking.

Traditional smart city applications have included the following:

- *Physical security*: Open-space security in streets, parks, stations, tunnels, bridges, trains, buses, ferries, and government buildings. Networked sensors (both fixed, portable and mobile [e.g. drones]) to support Internet Protocol (IP)-based surveillance video; license plate reading; gunshot detection; biohazard and radiological contamination monitoring; face recognition; and crowd monitoring and control. Initially, the IoT has been envisioned as supporting a large population of relatively low-bandwidth parameter-sensing devices, particularly in M2M environments, and generally in stationary locations (e.g. data-collecting meteorological weather stations, electric meters, industrial control, and the like). Nonetheless, video-oriented applications that have inbound streams ranging up to ultra-high definition (UHD) resolutions both in pixel density and in frame rate and video-oriented applications where the stream is outbound to the receiving devices (e.g. but not limited to smartphones) are emerging [4, 5]. Situational awareness also supports security objectives [66, 67] (as noted in Chapter 4).

- *Traffic, transportation, and mobility*: Optimized traffic flows, low congestion, low latency and high expediency, low noise, minimal waste of fuel and CO_2 emissions, safety. These applications include ITSs (e.g. smart mobility, vehicular automation, and traffic control): for example, (i) traffic monitoring, to assess traffic density and vehicle movement patterns, e.g. to adjust traffic lights to different hours of the day, special events and public safety (e.g. ambulances, police and fire trucks) and (ii) networked sensors to support traffic flow, driverless vehicles including driverless bus transit, and multi-modal transportation systems. For driverless vehicles, sensors will allow high-resolution mapping, telemetry data, traffic and hazard avoidance mechanisms. ITSs will be a critical element of urban planning and future smart cities, facilitating improved road and traffic safety, transportation efficiency, energy efficiency, and reduced environmental pollution [23]. Driverless vehicles are emerging as a technology. Advances in wireless communication techniques and location-aware sensor technologies are fueling the evolution of vehicle networks (VNs). In the next few years, driverless cars, buses, and trucks will begin to share the roadways with traditional vehicles. Additionally, one might see a significant expansion of autonomous transportation in fleet vehicles such as commercial trucks; fleets are the initial area where driverless vehicle technology is expected to be widely deployed – driverless public transit (trains and buses) may also shift toward autonomous transportation.

- *Power and other city-supporting utilities*: Reliable flow of electric energy, gas, and water; optimized waste-management and sewer; safe storage of fuel. The SG is an evolution of the electricity network that integrates the activities of power consumers, power generators, distribution grid, and devices connected to the grid (e.g. substations, transformers, etc.) [6, 7]; an SG encompasses all the various stages of power generation, distribution, and consumption. The goal of SG is to economically and efficiently deliver sustainable, reliable, and secure electricity supplies. M2M/IoT technology is designed for automated data exchange between devices, and thus has applicability to SGs; M2M communications takes place between two or more mechanistic entities that routinely avoid direct human intervention. With M2M technology, organizations track and manage assets; inventories; transportation fleets; oil and gas pipelines; mines; wide-spread infrastructure; natural phenomena such as weather conditions, farm production, forestry condition, water flows, and, as noted, SGs. Wireless communication is a staple of M2M. These wireless technologies range from unlicensed local (so called "fog") connectivity, to licensed 3G/4G/5G cellular, to low earth orbit (LEO) satellites. Utilities have started to gradually support M2M and SCADA (supervisory control and data acquisition) systems over wireless and satellite links; these connectivity technologies are applicable to the SG for city and rural environments, respectively, particularly for the transmission and distribution space (T&D) sector. The intelligent integration of information from actions of users connected to the electricity grid – consumers, generators, and the distribution grid – are performed by the SG. Efficient, sustainable, economic, and secure delivery of electricity supplies is the main goal of SG. The power industry is increasingly seeking to incorporate information and communications technology (ICT) in general and IoT in particular, into its operations, including at the edges; the thus-enhanced power grids are known as SGs. Three basic issues are of interest: monitoring of rural transmission systems, demand response (DR) support, and urban automated meter readers (AMRs). Well known examples of IoT usage in energy efficiency and interactions with the SG/advanced metering infrastructure (AMI) include the following: smart thermostats; smart appliances which can interwork with DR-based SG power management (with the use of appliance-level actuators); "plug-level" control of electrical outlets, where lower end devices can be turned on or off remotely; IoT-based light-emitting diode (LED) lighting and daylight sensors for "smart lighting" that not only allow intelligent centralized (and/or remote) control but also lower energy consumption while improving the residents' experience; and consumers' ability to generate green renewable power and sell it back to the SG.

- *Infrastructure and real estate management*: Monitor status and occupancy of spaces, buildings, roads, bridges, tunnels, railroad crossings, and street signals [8]. Networked sensors (possibly including drones) are used to provide real-time and historical trending data allowing city agencies to achieve enhanced visibility into the performance of resources, facilitating environmental and safety sensing, smart parking and smart parking meters, smart electric meters, and smart building functionality. Crowdsensing is just one of the applicable technologies. Crowdsensing allows a large population of mobile devices to measure phenomena of common interest over an extended geographic area, enabling "big data" collection, analysis, and sharing. It has major urban applications for the (active or passive) collection of traffic conditions, weather conditions, and even video images [16]. In recent years, the widespread availability of sensor-provided smartphones has enabled the possibility of harvesting large quantities of data in urban areas exploiting user devices, thus enabling a suite of urban crowdsensing applications.
- *Environmental monitoring*: Monitor outdoor temperature, humidity, air quality and the presence of contaminants or other environmental gases.
- *Pollution monitoring*: Monitor emission of dioxins, vaporized mercury, nanoparticles, radiation from factories, incinerators, and urban crematoria.
- *Water management*: To manage water usage, for example, for sprinklers, considering rain events.
- *Smart city lighting*: Control light intensity when area is empty or sparsely populated and/or when background light is adequate (e.g. depending on lunar phases, seasons, etc.), conversion to LED lighting and ensuing control via IoT for weather conditions, phases of the moon, seasons, traffic occupancy, and so on.
- *Flood abatement*: Flood and storm drainage control.
- *Waste management*: For example, for disposition of public containers or city-owned properties.
- *Intelligent Campus*: For example, external campus surveillance; internal and external surveillance; building Emergency Generator, Automatic Transfer Switch (ATS) and Digital Meter (DM – aka Smart Meter) monitoring and control; elevator monitoring and control; and HVAC monitoring and control. Other campus-related applications include remote door control, water leak detection, washing machine scheduler, smart parking, smart trash cans, light control, and emergency notification.
- *Logistics*: Supplying city dwellers with fresh food, supplies, goods, and other materials;
- *Livability*: QoL, expeditious access to services, efficient transportation, low delays, and safety.

Despite the plethora of (possible) applications, one of several challenges at the current time is that the smart cities IoT segment is fragmented into discrete vertical domains, multiple stakeholders, and disconnected information and ICT systems. A set of streamlined technical standards and a usable multi-service architecture that support a "plug and play" mode of expansion would greatly enhance deployment and broad-based penetration.

As stated and noted in Table 5.2, ML has become an important component to address the very large amounts of data collected by the multitude of sensors deployed in the environment and to draw meaningful on-point conclusions and actionable intelligence from the data. This chapter focuses specifically on smart buildings (e.g. as motivated in, but certainly not limited to, [11–13]).

5.2 Smart Building ML Applications

There is vast literature on this topic, as noted above. This section only provides one illustrative example (from the literature).

5.2.1 Basic Building Elements

An evolution of the smart city concept is the application of these concepts to commercial building environments, also possibly including multi-building campuses [11, 13, 14]. Commercial buildings have a wide gamut of monitoring, management, and resource optimization requirements and many of the applications for smart cities have applicability to building management; some of these applications include video surveillance, traffic/access control, surveillance, energy management (including lighting), indoor environmental and air quality/comfort control, and fire detection among others.

Energy management is a key consideration of (smart) buildings. The list that follows identifies typical elements and systems where energy is consumed, all of which benefit greatly from improved (IoT-based) sensing, automation, and management (list is not exhaustive):

- *Heating, ventilation, and air conditioning (HVAC) room*: Modular boilers, air compressors, and chillers;
- *Server room*: Uninterruptible power supplies (UPSs); computer room air conditioners (CRACs); telecom closets; racks and virtualized/blade servers;
- *Office space*: LEDs lighting; daylight sensors; thermostats (used in controlling HVAC systems and energy consumption); DR mechanisms;
- *Cooling system elements*: Rooftop units (RTUs), cooling towers, and heat pumps.

The need and desirability for energy consumption optimization spans commercial buildings, office buildings, campuses of integrated buildings, manufacturing plants, data centers, shopping malls, scientific research structures, and also residential buildings. According to the Commercial Buildings Energy Consumption Survey (CBECS), space heating was the most common end use in commercial buildings [68]. About 32% (2,167 TBTU) of energy was consumed in the United States for space heating (in recent years). Ventilation and lighting each accounted for 10% or more of total energy consumption. See Figure 5.11. HVAC in total (space heating, cooling, and ventilation) represents 52% of the consumption. CBECS is an independent, statistically representative source of national-level data on the characteristics and energy use of commercial buildings; they conducted the CBECS periodically since 1979, as required by Congress. The 2018 CBECS is the 11th iteration; respondents, such as building owners and managers, completed the survey for 6,436 buildings for the 2018 CBECS, representing 5.9 million buildings in the United States.

Building management systems (BMSs) have traditionally been used to manage several building-related functions. A BMS is a computerized platform that allows for monitoring and controlling a building's mechanical and electrical equipment. A BMS is typically used to manage loads and enhance efficiency, thus reducing the energy needed to illuminate, heat, cool and ventilate a building. BMSs interact with controls hardware in the various mechanical/electrical subsystems for real-time monitoring and controlling of the energy used; they are often used to implement DR arrangements. See Figure 5.12 for an illustrative example.

Although current-generation BMSs typically focus primarily on electrical consumption, in the future BMSs are expected to cover all energy sources supporting a building, also including natural gas, renewable energy, water usage, and steam systems. Indoor environmental and air quality capabilities are also important. Lately, BMSs have migrated to IP-based networking; this allows remote monitoring by a centralized operations center, such as cloud-based analytics. For example, smart in-building lighting not only allows intelligent centralized (and/or remote) control and improves the inhabitants' experiences but also lowers energy consumption. IoT will take the current BMS capabilities to the next level: inexpensive sensors are emerging, and user-friendly applications are becoming available, often as a Software-as-a-Service (SaaS) cloud-provided service [15]. HVAC optimization and intelligent lighting controls are just two of the key areas that are facilitated by the IoT.

5.2.2 Particle Swarm Optimization

Particle swarm optimization (PSO) may be used in some IoT-based applications. PSO is a natural world–inspired algorithm that is used to search for an optimal solution in the solution space utilizing very few hyperparameters; it is typically used to identify

Space heating accounted for close to one-third of end-use consumption in 2018

Major fuels consumption by end use, 2018
share of total

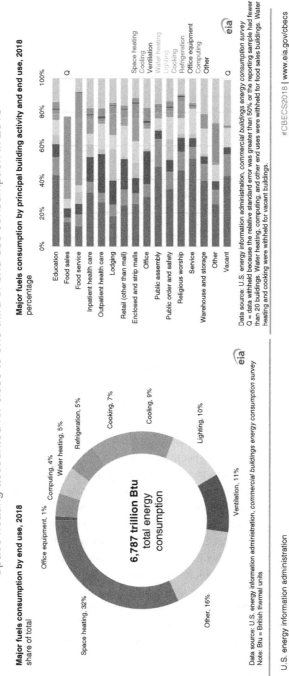

Office equipment, 1%
Computing, 4%
Water heating, 5%
Refrigeration, 5%
Cooking, 7%
Cooling, 9%
Lighting, 10%
Ventilation, 11%
Other, 16%
Space heating, 32%

6,787 trillion Btu
total energy
consumption

Data source: U.S. energy information administration, *commercial buildings energy consumption survey*
Note: Btu = British thermal units

U.S. energy information administration

Major fuels consumption by principal building activity and end use, 2018
percentage

0% 20% 40% 60% 80% 100%

Education
Food sales
Food service
Inpatient health care
Outpatient health care
Lodging
Retail (other than mall)
Enclosed and strip malls
Office
Public assembly
Public order and safety
Religious worship
Service
Warehouse and storage
Other
Vacant

Space heating
Cooling
Ventilation
Lighting
Water heating
Cooking
Refrigeration
Office equipment
Computing
Other

Data source: U.S. energy information administration, *commercial buildings energy consumption survey*
Q = data withheld because the relative standard error was greater than 50% or the reporting sample had fewer than 20 buildings. Water heating, computing, and other end uses were withheld for food sales buildings. Water heating and cooking were withheld for vacant buildings.

#CBECS2018 | www.eia.gov/cbecs

Figure 5.11 Commercial Buildings Energy Consumption Survey (CBECS). *Source*: U.S. Energy Information Administration/https://www.eia.gov/consumption/commercial/data/2018/pdf/CBECS%202018%20CE%202018%20Release%202%20Flipbook.pdf.

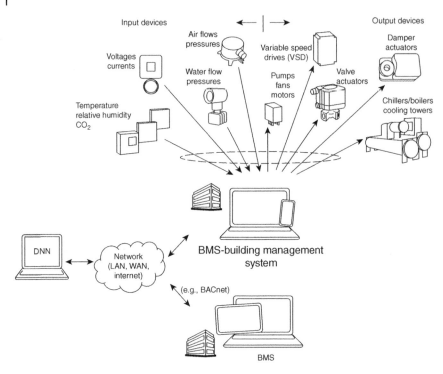

Figure 5.12 BMS complemented by an ML/DNN system.

the maximum or minimum of a function defined on a multidimensional vector space [69–71]. In PSO, only the objective function is required, and the methodology is not dependent on the gradient or any differential form of the objective.

The natural world–inspired concept derives from the belief that a school of fish or a flock of birds that moves in a group "can profit from the experience of all other members." Namely, while a bird flying is randomly searching for food, all birds in the flock can share their discovery and help the entire flock get the best hunt [72]. The concept is based on heuristics: *the best solution found by the flock is the best solution in the space.* Similar to the flock of birds looking for food, the algorithm starts with a number of random points on the plane (they are the "particles") and lets the particles look for the minimum point in random directions; at each step, every particle should search around the minimum point it ever found and also around the minimum point found by the entire swarm of particles. After a certain number of iterations, one considers the minimum point of the function as the minimum point ever explored by this swarm of particles. One often finds that the solution identified by PSO is reasonably close to the global optimal. Refer to [72] for a more inclusive explanation of the process.

5.2.3 Specific ML Example – Qin Model

Owing to the cumulative (high) cost of energy in a commercial building, there is an intrinsic goal to optimize the energy utilization efficiency of HVACs and other electrically-controlled occupant life support and convenience systems, including lighting systems. While traditional methods are available, the use of ML/DL to create an artificial intelligence control system (AICS) is advantageous.

Although there is a large literature on this topic, Ref. [73] provides an excellent description of matters related to DL use in building energy consumption optimization; this model and description is used liberally in this section as an illustrative application of DL to building energy management – in the rest of this section, we refer to this system as the Qin Model.[3] Some key terms related to energy management are included in the Glossary at the end of this chapter.

To address (smart) building energy optimization two phases are usually required: (i) an energy audit phase and (ii) an optimization phase. Preferably, the energy audit phase performs a zero-touch, objectively autonomous process that monitors and assesses energy losses and energy utilization efficiency without requiring an engineer to visit the structure being audited and optimized. To achieve this, a number of sensors can be deployed within or in proximity to the structure. These (IoT) sensors are equipped with communication links and report collected data to an audit computer that is programmed to perform a set of audit functions on the collected data and, thereby, compute both real-time energy utilization statistics and aggregated historical statistics. The optimization stage is performed by an optimization system, which may be physically implemented by the same computer that serves as the audit computer, or it may be physically implemented by a separate computer coupled by hardwired connection or by network connection to the audit computer (e.g. a local area network or a wide area network or Internet). The optimization computer implements an AI/ML algorithm to determine how the energy efficiency and performance of the structure may be improved. The results of the optimization algorithm are used to generate electronic control instructions that are sent to actuators in the building, causing the monitored energy consuming devices to adjust their settings and thereby improve the energy efficiency of the structure.

The ML algorithm comprises a learning-based model, such as a neural network (NN) model or set of models. The models are populated with data obtained from the structure, preferably as part of the audit stage, and these data define a nonlinear problem, which the optimization computer solves using a particle swarm optimizer.

---

3 The figures and the discussion in the remainder of this section are based directly on Ref. [73].

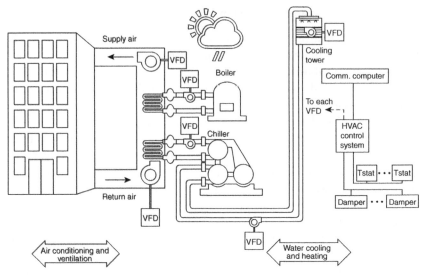

Figure 5.13 Block diagram of an exemplary commercial building structure.
Source: Reference [73]/U.S. Patent.

Figure 5.13 depicts an exemplary commercial building HVAC. Although the specifics of an HVAC system can vary, the basic elements typically include an (i) air conditioning and ventilation portion and (ii) a water cooling and heating portion. The air conditioning and ventilation portion comprises a forced air system that provides *supply air* to the structure through a *supply air ducts* and that receives return air from the structure through a return air duct. One or more air blowers cause the air to circulate from the supply air side, through the structure and, thus, through the return air side where the air is reconditioned before circulating back through the structure. A boiler system and a chiller system are utilized to provide both heating and cooling, respectively. Each of these includes heat exchanger coils positioned in the ductwork, intermediate between the supply air side and the return air side. In each of the boiler and chiller pump, a heat transfer fluid exists in respective closed loop paths flowing through the respective heat exchanger coils, using suitable pumps. In the case of the boiler system, heat energy is introduced into the structure using a suitable heat source, such as a natural gas burner, an electrical heating element, and so on. In the case of the chiller system, heat energy is extracted from the structure using a cooling tower. As illustrated in Figure 5.13, water (or other heat exchange fluid) from the chiller system is pumped using pump to the cooling tower. The cooling tower includes heat exchanger coil or radiator that transfers heat from the water or other heat exchange fluid into the atmosphere. A cooling fan increases the airflow through the radiator to increase transfer of heat energy to the atmosphere. As illustrated in the figure, each of the various pumps and blowers can be electronically controlled using variable frequency drive (VFD) controllers. These VFD controllers allow each of the pumps and blowers to be controlled not only on and off, but also at different variable speeds according to instructions from an HVAC control system. The HVAC control system is suitably coupled to each of the VFD controllers and may also be coupled to one or more thermostats deployed throughout the structure. The HVAC system may also be coupled to one or more electrically or pneumatically controlled dampers that can be controlled between fully open and fully closed and to positions in between, to regulate the forced airflow through air duct systems within the structure.

Without the benefit an optimization system, the HVAC would affect control over heating and cooling by controlling whether the chiller is on or off and by manipulating the settings of various dampers, with the goal of providing a constant temperature within a particular zone controlled by a particular thermostat. By way of example, in a typical HVAC control system to achieve a zone temperature of 72 °F, the HVAC control system might set the zone damper to a 15% open setting and command the chiller to be turned on. This might produce a supply air temperature from the chiller at 57 °F degrees at a static pressure of 350 Pa (Pascal), with the VFD controlling the supply air being set to 95%. When an optimization system is used to instruct the HVAC controller, a far more energy efficient control strategy becomes possible. Using *energy audit and optimization techniques*, the same structure might be controlled to achieve a zone temperature of 72 °F degrees (the same as in the example above), where the chiller system is commanded to be shut off, and the zone damper is set to 85.22% open. With the chiller off, the supply air temperature might be 64.4 °F degrees at a static pressure of 196.6 psi, with the VFD being set to 74.5%. Although the very same 72 °F degree zone temperature is achieved using the optimization system, far less energy is consumed, resulting in a 42% savings.

5.2.3.1 EnergyPlus™

EnergyPlus™ is a whole building energy simulation program that engineers, architects, and researchers use to model both energy consumption – for heating, cooling, ventilation, lighting, and plug and process loads – and water use in buildings. EnergyPlus™ is developed and funded by the US Department of Energy's (DOE) Building Technologies Office (BTO) and is managed by the National Renewable Energy Laboratory (NREL). It is developed in collaboration with NREL, various DOE National Laboratories, academic institutions, and private firms. DOE releases major updates to EnergyPlus twice annually. An updated version 22.2.0 with bug fixes was released on September 30, 2022. EnergyPlus is free, open-source, and cross-platform – it runs on the Windows, Mac OS X, and

Linux operating systems. Along with OpenStudio, EnergyPlus is part of BTO's building energy modeling program portfolio [74]. Some of the notable features and capabilities of EnergyPlus include:

- Integrated, simultaneous solution of thermal zone conditions and HVAC system response that does not assume that the HVAC system can meet zone loads and can simulate unconditioned and underconditioned spaces.
- Heat balance-based solution of radiant and convective effects that produce surface temperatures thermal comfort and condensation calculations.
- Sub-hourly, user-definable time steps for interaction between thermal zones and the environment implementing automatically varied time steps for interactions between thermal zones and HVAC systems. These allow EnergyPlus to model systems with fast dynamics while also trading off simulation speed for precision.
- Combined heat and mass transfer model that accounts for air movement between zones.
- Advanced fenestration models including controllable window blinds, electrochromic glazings, and layer-by-layer heat balances that calculate solar energy absorbed by window panes.
- Illuminance and glare calculations for reporting visual comfort and driving lighting controls.
- Component-based HVAC that supports both standard and novel system configurations.
- A large number of built-in HVAC and lighting control strategies and an extensible runtime scripting system for user-defined control.
- Functional mockup interface import and export for co-simulation with other engines.
- Standard summary and detailed output reports as well as user definable reports with selectable time-resolution from annual to sub-hourly, all with energy source multipliers.

EnergyPlus is a console-based program that reads input and writes output to text files. It ships with a number of utilities including IDF Editor for creating input files using a simple spreadsheet-like interface, EP-Launch for managing input and output files and performing batch simulations, and EP-Compare for graphically comparing the results of two or more simulations. Several comprehensive graphical interfaces for EnergyPlus are also available. DOE does most of its work with EnergyPlus using the OpenStudio software development kit and suite of applications.

5.2.3.2 Modeling and Simulation

Four different models are constructed in the Qin model: (i) a *thermal load model*, (ii) an *envelope model*, (iii) an *HVAC model*, and (iv) an *occupants model*. These models are implemented using a commercially available EnergyPlus™ computer-implemented energy simulation program. These four models take the form depicted by the equations below.

Thermal and Envelope Model

$$Q_A - U_A(T_A - T_O) - H_M(T_A - T_M) - C_A\frac{dT_A}{dt} = 0 \tag{5.1}$$

$$Q_M - H_M(T_M - T_A) - C_M\frac{dT_M}{dt} = 0$$

$$a = \frac{C_M C_A}{H_M}$$

$$b = \frac{C_{M(U_A + H_M)}}{H_M} + C_A \tag{5.2}$$

$$c = U^A$$

$$d = Q^M + Q^A + U^{A^T O}$$

$$a\frac{d^2 T^A}{dt^2} + b\frac{dT^A}{dt} + cT^A = d$$

$$g = \frac{Q^M}{H^M}$$

$$A^3 = \frac{r_1 C^A}{H^M} + \frac{U^A + H^M}{H^M} \tag{5.3}$$

$$A^3 = \frac{r_2 C^A}{H^M} + \frac{U^A + H^M}{H^M}$$

$$T^A = A^1 e^{r_1 t} + A^2 e^{r_2 t} + \frac{d}{c} \tag{5.4}$$

$$\frac{d^{T_{Ao}}}{dt} = \frac{H^M}{_cA^T{}_{Mo}} + \frac{U^A}{_cA^T{}_{Ao}} + \frac{U^A}{_cA^T{}_O} + \frac{Q^A}{C^A} \tag{5.5}$$

$$T^M = A^1 A^3 e^{r_1 t} + A^2 A^4 e^{r_2 t} + g + \frac{d}{c} \tag{5.6}$$

where

C_A: air heat capacity (Btu/°F. or Joules/°C.);
C_M: mass (of the building and its content) heat capacity (Btu/F or joules/°C.);
U_A: the gain/heat loss coefficient (Btu/F·hours or W/°C.) to the ambient;
H_M: the gain/heat loss coefficient (Btu/F·hours or W/°C.) between air and mass;
T_O: outdoor temperature (°F. or °C.);
T_A: air temperature inside the house (°F. or °C.);

T_M: mass temperature inside the house (°F. or °C.);
Q_A: heat added to the indoor air;
Q_M: heat added to the building mass.

Plug Load Model

$$P_{bui}(t) = P_{hvac}(t) + P_{plug}(t) + P_{light}(t)$$
$$Q_{bui}(t) = P_{bui}(t)\tan(\partial)$$

where

$P_{bui}(t)$: Active power of building at t time;
$P_{hvac}(t)$: Active power of HVAC load at t time;
$P_{plug}(t)$: Active power of plug load at t time;
$P_{light}(t)$: Active power of light load at t time;
$Q_{bui}(t)$: Reactive power of building at t time;
$\tan(\partial)$: Reactive power of building at t time;

HVAC Model

$$E(t_f) = \int_{t_0}^{t_f}\left(-P_{sup}^{(t)}\eta_{coo} + P_{hea}^{(t)}\right)\eta_{hea} + P_{fan}^{(t)}))dt$$
$$Er(t_f) = \int_{t_0}^{t_f}\left(-r^1 P_{sup}(t)\eta_{coo} + r^2 P_{hea}^{(t)}\eta_{hea}\right) + r^3 P_{fan}^{(t)}))dt$$

where

$E(t_f)$: The integral of power in HVAC system;
$P_{sup}(t)$: The cooling equal energy at t time;
η_{coo}: Cooling device efficiency;
$P_{hea}(t)$: The heating equal energy at t time;
η_{hea}: Heating device efficiency:
$P_{fan}(t)$: The air supply fan system equal energy at t time;
$E_Y(t_f)$: The integral of power in HVAC system under correction Y coefficient;
Y^1: Correction coefficient of cooling system;
Y^2: Correction coefficient of heating system;
Y^2: Correction coefficient of air supply fan system.

Occupants Model

$$H(x) = -\sum_{i=1}^{m} P(V_i)\log_2\left(P(V_i)\right)$$
$$H(Y \mid X) = \sum_{i=1}^{m} P(X = V_i)H(X \mid V_i)$$

where

$H(X)$: The information entropy of the set;
$P(V_i)$: The probability of getting the ith class value when randomly selecting one from the dataset X.
$H(Y|X = V_i)$: The entropy of Y among only those instances in which X has value V_i, named "specific conditional entropy."
$P(X = V_i)$: The probability of V_i in the dataset of X.

In their untrained form, the models express the fundamental thermal relationships by which heat is stored and propagated by conduction, convection, and radiation mechanisms; the models also express in statistical terms how occupants behave inside the structure as a function of time. The models are then populated with data reflecting the static conditions associated with the structure. This may be viewed as a first level of model population; then the models are further populated to reflect the dynamic behavior of the structure – this is done through simulation. The simulations are then compared with real operating data and the comparisons are used to calibrate the simulation models. The manner of performing these two levels of model population and the subsequent model calibration is shown in Figure 4.14.[4] As illustrated, the simulation models are typically expressed in an EnergyPlus™ form or other suitable form for computer analysis and are populated with model parameters reflecting properties of the actual structure. Such parameters include, for example, roof color (affecting its radiation rate), roof insulation value, wall insulation value, chiller efficiency, cooling system size, and floor area. The parameters also include certain operation parameters, such as temperature set points. Thus, the models now contain parameters that reflect the static properties or reflect generalized properties corresponding to a particular structure.

Next, the simulation models are operated on by a computerized simulation process that iteratively changes the initially assumed operating conditions and causes the models to calculate solutions or simulation output. Finally, to calibrate the simulation models, the simulation output is compared with real operation data and the results of the comparison are fed back to make adjustments to the parameters as initially chosen. In this way, the calibration process defines those parameters that are most influential and performs both a coarse grid calibration and a refined grid calibration of the simulation models. The simulation models can now be used to predict ranges of energy savings for real measures.

In addition to modeling the structure, the Qin model is also able to generate control instructions that may be fed back to the structure to induce energy

---

4 All figures in the rest of this section are from [73].

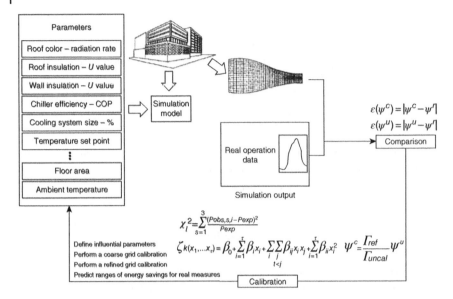

Figure 5.14 Calibration method by which the neural network may be trained.

optimization: The control instructions are fed back as control parameters that instruct the HVAC control system (and optionally other electronic control systems) to make changes to the HVAC component and other system settings to improve energy efficiency. The energy audit and optimization control system can either supply explicit commands to specific actuators within the structure (e.g. actuators that control heating and cooling equipment, adjust damper settings, change thermostat settings, and so on).

5.2.3.3 Energy Audit Stage

Referring to Figure 5.15, the structure under audit and optimization control includes energy consuming equipment, such as the HVAC; associated with the energy consuming equipment are a plurality of sensors and actuators. In general, the sensors (equipped with wired or wireless communication links) measure relevant conditions such as temperature, pressure, humidity, airflow, electric current, energy and heat flow, and the operating states of various components of the energy consuming equipment. The actuators perform functional changes to the energy consuming equipment, such as turning components on and off and changing variable settings, such as damper settings. In the illustration, these actuators and sensors are coupled to a communication computer that is programmed to collect and compile sensor readings and actuator instructions, and to communicate sensor readings to the audit computer and to receive actuator instructions

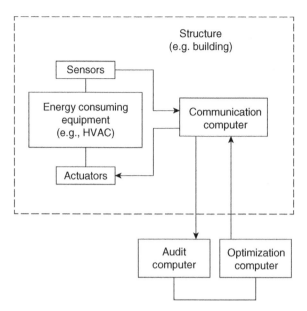

Figure 5.15 Block diagram illustrating optimization control system with an exemplary structure.

from the optimization computer. If desired, the communication computer may be implemented as a set of functions within a smart device, such as a smart power meter that meters electrical power being supplied to the structure.

To support the optimization stage, at least some of the actuators may be equipped with electronic control circuits, allowing them to be controlled by the optimization computer. The Qin optimization control technology system exploits these devices and structures as energy efficiency control measures, which the audit computer measures and the optimization computer adjusts settings, to achieve optimal energy efficiency. Although these energy efficiency control measures naturally depend on the nature of the structure, Table 5.3 provides an exemplary list of energy efficiency control measures that would be applicable to a reasonably sophisticated commercial building.

5.2.3.4 Optimization Stage

The optimization stage is performed by an optimization computer. The optimization computer implements an AI algorithm to determine how the energy efficiency and performance of the structure may be improved. In one implementation, the results of the optimization algorithm are used to generate electronic control instructions that are communicated to actuators disposed on or in the structure, causing the monitored energy consuming devices associated with that structure to

Table 5.3 List of energy efficiency control measures.

1. Outdoor and Exhaust Air Damper Faults and Control
2. Static Pressure Optimal Set
3. Optimal HVAC Schedules by Actual Status
4. Minimum Variable Air Volume (VAV) Terminal Box Damper Flow Reductions
5. Control Setpoint Dead Bands and Night Optimal Set
6. Chilled Water Differential Pressure Monitor
7. Chilled Water Temperature Optimal Set
8. Condenser Water Temperature Optimal Set
9. Water Differential Pressure Reset
10. Water Temperature Optimal Set
11. Supply Air Temperature Reset
12. Exhaust Fan Control
13. Fix Leaking Coil Valves
14. Fix Low Refrigerant Charge
15. Fix and Re-calibrate Faulty Sensors
16. Plant Optimal Low-load Shutdown
17. HVAC Optimal Start-Stop
18. Refrigerated Case Lighting Controls
19. Walk-in Refrigerator/Freezer Lighting Controls
20. Refrigeration Floating Head Pressure
21. Refrigeration Floating Suction Pressure
22. Optimize Defrost Strategy
23. Anti-sweat Heater Control
24. Evaporator Fan Speed Control
25. Occupancy Sensors for Thermostats and Room Lighting
26. Optimized Use of Heat Recovery Wheel
27. Lighting Occupancy Presence Sensors
28. Daylighting Controls
29. Exterior Lighting
30. Advanced Plug Load Controls
31. Night Purge
32. Advanced Rooftop Unit (RTU) Controls
33. Elevator Lighting
34. Cooling Tower Waterside Economizer
35. Cooling Tower Optimal Controls

Table 5.3 (Continued)

36. Demand Control Ventilation
37. Demand Response: Setpoint Changes
38. Demand Response: Pre-cool
39. Demand Response: Duty Cycle
40. Demand Response: Refrigeration
41. Demand Response: Lighting
42. Demand Response: Chilled Water Temperature Reset

adjust their settings and thereby improve the energy efficiency of the structure. The models are populated with data obtained from the structure, preferably as part of the audit stage, and these data define a nonlinear problem, which in the preferred implementation, the optimization computer solves using a particle swarm optimizer.

5.2.3.5 Model Construction

The audit computer uses parameter-based models and real-time control parameter data to capture relevant properties needed for the audit process. The parameter-based models are constructed using the EnergyPlus modeling tool described above and available from the US DOE's BTO, and managed by the NREL. The real-time control parameter data are obtained from actual measured values taken from the structure using the sensors discussed above.

5.2.3.6 EnergyPlus Models

In the Qin model, the EnergyPlus tool constructs and supplies the audit computer four models:

- A structure envelope model;
- A thermal model;
- A structure HVAC model; and
- A structure occupancy behavior model.

The structure envelope model captures the relevant properties about how the envelope of the structure is defined. The envelope defines the boundary between inside and outside of the structure (e.g. walls, roof, doors, windows, and so on). The thermal model captures the relevant properties related to how heat flows within the structure and through the structure. This includes how radiant heat from the sun is absorbed and transferred into the interior of the structure and how heat convectively flows through the envelope of the structure. The structure HVAC model captures the relevant properties concerning the operating parameters of the HVAC system within the structure. The HVAC model, thus, models the on/off states of HVAC system components, and their operational settings. Lastly, the structure occupancy behavior model captures properties about how building occupants move throughout the building both positionally and as a function of time. The structure occupancy behavior model would thus record when occupants enter or exit the building, whether windows have been opened, and the extent to which different rooms are occupied throughout the day and night. Collectively, the data captured in these four models represents a complex parameter space, describing how energy is utilized as a result of conditions external to the structure (e.g. weather and qualities of the structure envelope), and conditions internal to the structure (e.g. caused by occupant behavior) and as a result of the operation of the HVAC systems associated with the structure.

5.2.3.7 Real-Time Control Parameters

In addition to the model-based parameters (developed using EnergyPlus), the audit computer also captures real-time data from the sensors deployed within the structure, as described above. If desired, these real-time data can be aggregated into sets of historical control parameter data. The audit computer uses these two ingredients – the model-based parameters and the control parameter data – to train an NN that effectively learns how the control parameters within the actual physical structure correlate to the model-based parameters. The trained NN thus acquires an understanding of how control parameters that can be physically monitored relate to theoretical (model-based) parameters that are grounded in principles of thermodynamics. The trained NN model can predict how the HVAC systems within the structure will behave under certain weather and occupancy patterns, even if those have not actually been experienced before. The trained NN model can also predict how changes in certain control parameter settings (i.e. different HVAC equipment settings or operating conditions) will affect other systems when certain weather and occupancy patterns occur. The trained NN model empowers the audit computer to perform energy audits of the structure, without the need for an engineer to physically visit the structure site – this is possible because the trained NN is able to form associations from which the audit computer can draw conclusions about how the structure is operating from an energy efficiency standpoint.

5.2.3.8 Neural Networks in the Qin Model (DNN, RNN, CNN)

The Qin system uses a DNN having the general structure illustrated in Figure 5.16. The NN comprises a set of input layers, a set of middle (or hidden) layers and a set of output layers. The input layers receive DNN input parameters that correspond to at least some of the parameters used in the simulation models. Examples of such input parameters include (i) time: day, night, weekdays, and weekends;

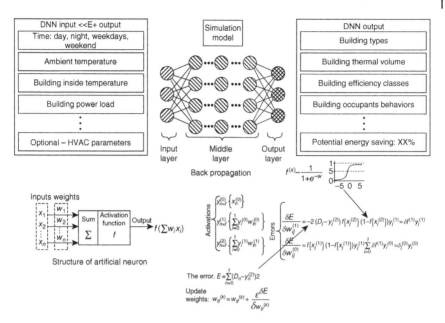

Figure 5.16 DNN used to implement the Qin optimization control system.

(ii) ambient temperature; (iii) structure (e.g. building) inside temperature; (iv) structure (e.g. building) power load; and (v) optional HVAC parameters or other optional operating parameters.

The NN is configured to effect backpropagation according to the equations shown in the figure. The output layer supplies the DNN output, which may reflect predictions made by the NN as to structure (e.g. building) types, thermal volume, efficiency classes; structure occupant behaviors; and potential energy saving. Although a number of different programming technologies can be used, a suitable DNN can also be implemented in Python using the Keras DL library. The process entails (i) importing the classes and functions from the Keras library, (ii) loading the dataset to be used, and (iii) encoding the classification variables and defining the DNN model. Once defined the DNN model is then trained and may then be used to perform evaluations and make predictions.

Referring to Figure 5.17, the NN is trained using the simulation output generated by the simulation model (as described in connection with Figure 5.14). The NN, so trained, is now able to make predictions relative to the four models established when developing the simulation model. These are the four key structure models used to perform energy audits, namely, the thermal load model, the envelope model, the HVAC model, and the occupant's model.

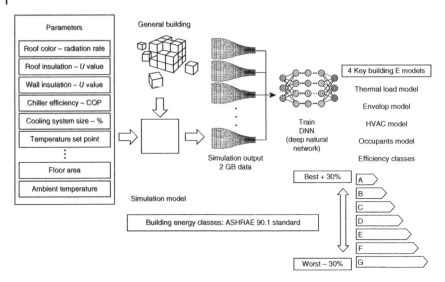

Figure 5.17 NN training using simulation output data.

In performing the energy audits, the trained NN is able to make predictions when real data is supplied at the DNN input. For example, in the exemplary configuration illustrated in Figure 5.18, the NN can predict that the structure type or building type is an "office" and that it has a thermal volume of 450,000 cubic feet. The NN is able to predict that the structure has a D class efficiency designation and that the occupants exhibit a behavior typical of office workers. In addition, the NN is able to predict that a potential energy savings of 25% is possible for the structure whose real data are supplied at the DNN input. The NN prediction that a 25% energy savings is possible has been made in an entirely automated fashion. The energy audit has required no audit engineers to visit the premises. All that was required was to supply the trained NN with real data from the structure. In the illustrated example, one week of real data was sufficient to produce the predictions illustrated.

The Qin model is implemented as a DNN using multiple disparate NN configurations, including a RNN and a CNN. A CNN is good at image recognition and the system uses the CNN to learn the overall shape of the performance curve of the structure. The focus of the CNN is, thus, on the shape of the performance data, as opposed to the actual data values themselves. Thus, the CNN is good at learning to recognize performance curves, which may be similar among structure types, even though different structures of that type may be quite different in size. The RNN is good at optimizing results and it is good at predicting individual output

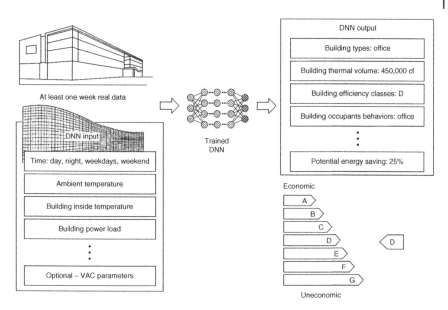

At least one week real data

DNN input

Time: day, night, weekdays, weekend

Ambient temperature

Building inside temperature

Building power load

⋮

Optional – VAC parameters

Trained DNN

DNN output

Building types: office

Building thermal volume: 450,000 cf

Building efficiency classes: D

Building occupants behaviors: office

⋮

Potential energy saving: 25%

Economic

A
B
C
D
E
F
G

D

Uneconomic

Figure 5.18 Trained NN performing the automated energy audit.

control parameters. When the RNN is fed with measured parameters, it is good at predicting how the control parameters need to be set at different times of the day.

The NN, comprising these CNN and RNN components, receives real-time data from parameter settings and sensors within the structure and performs predictions that control how the parameters will be controlled. Thus, the NN can be seen as the master controller that tells the individual subsystems within the structure how to perform. The NN also receives input from the PSO process and in this way the NN learns what the optimal behavior of the structure should be.

The NN is trained using two training mechanisms. (i) The NN is first trained using the simulation models. These models are constructed using the EnergyPlus tools to define a set of static models representing knowledge about how the structure functions from the standpoints of thermal loading, structure envelope, HVAC system operation and occupant behavior. (ii) The second training mechanism provides dynamic information about the structure. To perform the second training mechanism, the simulation models are fed with a series of time-varying conditions that cause the model to predict different responsive behaviors and these behaviors are used to further train the NN. The second training mechanism thus provides the NN with information about predicted energy consumption levels under time-varying dynamic conditions.

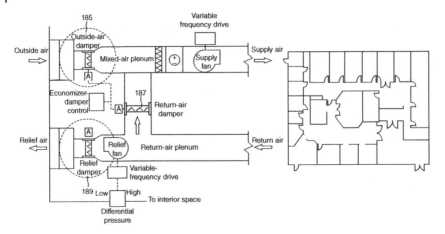

Figure 5.19 System block diagram of an economizer device.

By way of example, the second training mechanism might model the performance of an economizer such as that illustrated in Figure 5.19. The second training process generates simulation results showing economizer performance under different environmental conditions as damper and fan settings are changed.

5.2.3.9 Finding Inefficiency Measures

The NN can be used to find inefficiency measures, that is, to predict which parameters would lend themselves to being optimized. In Figure 5.19, the respective outside air damper, return air damper and exhaust or relief air damper are each set to different settings, which will differ, depending on the outside air temperature. Thus, when the outside air temperature is between 50° and 60°, the return air damper is nearly shut (between 0 and 10% open) while the outside air damper and exhaust damper are essentially open (between 90 and 95% open). However, when the outside air temperature rises above about 70°, the damper settings dramatically change, with the return air damper being set at about 75% open, while the outside air damper and exhaust air damper being set to about 25% open. The NN learns this behavior through the training process. Once trained, the NN can assess which damper operation strategies will produce efficient results and which do not.

5.2.3.10 Particle Swarm Optimizer

From the discussion above, the trained NN provides a useful tool in conducting an energy audit, automatically and without the need for a trained engineer to visit the structure site-which in some instances may not be physically possible. The ability to conduct such energy audits is itself valuable, however, the Qin model can do more than this. The optimization computer can examine all the control parameters and operating conditions and generate an optimal control parameter regime to optimize energy efficiency of the structure. To do this, the optimization computer uses a PSO algorithm. As discussed earlier, the PSO is a computational method that optimizes a problem by iteratively trying to improve a candidate solution with regard to a given measure of quality. It solves a problem by having a population of candidate solutions, called particles, and moving these particles around in the search-space according to simple mathematical formulae over the particle's position and velocity. Each particle's movement is influenced by its local best known position, but is also guided toward the best known positions in the search-space, which are updated as better positions are found by other particles. This is expected to move the swarm toward the best solutions.

In the Qin model, an exemplary NN might have on the order of 10–963 model parameters and 5–125 control parameters supplying the inputs. The network might have, for example, 15 adaptive layers, providing 15 outputs. Given the large number of different control parameters and condition permutations, the optimization problem spans a very large search space. Illustrated in Figure 5.20 the search space may be represented as a topographical map having local peaks and valleys. The candidate solutions are initially distributed across this space.

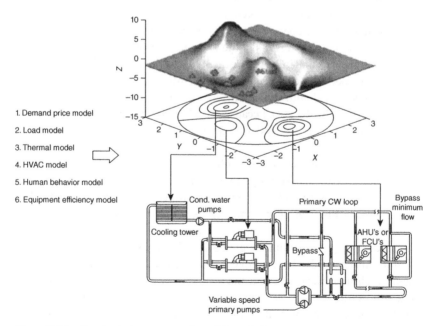

Figure 5.20 Topological representation of the solution space covering a set of control parameters.

The algorithm causes each candidate solution or particle to move, where the movement of each particle is influenced by its own local best known position, but also influenced by the best known positions discovered by other particles in the search space. After successive iterations, the algorithm causes the particles to swarm towards the optimal solution.

Figure 5.21 depicts an arrangement for configuring the PSO algorithm. In Figure 5.21, the system being optimized is an exemplary HVAC system comprising a water system and an air system. The particle swarm algorithm is shown at the lower left of the figure, which is also subject to the set of constraints.

5.2.3.11 Integration of Particle Swarm Optimization with Neural Networks

Figure 5.22 illustrates how the PSO is integrated with the NN. The operation parameters set points, represent the collection of controlled parameters within the HVAC system (or other system associated with the structure). These parameters are fed as inputs to the NN, which in turn supplies control instructions to operate the HVAC system. The operation parameters set points are supplied to the EnergyPlus simulation models, which generates simulation results that are fed to the PSO process. As illustrated diagrammatically in the figure, the results are mapped onto a topological space having points that correspond to specific operation parameters within the HVAC system. The particle swarm process finds the optimal settings for these operation parameters, to achieve the optimally efficient HVAC system. Notably, the results of the PSO are fed back as adjustments to the operation parameters set points. As a result of this feedback from the PSO process, the simulation models are changed to reflect the discoveries made by the PSO, and the NN is also supplied with the optimized operation parameters set points that have been improved by the PSO. Because the NN is controlling the HVAC system, the performance of the HVAC system is thus optimized by virtue of the feedback from the PSO process.

The PSO process takes as inputs a collection of operating parameters of the HVAC system at work. These might include air pressure set points, inside and outside temperature set points, air flow set point, cooling water flow set point, circulating water flow set point, and the like. The PSO process assesses the whole system energy consumption achieved for these set points and discovers the minimum energy consumption, giving a predefined set of constraints. These constraints may include, for example, occupant comfort parameters, such as temperature, humidity, ventilation, and the like. The particle swarm, in essence, examines how the system as a whole behaves when the individual control parameters change over their allowable ranges. To do this, the PSO process uses the results as generated by the simulation models. As illustrated in Figure 5.22, these simulation models generate results given the operation parameter set

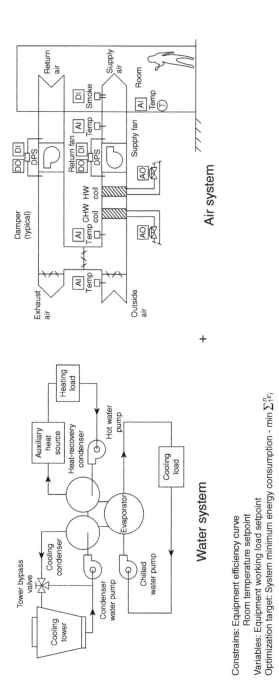

Air system

Water system

Constrains: Equipment efficiency curve
Room temperature setpoint
Variables: Equipment working load setpoint
Optimization target: System minimum energy consumption - $\min \sum_1^n \varepsilon_j$

Particle swarm optimization

Velocity of
particle i at time $k+1$

$$v_{k+1}^i = w v_k^i + c_1 \, rand \, \frac{(p_l^i - x_k^i)}{\Delta t} + c_2 \, rand \, \frac{(p_k - x_k^i)}{\Delta t}$$

Inertia factor
range: 0.4 to 1.4

Current
motion

Self confidence
range: 1.5 to 2

Particle memory
influence

Swarm confidence
range: 2 to 2.5

Swarm influence

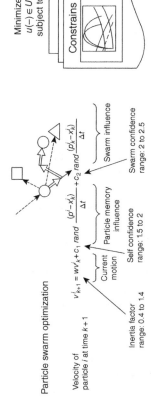

Minimize
$u(-) \in U$
subject to

Constrains

$E_y(t_f) + \int_{t_0}^{t_f} k \, u_{win}(t)^2 \, dt,$

$F(t, \dot{x}(t), x(t), u(t), y(t), \theta) = 0,$

$F_0(\dot{x}(t_0), x(t_0), u(t_0), y(t_0), \theta) = 0,$

$T_{air}^l(t) \leq T_{air}(t) \leq T_{air}^u(t),$

$0 \leq u_{win}(t) \leq u_{win}^u(t),$

$\dot{m}_{air}^l \leq \dot{m}_{air}(t) \leq \dot{m}_{air}^u,$

$P_{hea}(t) \geq 0,$

$T_{air}(t_0) = T_{air}(t_f),$

Figure 5.21 Performance optimization diagram.

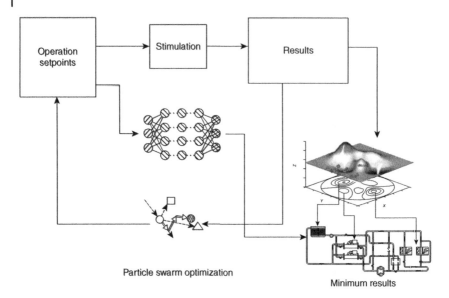

Particle swarm optimization

Minimum results

Figure 5.22 Data flow diagram illustrating how the output of the particle swarm optimization is fed back to further train the NN.

points as inputs. Thus, the optimizer is discovering the optimal operation parameter set point settings for a particular set of conditions (weather, structure occupancy, etc.).

5.2.3.12 Deep Reinforcement Learning

In conjunction with the techniques discussed above, a DRL process can be utilized. An example of DRL has been illustrated in Figure 5.23. In the DRL implementation a software agent responds to information from the environment, shown diagrammatically at and trains the dynamic DNN. The DNN supplies actions A to the environment, in feedback fashion. The DNN works in conjunction with a Q table that stores values for every state vs every action possible in the environment. This Q table is also mapped to another DNN; that DNN is used to drive the settings of coupled HVAC equipment and is also coupled to the simulation models which in turn regulate the environment (i.e. energy consumption) by the structure.

5.2.3.13 Deployments

There are a number of deployment scenarios. In particular, a single computer can perform both the functions of the audit computer and the optimization computer; here, the system interfaces with a BMS, which in turn controls the

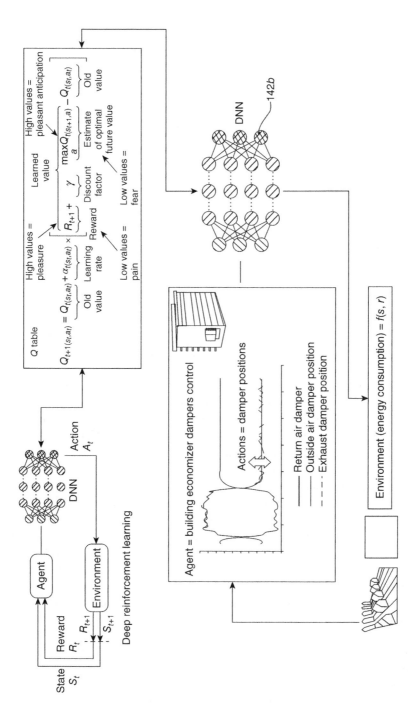

Figure 5.23 Data flow diagram illustrating how deep reinforcement learning optimization.

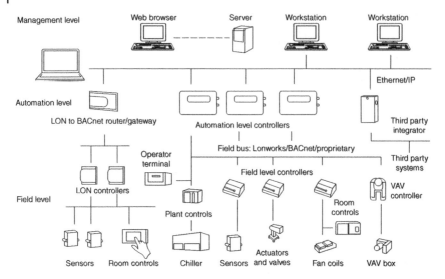

Figure 5.24 Networked implementation of the optimizer control system integrated into a BACnet interface.

systems within the structure. A suitable network (LAN, Internet, wireless link, cable, serial connection) is used to connect the computer running the DNN with the BMS.

In larger applications, it can be desirable to deploy the optimization control system on a computer that interfaces with the electronic controllers and actuators within the structure, in a manner that utilizes the structure's existing wiring and network topology. An example of this has been illustrated in Figure 5.24. Here, the sensors, actuators, room control thermostats, and other HVAC components such as the chiller, fan coils, variable air volume (VAV) box, and so on, are coupled to communicate over a LonWorks or BACnet Bus, as illustrated. Automation level controllers provide the local control over these HVAC components and these controllers are, in turn, responsive to DNN computer, which provides control strategies to be carried out by the automation level controllers. If needed, a router or gateway provides connectivity between the computer and certain HVAC devices that may not be accessible through the automation level controllers. In the example of Figure 5.24, the optimization control strategy, developed by the NNs and PSO functions, is supplied by the management computer to the automation level controllers as a control regimen which the controllers carry out. If certain components cannot be accessed through the automation level controllers, management computer can affect direct control over those components by sending instructions through the router or gateway.

Refer to [73] for specific operational applications and specific examples of control of elements (such as exhaust air dampers, and so on).

5.3 Example of a Commercial Product – BrainBox AI

This section provides a brief description of one useful *commercially-available off-the-shelf ML-based product* with which the authors are familiar, that can assist stakeholders reduce actual energy consumption by 15–25%, depending on conditions and geographic/environmental conditions. Section 5.3.1 provides some technical background and Section 5.3.2 provides some product capabilities, for illustrative purposes.[5]

5.3.1 Overview

As shown in Figure 5.11, total HVAC operation (space heating, cooling, and ventilation) equates to 52% of the aggregate energy consumption nationally and also for business stakeholders. Tools that can assist building managers and tenants of commercial buildings optimize energy use are highly valuable and directly impact the bottom line.

By way of introduction, some commercial HVAC systems use a constant air volume (CAV) approach while others use a VAV approach cited earlier. CAV systems supply air at a constant volume and variable temperature; the fresh air entering from outside is mixed with the return air to produce fresher air to the fan ventilation; the speed of the fan is fixed, and it is controlled by an (On/Off) switch. A VAV system can save more energy compared to a CAV: VAV systems enable energy-efficient air distribution by optimizing the amount and temperature of distributed air. VAV HVAC systems supply air at a variable temperature and airflow rate from an air handling unit (AHU) (see Figure 5.25 for an example, [75]). VAV systems use flow control to efficiently condition each building zone while maintaining required minimum flow rates. VAV systems can support varying heating and cooling needs of different building zones, thus, these systems are found in many commercial buildings.

For VAVs and/or CAVs the following ML methods have recently been proposed in the literature, as discussed in [76]: artificial neural network (ANN) clustering, ANN MLP (multi-layer perceptron), UrbanFM (a Context-Aware Neural Network Model – CANN), RNN-LSTM, RNN-NARX (Nonlinear autoregressive eXogenous – NARX), RNN-NNARX (Neural Nonlinear AutoRegressive eXogenous – NNARX), RNN-NNFL (Neural Network Feedback Linearization – NNFL), Regression Trees (RT), and RFs, among others. The ANN model has been widely

---

5 The material in this section was provided by BrainBox AI (https://brainboxai.com/en/). Headquarteredin Montreal, Canada. BrainBox AI brings sustainability to the built environment to significantly reduce energy consumption and costs. Since the launch in 2019, more than 100 MM/feet^2 of commercial building space has been impacted and optimized by the AI systems provided by the firm in 70 cities worldwide.

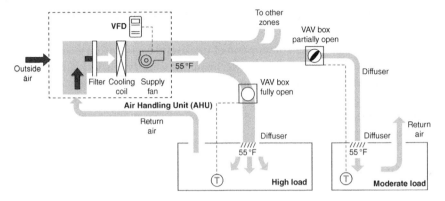

Figure 5.25 VAV HVAC system. *Source:* Reference [75]/U.S. Department of energy/https://www.pnnl.gov/projects/om-best-practices/variable-air-volume-systems/accessed May 02, 2023.

applied for several types of applications in HVAC sector, such as fault detection and diagnostics (FDD), thermal comfort approximation, and indoor air temperature (IAT) prediction. Many of these remain research endeavors and are not generally available *off-the-shelf products*.

5.3.2 LSTM Application – Technical Background

Accurate IAT predictions for building environments are challenging, especially for multi-zone buildings and for different types of HVAC systems. In particular, the nonlinearity of the buildings' thermal dynamics makes the IAT prediction difficult being that it is affected by complex factors such as (i) controlled and uncontrolled points, (ii) outside weather conditions, and (iii) occupancy schedule. ML models (MLMs) have been proposed and studied to predict IAT for multi-zone building including LSTMs based on direct multi-step prediction with a *direct sequence-to-sequence (direct-S2S)* approach [76]. Experimental results have indicated that the LSTM models outperform MLP models by reducing the prediction error by 50%, while at the same time ascertaining that temperature variations remain around the upper and lower boundaries of comfort (67 and 82°F, respectively, according to ASHRAE standard 55-2017). Thus, accurately modeling IAT is required. The IAT prediction can be achieved using physics-based or data-driven methods. Physics-based models are based on Fourier's law of heat conduction which is discretized into the finite difference. However, the IAT model varies from one zone to another, and this makes the IAT prediction modeling using a physics-based approach challenging and time-consuming. Data-driven models, on the other hand, are based on the data generated from a large number of sensors and thermostats already deployed. The main advantage of utilizing data-driven approaches is to eliminate the cost and time-consuming task to build IAT physics-based models. Besides, IAT model is easy to implement for a multi-zone system using ML techniques that can deal with nonlinearity in the system and incomplete or noisy data.

In recent years, researchers have developed ML-based tools to detect faults in the supply air temperature control loop in commercial buildings with VAV systems. Some researchers proposed a context-aware NNs model using human and environmental variables for approximating thermal comfort evaluation for HVAC systems; yet others developed a simple ANN-MISO (multi-input single-output) model to predict indoor temperature for different forecasting time steps; yet others developed a hybrid model predictive control (MPC)-based NN feedback-linearization model to predict IAT for CAV systems; an IAT prediction models of multi-zones using MISO structure; an optimal control of multi-zones VAV system; a framework named UrbanFM based on DNNs; a multi-zone modeling approach using the MLP-MIMO model; a multi-zone buildings using a different tuned model based on nonlinear autoregressive network with exogenous inputs models (refer to [76] for extensive literature).

However, prior work generally adopted a recursive prediction strategy and the typical NN method considers each input as an independent parameter, ignoring the time dependency between sequential values: in short, the temporal dependency among continuous variables is ignored. These limitations can be addressed by adopting a technique that takes into account the temporal relationship among input parameters and using a direct prediction approach to avoid the problem of error accumulation. A time series approach for multi-step prediction model using LSTM has been shown to be an accurate forecasting method for time series data.

RNNs are solutions for modeling sequence dependency, and LSTM network is a RNN that overcomes the problem of training a recurrent network with the architecture of learnable gates. As noted elsewhere in this text, LSTM is one of many variations of the RNN. The ability of LSTM to reduce the vanishing and exploding gradient problems efficiently makes such an approach more appropriate for contexts having a long-term dependency problem. Since the HVAC data are sequential and future outputs depend not only on the current values of inputs but also on the previous information, a model based on LSTM is a good choice to predict IAT in the HVAC system. The main advantage of LSTM is the use of gates to manage its own memory by choosing to update or not the information that goes through the cell. In fact, an LSTM network is able to learn long-term dependencies from an input sequence thanks to its internal memory cells. LSTMs have been found suitable for electric consumption, for predicting the photovoltaic power, for price forecasts, and also for emission factor prediction in the context of scheduling appliances use in the smart house domain; additionally, a few studies have investigated the usefulness of LSTM for IAT multi-step predictions in the HVAC system.

Reference [76] from BrainBox AI researchers discusses, among others, an LSTM-MIMO (multi-input multi-output) architecture that utilizes MIMO to predict IAT for all zones simultaneously. Only one model has proven advantageous and useful; this LSTM framework is based on a direct S2S multivariate multi-step time series prediction model, instead of the recursive model, which is helpful to better control the HVAC system's operation. (Note that an IAT prediction model used for a CAV system cannot be applied to a VAV system or vice versa without decreasing the performance.)

The IAT problem can be formulated as, given a time series of observations as input, predict a sequence of observations as output for a range of future time steps. To further improve the flexibility of the temperature forecasting methodology, an LSTM-based direct-S2S architecture is investigated to solve the defined problem. The framework considers a direct-S2S prediction instead of a recursive one (a recursive model increases the accumulation of prediction error throughout the prediction step ahead). Two direct-S2S models are developed in Ref. [76], one for a MISO architecture and one for a MIMO architecture. The direct-S2S for LSTM-MIMO is shown in Figure 5.26, which includes as input all n zones feature vectors, x_i; x_{i+1}; ...; x_n with all ℓ past time steps of each feature vector. All controlled variables of all n zones are included in the input vectors. Moreover, this architecture has n output prediction vectors, y_i; y_{i+1}; ...; y_n with p future steps each. The time steps, ℓ and p, have variable lengths. The main advantage of direct-S2S architecture is that it allows tuning ℓ to have the best prediction performance of p steps.

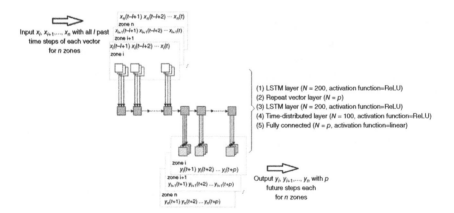

Figure 5.26 LSTM-MIMO architecture with multi-input multi-output to predict IAT for all zones simultaneously with only one model. *Source:* Adapted from Ref. [76].

As reported in [76], the performance of the multi-step IAT prediction model(s) was evaluated on LSTM and MLP models applied on real data collected from a hotel and retail store in Montreal with VAV and CAV system, respectively; the ensuing experimental results showed that the LSTM-MIMO models outperform MLP models by reducing the prediction error by 50%.

5.3.3   BrainBox AI Commercial Energy Optimization System

Technology overview

Using the theoretical machinery discussed in the previous section, *BrainBox AI* (https://brainboxai.com/en/) offers technology and software as a service that enhances existing HVAC equipment into predictive and self-adaptive HVAC systems using AI and cloud computing. See Figure 5.27. The following points highlight the features of the technology:

- Connects directly to a building's HVAC system;
- Transforms HVAC systems from reactive to preemptive or "self-driving";
- Utilizes existing data from building systems (e.g. BMS, access control systems) and third party sources (e.g. weather, occupancy) to drive decision-making;
- Autonomously drives a building's HVAC system by directly writing back to the controller in real-time, with no human intervention required;

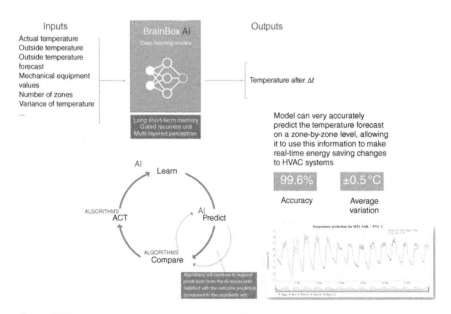

Figure 5.27   BrainBox AI system. *Source:* BrainBox AI.

- Makes decisions based on advanced DL models;
- Reduces number of service calls and increases Net Operating Income (NOI) – SaaS-based $/sq. ft. pricing model;
- Can save up to 25% in total energy costs in some environments;
- Can provide an improvement of up to 60% in some environments; and
- Can be net-cashflow positive relatively quickly.

Figure 5.28 illustrates some of the technical benefits of the BrainBox AI system compared with a traditional approach. A PID (proportional, integral, and derivative) loop is a control strategy (control loop) used in many types of process control systems. In building automation systems, HVACs in particular, PID loops are used to maintain precise control of temperature, pressure, and air flow (among other parameters). A PID loop consists of components, such as sensors, dampers, valves, and a loop controller. Figure 5.29 illustrates the business advantages of the AI system.

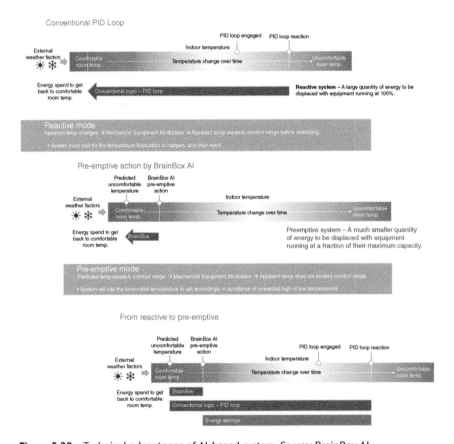

Figure 5.28 Technical advantages of AI-based system. *Source:* BrainBox AI.

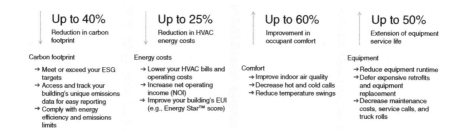

Up to 40%	Up to 25%	Up to 60%	Up to 50%
Reduction in carbon footprint	Reduction in HVAC energy costs	Improvement in occupant comfort	Extension of equipment service life

Carbon footprint
→ Meet or exceed your ESG targets
→ Access and track your building's unique emissions data for easy reporting
→ Comply with energy efficiency and emissions limits

Energy costs
→ Lower your HVAC bills and operating costs
→ Increase net operating income (NOI)
→ Improve your building's EUI (e.g., Energy Star™ score)

Comfort
→ Improve indoor air quality
→ Decrease hot and cold calls
→ Reduce temperature swings

Equipment
→ Reduce equipment runtime
→ Defer expensive retrofits and equipment replacement
→ Decrease maintenance costs, service calls, and truck rolls

Figure 5.29 Business benefits of AI-based system. *Source:* BrainBox AI.

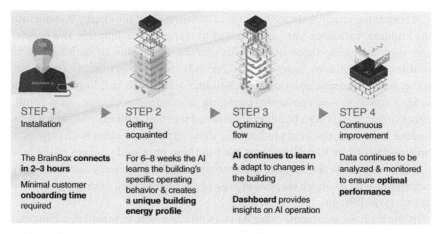

STEP 1	STEP 2	STEP 3	STEP 4
Installation	Getting acquainted	Optimizing flow	Continuous improvement
The BrainBox **connects in 2–3 hours** Minimal customer **onboarding time** required	For 6–8 weeks the AI learns the building's specific operating behavior & creates a **unique building energy profile**	**AI continues to learn** & adapt to changes in the building **Dashboard** provides insights on AI operation	Data continues to be analyzed & monitored to ensure **optimal performance**

Figure 5.30 Fully autonomous self-adaptive AI allows a stakeholder to move from reactive to preemptive HVAC operations management in four steps. *Source:* BrainBox AI.

Figure 5.30 depicts the simple implementation lifecycle that can be utilized to deploy the technology.

BrainBox AI Technology Background

BrainBox's DL engine and proprietary process autonomously optimizes existing HVAC control systems, maintaining a building's thermal equilibrium without human intervention. By (i) accurately predicting future conditions, (ii) evaluating the best possible configuration for the existing HVAC system, and then (iii) nstructing the control system to make the necessary adjustments, BrainBox AI allows the stakeholder to achieve real energy efficient gains in the built environment.

Energy efficiency must be embraced as a dynamic resource to achieve real efficiency gains in commercial building environments. Instead of using a static operating model of building operations and maintenance, stakeholders need to think about energy in terms of flow and modulate equipment performance dynamically in response to the ways in which the internal and external environments change over time. Effectively managing the thermal equilibrium of a building using dynamic modulation is a complex process that requires continuously optimizing energy flow to ensure occupant comfort and maximum energy efficiency. The first steps towards maintaining a building at equilibrium are to calculate its energy leak rate and determine the ideal HVAC system settings to ensure optimal levels for the power-to-thermal load relationship. However, both the energy leak rate and the power-to-thermal load relationship are themselves dynamic, which means that they change over time as the various factors that impact energy flow change over time, including occupancy, weather conditions, and demand.

Maintaining energy balance requires calculating all of the energy flowing into the building, including purchased energy and appliances, lighting, and people, and then balancing those inputs with thermal losses due to such factors as ventilation, infiltration, and drain water. But these factors are dynamic, not static, so the thermal equation for a building – and each building has its own unique equation – is constantly changing as conditions change. As a result, effectively maintaining a building at balanced thermal equilibrium also requires finding ways to accurately predict how these inputs and outputs change over time and to ensure that all systems react to these predicted changes in the best possible way. However, because of the complex nature of energy flow, manually determining a building's thermal energy equation and then managing the flow of energy is difficult.

In the built environment, using AI is an effective way to maintain a balanced thermal equation. One of the benefits of using AI in the building management space is that it can determine the thermal energy equation for a building in a fraction of the time required by individuals. The real value of AI arises when it is used to predict how the flow of energy will evolve over time and if it anticipates an unwanted thermal event in the future, to make adjustments to the HVAC system to eliminate that event before it happens. In addition to determining which adjustments are best for ensuring occupant comfort, AI can also evaluate the optimal HVAC system configuration required to achieve greater energy efficiency, thereby saving money and making buildings greener by reducing the load on the power grid.

Every built environment has its own unique set of habits and behaviors that affect the river of energy that flows through it – these characteristic behaviors are expressed in a collection of thermal energy equations which, once calculated, will not change over time for a building unless major renovations are carried out. Characterizing the river of energy by calculating a building's thermal energy equation is the first step towards optimization and requires calculating its energy leak rate and power-to-thermal load relationship.

The best way to characterize the collection of thermal equilibrium equations for a building is to first divide it up into zones or divisions and then derive the collection of equations for each zone. This requires collecting historical data from the zones and then looking for patterns in that data to isolate each zone's habits and behaviors. Once the specific energy leak rate and power-to-thermal load relationship has been calculated for each zone, the next step is to aggregate the characterizations of all the zones into a group of zones. While these tasks can be performed manually, an AI engine can perform them in a fraction of the time, accomplishing in a few minutes what it would otherwise take an engineer three weeks to do.

According to the laws of physics, thermal energy will always work to reach an equilibrium, so energy will always flow across a differential to balance the amount of energy on either side. If the energy levels are balanced in a system, then there is no flow. However, if the thermal energy on one side is higher than on the other, energy will start to flow at a certain rate across this differential until equilibrium is achieved.

In the built environment, there are barriers in place, including walls and windows, that work to slow down this rebalancing. The energy leak rate – hereafter referred to simply as the leak rate – is the speed at which energy moves across these barriers and measures the quantity of heat that a built environment loses or gains over a given period of time. It refers to the rate at which energy flows either into or out of a given zone. Each zone has its own unique leak rate that is based on how it was built – including the types of materials used, like insulation and windows – and on its operation and maintenance.

There are two critical factors to consider when calculating leak rate. The first is that every barrier has a unique leak rate that is dependent on its material composition and on how the building itself was built. Therefore, two identical windows installed in two different buildings will have different leak rates. The other factor is that leak rates are dynamic and vary according to changing conditions, including occupancy and weather, which means that the flow of energy in a zone is constantly changing over time. For example, while energy would leak out of a zone on a typical spring morning because it is colder outside than inside, the leak rate in that zone could reverse later in the day if it becomes hotter outside than inside, meaning that energy would start to flow inward through the walls and begin to heat the zone.

One of the keys to effective building management is to maintain the derived temperature of a conditioned space and ensure occupant comfort by continually compensating for the leak rate. This means that if energy is leaking out of a zone, then maintaining that zone at the derived temperature requires supplying thermal energy to that zone at the same rate at which it is leaving. If energy is leaking into a zone because the outside temperature is higher, then the stakeholder has to cool the air and withdraw thermal energy at the same rate at which it is entering.

Calculating the leak rate of a zone based on the materials used and also mapping the dynamic of its leak rate under all possible conditions gives the stakeholder a clear understanding of what is going to happen in that zone in terms of energy flow. Accurately calculating the leak rate is critical for maintaining a constant and stable temperature in a zone, because if the stakeholder knows how quickly energy is leaking in or out of a zone at any given time, then the stakeholder can adjust the HVAC system to produce and push the right amount of cold or hot air to compensate for the leak rate in real time and maintain that zone at a constant temperature.

Although maintaining a balanced thermal equilibrium in a zone can be done manually, the process of managing the flow of energy through every zone in a building is neither economically nor procedurally viable. Deploying AI in the building management space represents, through dynamic modulation and forecasting, an accurate and cost-effective way to maintain a balanced thermal equilibrium. Initially, AI can be used in combination with linear regression analysis tools to quickly and accurately calculate the leak rates for all zones within a building under all possible conditions. However, the real value of this technology becomes clear when AI (i) uses the leak rate behavior for a zone to predict what the temperature will be in that zone at different times in the future and then (ii) makes adjustments to the HVAC system in real time based on these predictions to eliminate any unwanted events from the zone's timeline.

Power-to-Thermal Load Relationship
The power-to-thermal load relationship refers to how efficiently the engine in an HVAC system converts power into thermal energy that it then delivers to the various zones in a building. Power is the input into the engine and is measured in kilowatts or cubic meters of gas; thermal load, measured as cubic meters of hot or cold air, is the amount of thermal energy delivered into the zones as output. In building management, it is the conversion relationship between the power that is fed into the engine and the air that is served that is interesting, and the power-to-thermal load relationship measures the level of efficiency of the engine in the HVAC system at maintaining the desired temperature within a building's zones.

Although the efficiency of the engine is determined to a large extent by the equipment that was purchased and how that equipment was installed, operators are able to adjust various settings on the engine to optimize its performance. For example, in addition to being able to open or close dampers, operators can also adjust the engine to serve more or less air/minute and to slow down the production of heat. Playing with these settings makes it possible to fine tune the engine and find the ideal configuration – or combination of settings – for achieving the most efficient power-to-thermal load relationship for a given zone.

But this relationship is dynamic in nature since the optimal power-to-thermal load relationship for a given zone is dependent on different internal and external conditions, such as occupancy and weather, that change over time. For instance, the ideal configuration for a specific zone with five occupants and with no sunlight during the winter will be drastically different when there are 50 occupants with full sunlight during the summer.

Therefore, maintaining balanced thermal equilibrium in a zone in the most efficient and effective way possible – which means ensuring occupant comfort, saving money, and reducing carbon footprint – requires calculating the ideal engine configuration under all possible conditions for each zone within a building. Also, while this would be costly and time consuming to achieve manually, managing this power-to-thermal load relationship in a building is a task that is perfectly suited for AI.

The process involves giving the AI engine all the relevant data concerning both the HVAC system's engine, including performance data, and the zones within the building, including historical data about occupancy levels and weather conditions. From a data perspective, the AI engine would not be concerned with the kind of HVAC system installed within the building; it would simply look at the available data and calculate the ideal configuration for the engine from an efficiency perspective for all zones under all possible given conditions.

For instance, to calculate the most efficient way to keep a zone in winter with five occupants at the desire temperature of 22°, the AI engine would compare the power-to-thermal load relationship to see if the engine would consume less power by changing the configuration of the system to maintain the desired temperature. The goal in this instance would be to analyze the available data to find the right combination of factors that would allow the engine to produce the thermal energy required to maintain occupant comfort in the most cost-effective and environmentally friendly manner.

Predicting Energy Flow in a Building
Having characterized all zones in a building, the next step is to ensure that each zone is reaching balanced thermal equilibrium at all times and under changing conditions. Accomplishing this requires predicting the future conditions within each zone according to selected variables and then, in response to those predictions, making adjustments in real time to specific pieces of HVAC equipment to ensure that each zone is always achieving thermal equilibrium.

In the building management space, AI is the most effective and efficient way to maintain equilibrium. Although the stakeholder could construct a simulation to model a building's behavior, the process is significantly more time-consuming and less accurate than using DL for optimization. In comparison to a 40–50% accuracy rate when using simulations to predict building behavior, research shows that AI is far more accurate (e.g. see [76]).

DL Methods in BrainBox AI System

The first step is to use DL and months of historical data that has been organized in advance to train the engine for each zone in a building. Using DL, it is possible to predict what will happen in a zone in the future by first calculating what the leak rate will be, and then using that projected leak rate to predict the temperature.

Then, the AI engine uses a zone's previously calculated leak rate behavior and power-to-thermal relationship to make predictions about its behavior.

Since these features typically do not change, this characterization can be used to calculate how the zone will react over time to changing conditions, such as occupancy and weather. BrainBox's method of DL uses the zone's leak rate behavior and power-to-thermal relationship and what the AI has learned based on the historical data to make predictions about the future value of different parameters. These parameters can include whatever variables building owners want to manage, including temperature levels, humidity levels, and the concentration levels of different gases. From these predictions regarding the future value of key variables, the AI is able to make decisions about how to best manage the thermal equilibrium for every zone in a building.

When the AI engine, based off of its predictions, sees an unwanted event in the future that will disrupt the balanced thermal equilibrium in a zone, the engine begins testing various micro-modifications to the HVAC system to figure out the effects that different combinations of adjustments will have on that zone's future conditions. The AI engine then uses the results of these tests to evaluate which combination of modifications to which pieces of equipment will, over time, most effectively eliminate the unwanted event. Once the right course of action is determined, algorithms then instruct the HVAC control system to make the modifications to the selected components to get the zone to the desired condition.

For example, consider a situation in which the engine predicts that the temperature on a particular floor in a building will be too high in three hours. In response to this prediction, the NN runs models on different micro-modifications to the HVAC system to see the impact that these possible modifications would have on the predicted temperature increase. Using this information, the AI decides on the best course of action – for instance, lowering the blinds and turning on the air conditioning – that will bring the floor to where it needs to be to ensure that the rise in temperature in three hours does not occur. It then instructs the HVAC control system in real time to make the desired modifications to create the right conditions on the floor.

In the next phase, the AI engine goes through the entire process again at a later time to check how the zone reacted to the modifications. Using the zone's thermal equation and the new conditions in the zone, it measures the results of the modifications on the specified variables and determines whether or not the instructions actually changed the future in the intended way and if these changes will, in fact, lead to the elimination of the unwanted event. The DL engine then tests all the possible combinations of different modifications that could now be applied to eliminate the predicted event. Once this is done, the AI again selects those adjustments that will most effectively eliminate the predicted event and instructs the HVAC control system in real time to make the required adjustments.

Continuing with the previous example of a predicted rise in temperature in three hours, the DL engine analyzes the new conditions on the floor and its learning based on historical data to see what effects, if any, the changes had on that predicted rise in temperature. Based on this information, the DL engine again calculates how effective different combinations of possible modifications will now be in eliminating the predicted rise in temperature. Once the AI determines the best course of action based on these calculations, it again instructs the HVAC control system to make the required modifications towards cancelling the predicted event. In this way, the AI, by continuously making adjustments and then reassessing the effectiveness of those adjustments, changes the future in real time and makes what the stakeholder want to happen a reality.

The above example only dealt with one variable, namely temperature. The BrainBox AI system can perform the same predictive work and system modifications on as many variables as the stakeholder wants to control in the conditioned space. The example also dealt with only one zone. BrainBox's solution performs these evaluations and modifications in real time for all the zones in a building based on the variables selected for each zone. In this way, by controlling the future in each zone, the aggregate result is that it is able to optimize the energy flow in the entire building without any human intervention.

In addition to working on multiple variables for the different zones in a building, BrainBox's AI engine can also look at energy spending and carbon emissions as part of its decision-making process when trying to figure out the best course of action for eliminating unwanted events in the future. Thus, in addition to being able to determine the right combination of modifications, BrainBox AI can also determine which modifications would be the most cost effective and which would most significantly reduce a building's carbon footprint. For example, from a cost analysis perspective, BrainBox's AI engine can determine the right moment during the timeline to make the required modifications to get the desired final result in the most cost-effective way. It does this by evaluating which modifications would cost less to create the desired conditions in the future, those that would need to be done immediately or in 20 minutes. It can also determine which modifications would be the greenest from an environmental perspective.

A. Basic Glossary of Key IoT (Smart Building) Terms and Concepts

This glossary is principally based on reference [77], which also included the original source of the definitions; the original sources are not identified here, but these sources can be identified in said reference; in particular, many of these terms are adapted from [78, 79]. Other references include [80, 81].

Note: Some of the terms are not explicitly called out in the chapter, but they are common terms in the context of HVACs and energy management.

Term	Definition/explanation/concept
Absolute humidity	The total amount of moisture contained in a cubic foot of air, measured in grains per cubic foot.
Absolute pressure	The total pressure on a surface, measured in pounds per square inch.
Actuator	A mechanized device of various sizes (from ultra-small to very large) that accomplishes a specified physical action, for example, controlling a mechanism or system, opening or closing a valve, starting some kind or rotary or linear motion, and initiating physical locomotion. It is the mechanism by which an entity acts on an environment. The actuator embodies a source of energy, such as an electric current (battery, solar, or motion), a hydraulic fluid pressure, or a pneumatic pressure; the device converts that energy into some kind of action or motion on external command.
Advanced meter infrastructure (AMI) system	An infrastructure that contains electric meters capable of two-way communications with a centralized grid control system. These are meters that can receive signals, including the cost of electricity and status of the grid, track electricity usage on a short-term basis, and automatically report the meter readings back to the utility. The electric information service infrastructure between the end user or end device and the electric company. A system for implementing smart grid (SG) and a principal means of realizing demand response. To communicate between physical service layers, some combinations and transformations of the protocols are required.
Air conditioning airflow efficiency (ACAE)	The amount of heat removed per standard cubic foot of airflow per minute.

Term	Definition/explanation/concept
Air handling unit (AHU)	A device used to condition and circulate air as part of an HVAC system. Typically, the air handler is a metal box housing a blower, heating and/or cooling elements, filter racks or chambers, sound attenuators, and dampers. Air handlers connect to a ductwork that distributes the conditioned air throughout the building.
Air mixing	The unintended mixing of cold and hot air.
Air space	Space where the air space below a raised floor or above a suspended ceiling is used to recirculate data center/telecom room environmental air.
Air vent valve	Valves connected to the top of the water box or connecting pipe used to vent trapped air.
Air, bypass	Air diverted around a cooling coil in a controlled manner for the purpose of avoiding saturated discharge air. On room scale equipment, bypass air can also refer to the supply air that "short-cycles" around the load and returns to the air handler without producing effective cooling at the load.
Air, conditioned	Air that is treated to control its temperature, relative humidity, purity, pressure, and movement.
Air, return	Air extracted from a space and totally or partially returned to an air conditioner, furnace, or other heat source.
Air/liquid cooling	Cooling: removal of heat. Air cooling: direct removal of heat at its source using air. Liquid cooling: direct removal of heat at its source using a liquid (usually water, water/glycol mixture, Fluorinert^{TM} or refrigerant).
Air-cooled system	(data center) System where conditioned air is supplied to the inlets of the rack/cabinet for convective cooling of the heat rejected by the components of the electronic equipment within the rack. Within the rack, the transport of heat from the actual source component within the rack itself can be either liquid or air based, but the heat rejection media from the rack to the terminal cooling device outside of the rack is air.
Airside economizer	An economizer that directs exterior air into the data center when the air temperature is at or below the cooling set point.
American Society of Heating, Refrigerating and Air-Conditioning Engineers (ASHRAE)	A well-known international technical society organized to advance the arts and sciences of air management.
Atmospheric pressure	The pressure exerted on the earth's surface by the weight of atmosphere above it.

(Continued)

(Continued)

Term	Definition/explanation/concept
BACnet (protocol)	A widely-deployed communication protocol for building automation and control networks developed in the late 1980s that utilizes the ASHRAE, ANSI, and ISO 16484-5 standards protocol. It supports communication of building automation and control systems for applications such HVAC, lighting control, access control, and fire detection systems.
Base load generation	(electric utility industry) Those generating facilities within a utility system that are operated to the greatest extent possible to maximize system mechanical and thermal efficiency and minimize system operating costs.
Base rate	(electric utility industry) That part of the total electric rate covering the general costs of doing business unrelated to fuel expenses.
Benchmark	A reference point.
Boiling point	The temperature at which the addition of any heat will begin a change of state from a liquid to a vapor.
BTU	British thermal unit; one BTU is the amount of heat required to raise the temperature of one pound of water 1 at 60°F.
Building automation system (BAS)	Centralized building control. Usually utilized for the purpose of monitoring and controlling environment, lighting, power, security, fire safety, and elevators.
Building management systems (BMSs)	A computerized platform that allows for monitoring and controlling a building's mechanical and electrical equipment.
Busy hour drain	The actual average current drawn by a circuit or group of circuits during the busy hour of the season (excluding power fail/battery discharge events). Busy hour drains for DC power plants are measured by power monitor devices.
Bypass	A piping detour around a component.
Bypass airflow	Conditioned air that does not reach computer equipment, escaping through cable cut-outs, holes under cabinets, misplaced perforated tiles or holes in the computer room perimeter walls.
Capacity	(electric utility industry) The load for which a generating unit, generating plant or other electrical apparatus is rated either by the user or by the manufacturer.
Carbon footprint	A term to describe a calculation of total carbon emissions of some system, namely, to describe energy consumption. It is the amount of greenhouse gases (GHG) produced to support a given activity (or piece of equipment), typically expressed in equivalent tons of carbon dioxide (CO_2).

Term	Definition/explanation/concept
Carbon neutral	Calculating total carbon emissions, reducing them where possible, and balancing remaining emissions with the purchase of carbon offsets.
Central air conditioner	A central air-conditioner model consists of one or more factory-made assemblies that normally include an evaporator or cooling coil(s), compressor(s), and condenser(s). Central air conditioners provide the function of air-cooling, and may include the functions of air circulation, air cleaning, dehumidifying, or humidifying. A split system is a system with components located both inside and outside of a building. A 'single package unit' is a system that has all components completely contained in one unit.
Charge (HVAC context)	The refrigerant and lithium bromide contained in a sealed system; the total of refrigerant and lithium bromide required.
Chilled water system	A type of air-conditioning system that has no refrigerant in the unit itself. The refrigerant is contained in a chiller, which is located remotely. The chiller cools water, which is piped to the air conditioner to cool the space. An air or process conditioning system containing chiller(s), water pump(s), a water piping distribution system, chilled-water cooling coil(s), and associated controls. The refrigerant cycle is contained in a remotely located water chiller. The chiller cools the water, which is pumped through the piping system to the cooling coils.
Close-coupled cooling	Cooling technology that is installed adjacent to server racks and enclosed to direct airflow directly to the rack without mixing with data center air.
Cloud computing	The latest term to describe a grid/utility computing service – such service is provided in the network. From the perspective of the user, the service is virtualized. In turn, the service provider will most likely use virtualization technologies (virtualized computing, virtualized storage, etc.) to provide the service to the user.
Clusters	Aggregating of processors in parallel-based configurations, typically in local environment (within a data center); all nodes work cooperatively as a single unified resource. Resource allocation is performed by a centralized resource manager and scheduling system.
CO_2 equivalent	A metric used to equate the impact of various greenhouse gas emissions on the environment's global warming potential based on CO_2.
Coefficient of Effectiveness (CoE)	Uptime Institute metric based on the Nash-Sutcliffe model efficiency coefficient.

(Continued)

(Continued)

Term	Definition/explanation/concept
Coefficient of Performance (CoP)	A measure of efficiency in the heating mode that represents the ratio of total heating capacity (BTU) to electrical input (also in BTU). The ratio of the cooling load, in kW, to power input at the compressor, in kW. The ratio of the rate of heat removal to the rate of energy input, in consistent units, for a complete cooling system or factory-assembled equipment, as tested under a nationally recognized standard or designated operating conditions. The definition is similar to EER, but the units are different.
Cold spot	An area where ambient air temperature is below acceptable levels. Typically caused by cooling equipment capacity exceeding heat generation.
Cold Supply Infiltration Index (CSI)	Term that quantifies the amount of hot air mixing with cold inlet air before entering the rack.
Commissioning	A quality control process required by most green building certification programs. This process incorporates verification and documentation to ensure that building systems and assemblies are planned, designed, installed, tested, operated, and maintained to meet specified requirements.
Computational fluid dynamics (CFD)	Scientific calculations applied to airflow analysis.
Computer room air handler (CRAH)	Computer room air handler that uses chilled water to cool air.
Condensation point	The temperature at which the removal of any heat will begin a change of state from a vapor to a liquid.
Condenser	A device in which the superheat and latent heat of condensation is removed to effect a change of state from vapor to a liquid. Some sub-cooling is also usually accomplished.
Condenser water	Water that removes the heat from the lithium bromide in the absorber and from condenser vapor. The heat is rejected to atmosphere by a cooling tower.
Conditioned air	Air treated to control its temperature, relative humidity, purity, and movement.
Constant air volume (CAV) HVAC	An HVAC system that supplies air at a constant volume and variable temperature; the fresh air entering from outside is mixed with the return air to produce fresher air to the fan ventilation; the speed of the fan is fixed, and it is controlled by an (On/Off) switch.
Constant volume (CV) air conditioner	An air conditioner that supplies a constant rate of airflow to a space at reasonably high velocity and removes an equal amount of the air from the space (return air).

Term	Definition/explanation/concept
Contactor	A switch that can repeatedly cycle, making and breaking an electrical circuit: a circuit control. When sufficient current flows through the coil built into the contactor the resulting magnetic field causes the contacts to be pulled in or closed.
Control	Any component that regulates the flow of fluid, or electricity.
Control device	Any device that changes the energy input to the chiller/heater when the building load changes and shuts it down when the chiller is not needed.
Cooling capacity	The cooling capacity is the quantity of heat in BTUs that an air conditioner or heat pump is able to remove from an enclosed space during a one-hour period.
Cooling tower	Heat-transfer device where atmospheric air cools warm water, generally by direct contact (heat transfer and evaporation).
Cooling, air	Conditioned air is supplied to the inlets of the rack/cabinet/server for convective cooling of the heat rejected by the components of the electronic equipment within the rack. Within the rack, the transport of heat from the actual source component within the rack itself can be either liquid or air based, but the heat rejection media from the rack to the building cooling device outside the rack is air. The use of heat pipes or pumped loops inside a server or rack where the liquid remains is still considered air cooling.
Cooling, liquid	Conditioned liquid is supplied to the inlets of the rack/cabinet/server for thermal cooling of the heat rejected by the components within the rack.
Computer room air conditioning (CRAC)	Computer room air conditioner (pronounced crack) that uses a compressor to mechanically cool air. A modular packaged environmental control unit designed specifically to maintain the ambient air temperature and/or humidity of spaces that typically contain telecom/datacom/data center equipment. These products can typically perform all (or a subset) of the following functions: cool, reheat, humidify, dehumidify. CRAC units designed for data and communications equipment room applications typically meet the requirements of ANSI/ASHRAE Standard 127-2001, Method of Testing for Rating Computer and Data Processing Room Unitary Air-Conditioners.
Critical load	Computer equipment load delivered by PDU output.
Cubic feet per minute (CFM)	An airflow volume measurement.
Cutout	An open area in a raised floor that allows airflow or cable feeds.
Data grid	A kind of Grid Computing grid used for housing and providing access to data across multiple organizations; users are not focused on where this data is located as long as they have access to the data.

(Continued)

(Continued)

Term	Definition/explanation/concept
Daylighting	Using various design methods, such as windows and skylights, to reduce the building's reliance on electric lighting. Numerous studies have highlighted the productivity benefits of natural lighting for building occupants.
Dead band	An HVAC energy saving technique whereby sensitivity set points of equipment are set more broadly to improve coordination of the equipment and avoid offsetting behaviors, also dead band control strategy.
Degree-day	A unit measuring the extent to which the outdoor mean (average of maximum and minimum) daily dry-bulb temperature falls below or rises above an assumed base. The base is normally taken as 65 °F for heating and for cooling unless otherwise designated. One degree-day is counted for each degree below (deficiency heating) or above (excess cooling) the assumed base, for each calendar day on which such deficiency or excess occurs.
Delta T	Delta temperature, the spread between the inlet and outlet air temperatures of air conditioning equipment, measured as the maximum achievable difference between inlet (return) and outlet (supply) temperatures.
Demand	(electric utility industry) The rate at which electric energy is delivered to or by a system, part of a system or a piece of equipment. It is expressed usually in kilowatts at a given instant or averaged over any designated period of time. The primary source of "demand" is the power-consuming equipment of customers.
Demand charge	(electric utility industry) That part of the charge for electric service based on the electric capacity (kW) consumed and billed on the basis of billing demand under an applicable rate schedule.
Demand interval	(electric utility industry) The period of time during which the electric energy flow is averaged in determining demand, such as 60, 30, 15-minutes, or instantaneous. Electric utilities typically use 15-minutes demands for most demand-billed rate classes.
Demand, average	(electric utility industry) The demand on, or the power output of, an electric system or any of its parts over any interval of time, as determined by dividing the total number of kilowatt-hours by the number of units of time in the interval.
Demand, billing	(electric utility industry) The demand on which billing to a customer is based, as specified in a rate schedule or contract. It may be based on the contract year, a contract minimum or a previous maximum and, therefore, does not necessarily coincide with the actual measured demand of the billing period.

Term	Definition/explanation/concept
Demand, instantaneous peak	(electric utility industry) The demand at the instant of greatest load, usually determined from the readings of indicating or graphic meters.
Demand, maximum	(electric utility industry) The greatest demand that occurred during a specified period of time such as a billing period.
Dew point temperature (DPT)	The temperature at which a moist air sample at the same pressure would reach water vapor saturation. At this saturation point, water vapor begins to condense into liquid water fog or solid frost, as heat is removed. The temperature at which air reaches water vapor saturation, typically used when examining environmental conditions to ensure they support optimum hardware reliability.
Direct expansion (DX) system	An air conditioning system where the cooling effect is obtained directly from the refrigerant. It typically incorporates a compressor; almost invariably the refrigerant undergoes a change of state in the system.
Dispatch, dispatching	(electric utility industry) The operating control of an integrated electric system to (i) assign generation to specific generating plants and other sources of supply to effect the most reliable and economical supply as the total of the significant area loads rises or falls; (ii) control operations and maintenance of high-voltage lines, substations and equipment, including administration of safety procedures; (iii) operate the interconnection; and (iv) schedule energy transactions with other interconnected electric utilities.
Distribution	(electric utility industry) The act or process of delivering electric energy from convenient points on the transmission system (usually a substation) to consumers. The network of wires and equipment that distributes, transports or delivers electricity to customers. The delivery of electric energy to customers on the distribution service. Electric energy is carried at high voltages along the transmission lines. For consumers needing lower voltages, it is reduced in voltage at a substation and delivered over primary distribution lines extending throughout the area where the electricity is distributed. For users needing even lower voltages, the voltage is reduced once more by a distribution transformer or line transformer. At this point, it changes from primary to secondary distribution.
Dry-bulb temperature (DBT)	The temperature of an air sample, as determined by an ordinary thermometer, the thermometer's bulb being dry. The SI unit is Kelvin; in the US unit is Fahrenheit.
Economizer, air	A ducting arrangement and automatic control system that allow a cooling supply fan system to supply outdoor (outside) air to reduce or eliminate the need for mechanical refrigeration during mild or cold weather.

(*Continued*)

(Continued)

Term	Definition/explanation/concept
Economizer, water	A system by which the supply air of a cooling system is cooled directly or indirectly or both by evaporation of water or by other appropriate fluid (to reduce or eliminate the need for mechanical refrigeration).
Efficiency	The amount of usable energy produced by a machine, divided by the amount of energy supplied to it, i.e. the ratio of the *output useful energy* to the *input energy* of any system.
Efficiency, HVAC system	The ratio of the useful energy output to the energy input, in consistent units, expressed in percent.
Electricity service	The network of generating plants, wires and equipment needed to produce or purchase electricity (generation) and to deliver it to the local distribution system (transmission). Priced in cents per kilowatt-hour for energy-only customers and in dollars per kilowatt and in cents per kilowatt-hour for demand-billed customers.
Energy charge	That part of the charge for electric service based on the electric energy (kWh) consumed or billed.
Energy costs	Costs, such as fuel, related to and vary with energy production or consumption.
Energy efficiency	Designing a system, a data center, a network node, a building, and so on, to use less energy for the same or higher performance as compared to conventional approaches. All data center subsystems, network subsystems, and buildings subsystems (e.g. HVAC, lighting, and so on) can contribute to higher energy efficiency.
Energy efficiency ratio (EER)	EER is a measure of efficiency in the cooling mode that represents the ratio of total cooling capacity (BTU/hour) to electrical energy input (Watts). It is measured by ratio of BTU/hour of cooling or heating load, to watts of electrical power input.
Energy, off-peak	Energy supplied during periods of relatively low system demand as specified by the supplier. E.g. from 9 p.m.–7 a.m., Monday through Friday, all holidays, and all weekends.
Energy, on-peak	Energy supplied during periods of relatively high system demand as specified by the supplier.
Equilibrium	When refrigerant (water) molecules leave the solution at the same rate that they are being absorbed, the solution is said to be in equilibrium.
Equilibrium chart	A pressure-temperature concentration chart that can be used to plot solution equilibrium at any point in the absorption cycle.

Term	Definition/explanation/concept
Equipment room	Data center or networking/telecom rooms – including carrier central office room and points-of-presence – that houses computer and/or telecom equipment.
Evacuate	To remove, through the use of the vacuum pump, all non-condensables from a machine.
Evaporative condenser	Condenser where the removal of heat from the refrigerant is achieved by the evaporation of water from the exterior of the condensing surface, induced by the forced circulation of air and sensible cooling by the air.
Evaporator	Heats and vaporizes refrigerant liquid from the condenser, using building system water.
Extreme learning machines (ELM)	Feed-forward NNs used for classification, regression, clustering, sparse approximation, compression and feature learning with a single layer or multiple layers of hidden nodes, where the parameters of hidden nodes (not just the weights connecting inputs to hidden nodes) need to be tuned. These hidden nodes can be randomly assigned and never updated (i.e. they are random projection but with nonlinear transforms), or can be inherited from their ancestors without being changed. In most cases, the output weights of hidden nodes are usually learned in a single step, which practically corresponds to learning a linear model [80].
Fan	Device for moving air. Various systems include: • *Airfoil fan*: shaped blade in a fan assembly to optimize flow with less turbulence. • *Axial fan*: fan that moves air in the general direction of the axis about which it rotates. • *Centrifugal fan*: fan in which the air enters the impeller axially and leaves it substantially in a radial direction. • *Propeller fan*: fan in which the air enters and leaves the impeller in a direction substantially parallel to its axis.
Freezing point	The temperature at which the removal of any heat will begin a change of state from a liquid to a solid.
Fuel cost adjustments	(electric utility industry) A provision in a rate schedule that provides for an adjustment to the customer's bill if the cost of fuel at the supplier's generating stations varies from a specified unit cost.
Gauge pressure	A fluid pressure scale in which atmospheric pressure equals zero pounds and a perfect vacuum equals 30 in mercury.
Generation, generating plant electric power	The large-scale production of electricity in a central plant. A power plant consists of one or more units. Each unit includes an individual turbine generator. Turbine generators (turbines directly connected to electric generators) use steam, wind, hot gas or falling water to generate power.

(Continued)

(Continued)

Term	Definition/explanation/concept
Gigawatt (gW)	One gigawatt equals one billion (1,000,000,000) watts, one million (1,000,000) kilowatts, or one thousand (1,000) megawatts.
Gigawatt-hours (gWh)	One gigawatt-hour equals one billion (1,000,000,000) watt-hours, one million (1,000,000) kilowatt-hours, or one thousand (1,000) megawatt-hours.
Green	Refers to the tangible or physical attributes of a product or a property, in the context of energy efficiency/carbon footprint and/or eco-friendliness.
Green buildings	The implementation of design, construction, and operational strategies that reduce a building's environmental impact during both construction and operation, and that improve its occupants' health, comfort, and productivity throughout the building's life cycle. "Green buildings'" is part of a larger trend toward *sustainable design*. (Some call these "smart buildings").
Green power, green pricing	Optional service choices that feature renewable fuels such as wind or solar, usually priced at some form of premium.
Greenhouse gases (GHG)	Specific gases that absorb terrestrial radiation and contribute to the global warming or greenhouse effect. Six greenhouse gases are listed in the Kyoto Protocol: carbon dioxide, methane, nitrous oxide, hydrofluorocarbons, perfluorocarbons, and sulfur hexafluoride.
Greening	Refers to minimizing energy consumption, maximizing energy use efficiency, and using, whenever possible, renewable energy sources. Use of eco-friendly components and consumables also plays a role. Succinctly: energy efficiency and renewable energy initiatives.
Heat exchanger	Any device for transferring heat from one fluid to another. A device used to transfer heat energy, typically used for removing heat from a chilled liquid system. Namely, a device used to transfer heat between two physically separated fluids. • Counterflow heat exchanger: heat exchanger in which fluids flow in opposite directions approximately parallel to each other. • Cross-flow heat exchanger: heat exchanger in which fluids flow perpendicular to each other. • Parallel-flow heat exchanger: heat exchanger in which fluids flow approximately parallel to each other and in the same direction. • Plate heat exchanger or plate liquid cooler: thin plates formed so that liquid to be cooled flows through passages between the plates and the cooling fluid flows through alternate passages.

Term	Definition/explanation/concept
Heat of condensation	The latent heat energy liberated in the transition from a gaseous to a liquid state.
Heat pump	A heat pump model consists of one or more factory-made assemblies that normally include an indoor conditioning coil(s), compressor(s), and outdoor coil(s), including means to provide a heating function. Heat pumps shall provide the function of air heating with controlled temperature, and may include the functions of air-cooling, air circulation, air cleaning, dehumidifying, or humidifying.
Heat transfer	The three methods of heat transfer are conduction, convection, and radiation.
Heating Seasonal Performance Factor (HSPF)	HSPF is a measure of a heat pump's energy efficiency over one heating season. It represents the total heating output of a heat pump (including supplementary electric heat) during the normal heating season (in BTU) as compared to the total electricity consumed (in watt-hours) during the same period.
High temp generator	The section of a chiller where heat is applied to the lithium bromide solution to separate water vapor.
High-flow constraint day	(gas utility industry) A day when the utility expects natural gas demand to exceed its available deliverable supply of gas for gas sales and service needs. On such a day, the utility will interrupt Interruptible customers and/or require customers using third-party natural gas supplies to use no more than their daily confirmed pipeline deliveries to avoid incurring pipeline penalties and assure that adequate supplies are available for Firm Sales Service needs.
High-performance building	A building that uses significantly less energy than a conventional building. Such buildings also feature high water efficiency and superior indoor air quality.
High-performance data center (HPDC)	A data center with above average kW loading, typically greater than 10 kW/rack.
Horizontal displacement (HDP)	An air-distribution system used predominantly in telecommunications central offices in Europe and Asia; typically, this system introduces air horizontally from one end of a cold aisle.
Hot spot	An area, typically related to a rack or set of racks, where ambient air temperature is above acceptable levels. Typically caused by heat generation in excess of cooling equipment capacity.

(Continued)

(Continued)

Term	Definition/explanation/concept
Humidity	Water vapor within a given space/volume. • Absolute humidity: The mass of water vapor in a specific volume of a mixture of water vapor and dry air. • Relative humidity: Ratio of the partial pressure or density of water vapor to the saturation pressure or density, respectively, at the same dry-bulb temperature and barometric pressure of the ambient air.
Humidity ratio	The proportion of mass of water vapor per unit mass of dry air at the given conditions (DBT, WBT, DPT, RH, etc.). The humidity ratio is also known as moisture content or mixing ratio.
HVAC (heating, ventilation, and air conditioning)	Heating, ventilation, and air conditioning system, the set of components used to condition interior air including heating and cooling equipment as well as ducting and related airflow devices. HVAC systems provide heating, cooling, humidity control, filtration, fresh air makeup, building pressure control, and comfort control. CRACs are HVAC systems specifically designed for data centers applications.
Inlet air	The air entering the referenced equipment. For air conditioning equipment this is the heated air returning to be cooled, also called return air. For racks and servers this is the cooled air entering the equipment.
Input rate	The quantity of heat or fuel supplied to an appliance, expressed in volume or heat units per unit time, such as cubic feet per hour or BTU per hour.
In-row cooling	Data center cooling technology installed between server racks in a row that delivers cooled air to equipment more efficiently.
Integrated design	The main method used by green builders to design high-performance buildings on conventional budgets. This is accomplished by incorporating efficient building system design that reduces the anticipated energy use of the building so that smaller building systems can be installed.
Integrated energy efficiency ratio (IEER)	A measure that expresses cooling part-load EER efficiency for commercial unitary air-conditioning and heat pump equipment on the basis of weighted operation at various load capacities.
Interruptible customer	(gas utility industry) A customer receiving service under rate schedules or contracts which permit interruption of service on short notice because of insufficient gas supply or capacity to deliver that supply.

Term	Definition/explanation/concept
Interruptible service	(gas utility industry) Low-priority service offered to customers under schedules or contracts which anticipate and permit interruption on short notice, generally in peak-load seasons, by reason of the claim of firm service customers and higher priority users. Gas is available at any time of the year if the supply is sufficient and the supply system is adequate.
Inverted, inverted block rate design	A rate design for a customer class for which the unit charge for electricity increases as usage increases.
Kilowatt (kW)	One kilowatt equals 1,000 watts.
Kilowatt-hour (kWh)	This is the basic unit of electric energy equal to one kilowatt of power supplied to or taken from an electric circuit steadily for one hour. One kilowatt-hour equals 1,000 Wh.
kVA	Kilovolt amperes = voltage × current (amperage).
kWc	Kilowatts of cooling, alternate unit of measurement for the cooling capacity of a CRAH.
Latent cooling capacity	Cooling capacity related to wet bulb temperature and objects that produce condensation.
Latent heat	Heat that produces a change of state without a change in temperature; i.e. ice to water at 32 °F or water to steam at 212 °F.
Latent heat of condensation	The amount of heat energy in BTUs that must be removed to change the state of one pound of vapor to one pound of liquid at the same temperature.
Latent heat of vaporization	The amount of heat energy in BTUs required changing the state of one pound of a liquid to one pound of vapor at the same temperature.
LEED™ (Leadership in Energy and Environmental Design)	A third-party certification program operated by the US Green Building Council (USGBC). LEED™ is the primary US benchmark for the design, construction, and operation of high-performance green buildings.
Liquid cooled system	Conditioned liquid (e.g. water, etc., usually above dew point) is channeled to the actual heat-producing electronic equipment components and used to transport heat from that component where it is rejected via a heat exchanger (air to liquid or liquid to liquid) or extended to the cooling terminal device outside of the rack.
Liquid cooling	A general term used to refer to cooling technology that uses a liquid circulation system to evacuate heat as opposed to a condenser, most commonly used in reference to specific types of in-row or close-coupled cooling technologies.

(Continued)

(Continued)

Term	Definition/explanation/concept
Load	The kW consumption of equipment, typically installed in a rack. Also, the heat level a cooling system is required to remove from the data center environment.
Load curve	A curve on a chart showing power (kilowatts) supplied, plotted against time of occurrence, and illustrating the varying magnitude of the load during the period covered.
Load factor	The ratio of the average load in kilowatts supplied during a designated period to the peak or maximum load in kilowatts occurring in that period. Load factor, in percent, also may be derived by multiplying the kilowatt-hours (kWh) in the period by 100 and dividing by the product of the maximum demand in kilowatts and the number of hours in the period.
Load level	For any benchmark which submits various amounts of work to a system under test (SUT), a load level is one such amount of work.
Load management	(electric utility industry) Economic reduction of electric energy demand during a utility's peak generating periods. Load management differs from conservation in that load-management strategies are designed to either reduce or shift demand from on-peak to off-peak times, while conservation strategies may primarily reduce usage over the entire 24-hour period. Motivations for initiating load management include the reduction of capital expenditure (for new power plants), circumvention of capacity limitations, provision for economic dispatch, cost of service reductions, system efficiency improvements or system reliability improvements. Actions may take the form of normal or emergency procedures. Many utilities encourage load management by offering customers a choice of service options with various price incentives.
Load shifting	(electric utility industry) Involves moving load from on-peak to off-peak periods. Popular applications include use of storage water heating, storage space heating, cool storage and customer load shifts to take advantage of time-of-use or other special rates.
Loop/looped	(electric utility industry) An electrical circuit that provides two sources of power to a load or to a substation so that if one source is de-energized the remaining source continues to provide power.
Loss (losses)	The general term applied to energy (kilowatt-hours) and power (kilowatts) lost or unaccounted for in the operation of an electric system. Losses occur primarily as energy transformations from kilowatt-hours to waste heat in electric conductors and apparatus.

Term	Definition/explanation/concept
M2M (machine-to-machine)	Term used to refer to machine-to-machine communication, i.e. automated data exchange between machines. ("Machine" may also refer to virtual machines such as software applications.) M2M is an enabler of the Internet of Things (IoT).
M2M service provider's domain	Domain that includes the Network Application Domain and any standardized systems under the control of the M2M Service Provider which interact with the M2M Service Capabilities.
M2M system	Comprises Network Application Domain, M2M Devices Domain, and any interfaces or networks required to connect those entities.
Machine (host)	A node that cannot send datagrams not created by itself. A machine (host) is both the source and destination of Internet Protocol (IP) traffic and will discard traffic that is not specifically addressed to it.
M2M communication	Communication between remotely deployed (generally low-end) devices with specific responsibilities and requiring little or no human intervention, which are all connected to an application server via the mobile network data communications.
Machine-type communications (MTC)	M2M system communication as described by the 3rd Generation Partnership Project (3GPP)
Makeup Air Handler (MAH)	Device used to manage humidity control for the entire data center. A MAH is a larger air handler that conditions 100% outside air. Synonymous with MAU (makeup air unit). The MAH filters and conditions the makeup air, either adding or removing water vapor.
Makeup Air Unit (MAU)	A large air handler that conditions 100% outside air. Synonymous with MAH.
Maximum temperature rate of change	An ASHRAE standard established to ensure stable air temperatures. The standard is 9 °F/h.
M-Bus	The M-Bus ("Meter-Bus") is a European standard for remote reading of gas and electric meters; it is also usable for all other types of consumption meters. It is specified as follows: • EN 13757-2 (physical and link layer) • EN 13757-3 (application layer) • Note: The frame layer uses IEC 870 and the network (packet layer) is optional. A radio variant of M-Bus (Wireless M-Bus) is also specified in EN 13757-4.

(Continued)

(Continued)

Term	Definition/explanation/concept
Media access control (MAC)	In a computer networking environment MAC is the media access and control of the Ethernet IEEE 802 standard and protocols.
Medical body area network system (MBANS)	Low power radio system used for the transmission of non-voice data to and from medical devices for the purposes of monitoring, diagnosing and treating patients as prescribed by duly authorized healthcare professionals.
Megawatt (MW)	One megawatt equals one million (1,000,000) watts.
Megawatt-hour (mWh)	One megawatt-hour equals one million (1,000,000) watt-hours.
MERV	Minimum efficiency reporting value, ASHRAE 52.2, for air filtration measured in particulate size.
NEBS™	Defines a set of physical, environmental, and electrical requirements for a central office (CO) of a carrier; specs can be used by other entities.
Network	(electric utility industry) A system of transmission or distribution lines so cross-connected and operated as to permit multiple power supply to any principal point on it.
Nominal cooling capacity	The total cooling capacity of air conditioning equipment, includes both latent and sensible capacities. Owing to humidity control in data centers, the latent capacity should be deducted from nominal capacity to determine useful capacity.
Noncondensable gas	Air or any gas in the machine that will not liquefy under operating pressures and temperatures.
Overcooling	A situation where air is cooled below optimum levels. Typically used in reference to rack inlet temperatures.
Particle swarm optimization (PSO)	Natural-world-inspired algorithm that is utilized to search for an optimal solution in the solution space utilizing a heuristic: *the best solution found by the flock is the best solution in the space.*
Peak day	(gas utility industry) The maximum daily quantity of gas distributed through the utility's system.
PID (proportional, integral, and derivative) loop	A control strategy (control loop) used in many types of process control systems. In building automation systems, HVACs in particular, PID loops are used to maintain precise control of temperature, pressure, and air flow (among other parameters). A PID loop consists of components, such as sensors, dampers, valves, and a loop controller
Plenum	A receiving chamber for air used to direct air flow.
Pole	A row of power receptacles with power supplied from a PDU.

Term	Definition/explanation/concept
Post-occupancy evaluation (POE)	A process for evaluating the performance of a building once the building has been completed. This evaluation particularly focuses on energy and water use, occupant comfort, indoor air quality, and the proper operation of all building systems. The results of this evaluation can often lead to operational improvements.
Power distribution unit (PDU)	The junction point between the UPS and the cabinets containing equipment.
Power usage effectiveness (PUE)	A measure of data center energy efficiency calculated by dividing the total data center energy consumption by the energy consumption of the IT computing equipment.
Power, firm	(electric utility industry) Power or power-producing capacity intended to be available at all times during the period covered by a commitment, even under adverse conditions.
Power, interruptible	(electric utility industry) Power made available under agreements that permit curtailment or cessation of delivery by the supplier.
Power, nonfirm	(electric utility industry) Power or power-producing capacity supplied or available under an arrangement that does not have the guaranteed continuous availability feature of firm power. Power supplied based on the availability of a generating unit is one type of such power.
Pressure differential	The difference in pressure between two locations in the data center used to analyze air flow behaviors.
Primary distribution, Primary distribution feeder	A primary voltage distribution circuit – usually considered to be between a substation or point of supply and the distribution transformers – that supplies lower voltage distribution circuits or consumer service circuits.
Primary voltage	The voltage of the circuit supplying power to a transformer is called the primary voltage, as opposed to the output voltage or load-supply voltage which is called secondary voltage. In power supply practice, the primary is almost always the high-voltage side and the secondary is the low-voltage side of a transformer, except at generating stations.
Production	(electric utility industry) The act or process of generating electric energy.
Psychrometric chart	A graph of the properties of air (temperature, relative humidity, etc.), used to determine how these properties vary as the amount of moisture in the air changes.
PU	Packaged unit, an air handler designed for outdoor use.
Pump	Machine for imparting energy to a fluid, causing it to do work.

(Continued)

(Continued)

Term	Definition/explanation/concept
Raised floor	Also known as access floor. A platform with removable panels (tiles) where equipment is installed, with the intervening space between it and the main building floor used to house the interconnecting cables. Often (but not always) the space under the floor is used as a means for supplying conditioned air to the information technology equipment and the room. The floors utilize pedestals to support the floor panels (tiles). The cavity between the building floor slab and the finished floor can be used as an air distribution plenum to provide conditioned air throughout the raised floor area.
Rate level	The electric price a utility is authorized to collect.
Rate structure	The design and organization of billing charges to customers.
Rates, demand	(electric utility industry) Any method of charge for electric service that is based on, or is a function of, the rate of use, or size, of the customer's installation or maximum demand (expressed in kilowatts) during a given period of time like a billing period.
Rates, flat	(electric utility industry) The price charged per unit is constant, does not vary because of an increase or decrease in the number of units.
Rates, seasonal	(electric utility industry) Rates vary depending on the time of year. Charges are generally higher during the summer months when greater demand levels push up costs for generating electricity. Typically, there are summer and winter seasonal rates. Summer rates are effective from June 1 through September 30; all other times of the year winter rates are effective.
Rates, step	(electric utility industry) A certain specified price per unit is charged for the entire consumption, the rate or price depending on the particular step within which the total consumption falls.
Rates, time-of-use	(electric utility industry) Prices for electricity that vary depending on what time of day or night a customer uses it. Time-of-use rates are designed to reflect the different costs an electric company incurs in providing electricity during peak periods when electricity demand is high and off-peak periods when electricity demand is low. A power company may have two time periods defined for its time-of-use services: on-peak and off-peak. On-peak periods are defined as 10 a.m. through 9 p.m., Monday through Friday, excluding holidays. All other periods are off-peak. Whether customers benefit from time-of-use rates depends on the percentage of total consumption used during on-peak periods. Generally, customers who use less than 30–36% of their total consumption during on-peak periods may benefit from these rates. However, individual analysis of electricity usage habits is required to see if a time-of-use service would be of potential value.

Term	Definition/explanation/concept
Recirculation	Chilled airflow returning to cooling units without passing through IT equipment, also referred to as short cycling.
Recirculation air handler (RAH)	A device that circulates air but does not cool the air.
Refrigerant	Any substance that transfers heat from one place to another, creating a cooling effect. The medium of heat transfer that picks up heat by evaporating at a low temperature and pressure and gives up heat when condensing at a higher temperature and pressure. Water is a refrigerant in absorption machines.
Relative humidity (RH)	(i) Ratio of the partial pressure or density of water vapor to the saturation pressure or density, respectively, at the same dry-bulb temperature and barometric pressure of the ambient air. (ii) Ratio of the mole fraction of water vapor to the mole fraction of water vapor saturated at the same temperature and barometric pressure-at 100% relative humidity, the dry-bulb, wet-bulb, and dew-point temperatures are equal. RH is dimensionless, and is usually expressed as a percentage.
Relief valve	A valve that opens before a dangerously high pressure is reached.
Renewable energy	Energy generated from natural resources that are inexhaustible. Renewable energy technologies include solar power, wind power, hydroelectricity and micro hydro, biomass, and biofuels.
Reserve margin	(electric utility industry) The difference between net system capability and system maximum load requirements (peak load or peak demand).
Return air	The heated air returning to air conditioning equipment
Rooftop unit (RTU)	An air handler designed for outdoor use mounted on a rooftop. A typical application of a PU.
Room load capacity	The point at which the equipment heat load in the room no longer allows the equipment to run within the specified temperature requirements of the equipment. The load capacity is influenced by many factors, the primary one being the room's theoretical capacity. Other factors, such as the layout of the room and load distribution, also influence the room load capacity.
Scheduling	(gas utility industry) Is a process by which nominations are first consolidated by receipt point, by contract, and verified with upstream/downstream parties. If the verified capacity is greater than or equal to the total nominated quantities, all nominated quantities are scheduled. If verified capacity is less than nominated quantities, nominated quantities will be allocated according to scheduling priorities.

(Continued)

(Continued)

Term	Definition/explanation/concept
Seasonal Energy Efficiency Ratio (SEER)	SEER is a measure of equipment energy efficiency over the cooling season. It represents the total cooling of a central air-conditioner or heat pump (in BTU) during the normal cooling season as compared to the total electric energy input (in watthours) consumed during the same period. The values for SEER, are determined through averaging readings of different air conditions, to represent air conditioner efficiency throughout the season.
Sensible cooling capacity	Cooling capacity related to dry bulb temperature and objects that do not produce condensation.
Sensitivity	An equipment setting that bounds the set point range and triggers a change in device function when exceeded. Most commonly referring to CRAC/CRAH temperature and humidity set points.
Sensor network/ wireless sensor network (WSN)	A sensor network is an infrastructure comprised of sensing (measuring), computing, and communication elements that gives the administrator the ability to instrument, observe, and react to events and phenomena in a specified environment. The administrator typically is some civil, government, commercial, or, industrial entity. Typically, the connectivity is by wireless means, hence the term WSN. Applications include, but are not limited to, data collection, monitoring, surveillance, and medical telemetry. There are four basic components in a sensor network: (i) an assembly of distributed or localized sensors; (ii) an interconnecting network (usually but not always wireless-based); (iii) a central point of information clustering; and (iv) a set of computing resources at the central point (or beyond) to handle data correlation, event-trending, querying, and data mining.
Sensors	Active devices that measure some variable of the natural or man-made environment (for example, a building, an assembly line). The technology for sensing and control includes electric and magnetic field sensors; radio-wave frequency sensors; optical-, electro-optic-, and infrared-sensors; radars; lasers; location/navigation sensors; seismic and pressure-wave sensors; environmental parameter sensors (for example, wind, humidity, heat, etc.); and, biochemical Homeland Security-oriented sensors. Sensors can be described as "smart" inexpensive devices equipped with multiple on-board sensing elements: they are low-cost, low-power, untethered multifunctional nodes that are logically homed to a central sink node. Sensors are typically internetworked via a series of multi-hop short-distance low-power wireless links (particularly within a defined "sensor field"); they typically utilize the Internet or some other network for long-haul delivery of information to a point (or points) of final data aggregation and analysis.

Term	Definition/explanation/concept
Service drop	(electric utility industry) The overhead conductors between the electric supply, such as the last pole, and the building or structure being served.
Service entrance	(electric utility industry) The equipment installed between the utility's service drop, or lateral, and the customer's conductors. Typically consists of the meter used for billing, switches and/or circuit breakers and/or fuses, and a metal housing.
Service lateral	(electric utility industry) The underground service conductors between the street main and the first point of connection to the service entrance conductors.
Short cycling	Chilled airflow returning to cooling units without passing through IT equipment, also referred to as recirculation.
Single-phase service	(electric utility industry) Service where the facility (e.g. house, office, warehouse, barn) has two energized wires coming into it. Typically serves smaller needs of 120 V/240 V. Requires less and simpler equipment and infrastructure to support and tends to be less expensive to install and to maintain.
Smart grid (SG)	An electricity network that can intelligently integrate the actions of all users connected to it – consumers, generators, and those that do both –to efficiently deliver sustainable, economic and secure electricity supplies.
Solution pump	Pump that recirculates lithium bromide solution in the absorption cycle.
SPEC	Standard Performance Evaluation Corporation (SPEC) is an organization that develops standardized benchmarks and publishing reviewed results.
Specific heat	The amount of heat necessary to change the temperature of one pound of a substance 1 °F.
Specific volume	The volume per unit mass of the air sample. The SI units are cubic meters per kilogram of dry air; other units are cubic feet per pound of dry air. Specific volume is also called inverse density.
Spiking neural networks (SNNs)	ANNs that incorporate the concept of time into their operating model; the concept is that neurons in the SNN do not transmit information at each propagation cycle (as it happens with typical MLPs), but rather transmit information only when a membrane potential – an intrinsic quality of the neuron related to its membrane electrical charge – reaches a specific value, called the threshold. When the membrane potential reaches the threshold, the neuron fires, and generates a signal that travels to other neurons which, in turn, increase or decrease their potentials in response to this signal. A NN that fires at the moment of threshold crossing is also called a spiking neuron model [81].

(Continued)

(Continued)

Term	Definition/explanation/concept
Spill point	When the evaporator pan overflows into the absorber.
Standard cubic feet of air per minute (SCFM)	A measure of free air flow into the intake filter on an air compressor. It is the volumetric flow rate of a gas representing a precise mass flow rate corrected to "standardized" conditions of temperature, pressure and relative humidity. SCFM is used to designate flow in terms of some base or reference pressure, temperature and relative humidity. Many standards are used, the most common being the Compressed Air & Gas Institute (CAGI) and the American Society of Mechanical Engineers (ASME) standards.
Step-down	(electric utility industry) To change electricity from a higher to a lower voltage.
Step-up	(electric utility industry) To change electricity from a lower to a higher voltage.
Strainer	Filter that removes solid particles from the liquid passing through it.
Sub-cooling	Cooling of a liquid, at a constant pressure, below the point at which it was condensed.
Sub-floor	The open area underneath a raised computer floor, also called a sub-floor plenum.
Submetering	(electric utility industry) Remetering of purchased energy by a customer for distribution to his tenants through privately owned or rented meters.
Substation	(electric utility industry) An assemblage of equipment for the purposes of switching and/or changing or regulating the voltage of electricity. Service equipment, line transformer installations or minor distribution and transmission equipment are not classified as substations.
Summer peak	(electric utility industry) The greatest load on an electric system during any prescribed demand interval in the summer (or cooling) season.
Superheat	The heat added to vapor after all liquid has been vaporized.
Supervisory control and data acquisition (SCADA)	A legacy, but widely-deployed system used to monitor and control a plant or equipment in industries such as but not limited to energy, oil and gas refining, water and waste control, transportation, and telecommunications. A SCADA system encompass the transfer of data between a SCADA central host computer and a number of remote terminal units (RTUs) and/ or programmable logic controllers (PLCs), and the central host and the operator terminals.
Supply air	The cooled airflow emitted from air conditioning equipment.

Term	Definition/explanation/concept
Sustainable	Refers not only to green physical attributes, as in a building, but also business processes, ethics, values, and social justice.
Sustainable design	System design where the objective is to create places, products, and services in a way that reduces the use of nonrenewable resources, minimizes environmental impact, and relates people with the natural environment.
Tariff	A schedule of prices or fees.
Temperature	A measurement of heat intensity.
Temperature, dew-point	The temperature at which water vapor reaches the saturation point (100% relative humidity).
Temperature, dry-bulb	The temperature of air indicated by a thermometer.
Temperature, wet-bulb	The temperature indicated by a psychrometer when the bulb of one thermometer is covered with a water-saturated wick over which air is caused to flow at approximately 4.5 m/s (900 ft/min) to reach an equilibrium temperature of water evaporating into air, where the heat of vaporization is supplied by the sensible heat of the air.
Therm	The quantity of heat energy which is equivalent to one hundred thousand (100,000) BTU.
Thermal effectiveness	Measure of the amount of mixing between hot and cold airstreams before the supply air can enter the equipment and before the equipment discharge air can return to the air handling unit.
Three-phase service	(electric utility industry) Service where the facility (e.g. manufacturing plant, office building, warehouse, barn) has three energized wires coming into it. Typically serves larger power needs of greater than 120V/240V. Usually required for motors exceeding 10 horsepower or other inductive loads. Requires more sophisticated equipment and infrastructure to support and tends to be more expensive to install and maintain.
Ton (refrigeration)	The amount of heat absorbed by melting one ton of ice in 24 hours. Equal to 288,000 BTU's/day, 12,000 BTU's/h or 200 BTU's/min.
Transformer	(in electric utility industry context) An electromagnetic device for changing the voltage level of alternating-current electricity.
Transmission	(electric utility industry) The act or process of transporting electric energy in bulk from a source or sources of supply to other principal parts of the system or to other utility systems.
Trimming the machine	Mechanisms for automatic adjustment of the generator solution flow to deliver the correct solution concentration at any operating condition and providing the right evaporator pan water storage capacity at design conditions.

(Continued)

(Continued)

Term	Definition/explanation/concept
Uninterruptible power supply (UPS)	Device used to supply short-term power to computing equipment for brief outages or until an alternate power source, such as a generator, can begin supplying power.
US Green Building Council (USGBC)	A nonprofit trade organization that promotes sustainability in the way buildings are designed, built, and operated. The USGBC is best known for the development of the Leadership in Energy and Environmental Design (LEED™) rating system.
Vacuum	A pressure below atmospheric pressure. A perfect vacuum is 30-in. HG.
Variable air volume (VAV) AC	ACs where both the supply airflow and return airflow rates vary according to the thermal demands in the space.
Variable air volume (VAV) HVAC	An HVAC system that enables energy-efficient air distribution by optimizing the amount and temperature of distributed air. VAV systems supply air at a variable temperature and airflow rate from an air handling unit (AHU).
Variable frequency drive (VFD)	An electronic controller that adjusts the speed of an electric motor by regulating the power being delivered. VFDs provide continuous control, matching motor speed to the specific demands of the work being performed. VFDs allow operators to fine-tune processes while reducing costs for energy and equipment maintenance. For example, by lowering fan or pump speed by 15–20%, shaft power can be reduced by as much as 30%. VFDs reduce the capacity of centrifugal chillers and, thus, saving energy. They include electronic control of the fan motor's speed and torque to continually match fan speed with changing building-load conditions. Electronic control of the fan speed and airflow can replace inefficient mechanical controls, such as inlet vanes or outlet dampers.
Variable-capacity cooling	The application of newer technologies, such as digital scroll compressors and variable frequency drives in CRACs, which allow high efficiencies to be maintained at partial loads. Digital scroll compressors allow the capacity of room air conditioners to be matched exactly to room conditions without turning compressors on and off. Typically, CRAC fans run at a constant speed and deliver a constant volume of air flow; converting these fans to variable frequency drive fans allows fan speed and power draw to be reduced as load decreases.
Waterside economizer	An economizer that redirects water flow to an external heat exchanger when the exterior ambient air temperature is at or below a temperature required to chill water to a given set point, simultaneously shutting down the mechanical chiller equipment.

Term	Definition/explanation/concept
Watts per square foot (WPSF)	A measure used to normalize power and heat density on a per-cabinet basis.
Wet-bulb temperature (WBT)	The temperature of an air sample after it has passed through a constant-pressure, ideal, adiabatic saturation process. Effectively, this is after the air has passed over a large surface of liquid water in an insulated channel.
Wireless M-BUS	The Wireless M-BUS standard (EN 13757-4:2005) specifies communications between water, gas, heat, and electric meters and is becoming widely accepted in Europe for smart metering or advanced metering infrastructure (AMI) applications. Wireless M-BUS is targeted to operate in the 868 MHz band (from 868 to 870 MHz); this band enjoys good tradeoffs between RF range and antenna size.
Wireless sensor network (WSN)	An infrastructure comprised of sensing (measuring), computing, and communication elements that gives the administrator the ability to instrument, observe, and react to events and phenomena in a specified environment. See sensor network above.
ZigBee smart energy	A leading standard for interoperable products that monitor, control, inform and automate the delivery and use of energy and water. It helps create greener homes by giving consumers the information and automation needed to easily reduce their consumption.

References

1 Minoli, D. (2007). *Wireless Sensor Networks (Coauthored with K. Sohraby and T. Znati)*. Wiley.

2 Minoli, D. (2013). *Building the Internet of Things with IPv6 and MIPv6*. Wiley.

3 D. Minoli, B. Occhiogrosso, "Practical Aspects for the Integration of 5G Networks and IoT Applications in Smart Cities Environments", in Special Issue titled "Integration of 5G Networks and Internet of Things for Future Smart City", Wireless Communications and Mobile Computing, vol. 2019, Article ID 5710834, 30 pages, Hindawi & Wiley, Aug. 2019. https://doi.org/10.1155/2019/5710834.

4 D. Minoli, B. Occhiogrosso, "Mobile IPv6 Protocols and High Efficiency Video Coding for Smart City IoT Applications", in CEWIT2017 Conference, Stony Brook University, New York, Nov. 8–9, 2017.

5 D. Minoli, B. Occhiogrosso, *et al*, "Multimedia IoT Systems and Applications", in Global IoT Summit, GIoTS-2017, Organized by Mandat International, IEEE IoT TsC (Transactions on Service Computing), The IoT Forum and IPv6 Forum (Collocated with the IoT Week), Geneva, Switzerland, June 6–9, 2017.

6 Minoli, D., Occhiogrosso, B. et al. (2017). A Review of Wireless and Satellite-based M2M/IoT Services in Support of Smart Grids. *Mobile Networks and Applications Journal* 23: 881–895.

7 D. Minoli, B. Occhiogrosso, *et al*, "A Review of Wireless and Satellite-based M2M Services in Support of Smart Grids", in 1st EAI International Conference on Smart Grid Assisted Internet of Things (SGIoT 2017), Sault Ste, Marie, ON, Canada, July 11–13, 2017.

8 D. Minoli, B. Occhiogrosso, "Internet of Things (IoT)-Based Apparatus and Method for Rail Crossing Alerting of Static Or Dynamic Rail Track Intrusions", in JRC2017-2304, Proceedings of JRC (Joint Rail Conference), Philadelphia, PA, Apr. 4–7, 2017.

9 Minoli, D. and Finco, L. (2018). Implementing the IoT for Renewable Energy, Chapter 15. In: *Internet of Things A to Z: Technologies and Applications* (ed. Q. Hassan). IEEE Press/Wiley. ISBN-13: 978-1119456742.

10 Minoli, D. and Occhiogrosso, B. (2018). The Emerging 'Energy Internet of Things', Chapter 14. In: *Internet of Things A to Z: Technologies and Applications* (ed. Q. Hassan). IEEE Press/Wiley. ISBN-13: 978-1119456742.

11 Minoli, D. and Occhiogrosso, B. (2020). IoT-driven Advances in Commercial and Industrial Building Lighting and in Street Lighting. In: *Industrial IoT: Challenges, Design Principles, Applications, and Security* (ed. I. Butun). Springer. ISBN: 978-3-030-42500-5. https://www.springer.com/gp/book/9783030424992.

12 D. Minoli, B. Occhiogrosso, "Ultrawideband (UWB) Technology for Smart Cities IoT Applications", in 2018 IEEE International Smart Cities Conference (ISC2) – IEEE ISC2 2018 – Buildings, Infrastructure, Environment Track, Kansas City, Sept. 16–19, 2018.

13 D. Minoli, B. Occhiogrosso, K. Sohraby, "IoT Considerations, Requirements, and Architectures for Smart Buildings – Energy Optimization and Next Generation Building Management Systems", *IEEE Int. Things J.*: 2017, Volume: 4, Issue: 1, Pages: 269 – 283. doi: https://doi.org/10.1109/JIOT.2017.2647881.

14 D. Minoli, B. Occhiogrosso, "IoT Applications to Smart Campuses and a Case Study", in EAI (European Alliance for Innovation) Endorsed Transactions on Smart Cities, European Union Digital Library, ISSN: 2518-3893, Vol. 2. Published Dec. 19, 2017, http://eudl.eu/doi/10.4108/eai.19-12-2017.153483.

15 Minoli, D. and Occhiogrosso, B. (2018). Internet of Things Applications for Smart Cities, Chapter 12. In: *Internet of Things A to Z: Technologies and Applications* (ed. Q. Hassan). IEEE Press/Wiley. ISBN-13: 978-1119456742.

16 Minoli, D., Occhiogrosso, B., and Wang, W. (2020). MIPv6 in Crowdsensing Applications for SIoT Environments, Chapter 3. In: *Towards Social Internet of Things: Enabling Technologies, Architectures and Applications*, Part of the Studies in Computational Intelligence book series (SCI), vol. 846 (ed. A. Hassanien, R. Bhatnagar, N. Khalifa, and M. Taha), 31–49. Springer. https://link.springer.com/chapter/10.1007/978-3-030-24513-9_3.

17 Minoli, D. and Occhiogrosso, B. (2020). Constrained Average Design Method for QoS-based Traffic Engineering at the Edge/Gateway Boundary in VANETs and Cyber-Physical Environments. In: *Managing Resources for Futuristic Wireless Networks* (ed. M. Rath). IGI Global. https://www.igi-global.com/chapter/ constrained-average-design-method-for-qos-based-traffic-engineering-at-the-edgegateway-boundary-in-vanets-and-cyber-physical-environments/262549.

18 Minoli, D., Occhiogrosso, B. et al. (2018). Security Considerations for IoT Support of E-Health Applications, Chapter 16. In: *Internet of Things: Challenges, Advances and Applications*, Chapman & Hall/CRC Computer and Information Science Series (ed. Q. Hassan, A.R. Khan, and S.A. Madani). CRC Press, Taylor & Francis. ISBN: 9781498778510.

19 D. Minoli, B. Occhiogrosso, *et al*, "IoT Security (IoTSec) Considerations, Requirements, and Architectures", in 2017 14th IEEE Annual Consumer Communications & Networking Conference (CCNC), Las Vegas, NV, USA, Jan. 8–11, 2017. IEEE catalog number: CFP17CCN-CDR, ISBN: 978-1-5090-6195-2, ISSN: 2331-9860.

20 D. Minoli, B. Occhiogrosso, *et al*, "IoT Security (IoTSec) Mechanisms for e-Health and Ambient Assisted Living Applications", in The Second IEEE/ACM International Workshop on Safe, Energy-Aware, & Reliable Connected Health (SEARCH 2017) (Collocated with CHASE 2017, Conference on Connected Health: Applications, Systems, and Engineering Technologies), Philadelphia, July 2017.

21 Minoli, D., Occhiogrosso, B. et al. (2018). IoT Considerations, Requirements, and Architectures for Insurance Applications, Chapter 17. In: *Internet of Things: Challenges, Advances and Applications*, Chapman & Hall/CRC Computer and Information Science Series (ed. Q. Hassan, A.R. Khan, and S.A. Madani). CRC Press, Taylor & Francis. ISBN: 9781498778510.

22 S. Shea, "Machine to Machine (M2M)", Techtarget.com, https://www.techtarget. com/iotagenda/definition/machine-to-machine-M2M. Accessed Feb. 27, 2023.

23 T. Yuan, W. Da Rocha Neto, *et al*, "Machine Learning For Next-Generation Intelligent Transportation Systems: A Survey", *Trans. Emerg. Telecommun. Technol.*,2021 doi: https://doi.org/10.1002/ett.4427.

24 S. Azad, F. Sabrina, S. Wasimi, "Transformation of Smart Grid using Machine Learning", in 2019 29th Australasian Universities Power Engineering Conference (AUPEC), Nadi, Fiji, 2019, pp. 1–6, https://doi.org/10.1109/ AUPEC48547.2019.211809. Accessed Nov. 26–29, 2019.

25 E. Hossain, I. Khan, *et al*, "Application of Big Data and Machine Learning in Smart Grid, and Associated Security Concerns: A Review," in *IEEE Access*, vol. 7, pp. 13960–13988, 2019, doi: https://doi.org/10.1109/ACCESS.2019.2894819.

26 Kotsiopoulos, T., Sarigiannidis, P. et al. (2021). Machine Learning and Deep Learning in Smart Manufacturing: The Smart Grid Paradigm. *Comput. Sci. Rev.* 40: 100341. ISSN: 1574-0137, https://doi.org/10.1016/j.cosrev.2020.100341.

27 M. Ozay, I. Esnaola, *et al*, "Machine Learning Methods for Attack Detection in the Smart Grid," in *IEEE Trans. Neural Netw. Learn. Syst.*, vol. 27, no. 8, pp. 1773–1786, 2016, doi: https://doi.org/10.1109/TNNLS.2015.2404803.

28 Xu, C., Liao, Z. et al. (2022). Review on Interpretable Machine Learning in Smart Grid. *Energies* 15, 2022 (12): 4427. https://doi.org/10.3390/en15124427.

29 Luo, X.J., Oyedele, L.O. et al. (2020). Comparative Study of Machine Learning-Based Multi-Objective Prediction Framework for Multiple Building Energy Loads. *Sustainable Cities and Society* 61: 102283. ISSN: 2210-6707, https://doi.org/10.1016/j.scs.2020.102283.

30 A. K. Gopalakrishna, T. Özçelebi, *et al*, "Exploiting Machine Learning for Intelligent Room Lighting Applications", in 2012 6th IEEE International Conference Intelligent Systems, Sofia, Bulgaria, 2012, pp. 406–411, https://doi.org/10.1109/IS.2012.6335169. Accessed Sept. 6–8, 2012.

31 Shook, P. (2022). Predicting the Impact of Utility Lighting Rebate Programs on Promoting Industrial Energy Efficiency: A Machine Learning Approach. *Environments* 9 (8): 100. https://doi.org/10.3390/environments9080100, (This article belongs to the Special Issue Improving the Sustainability of Industrial Operations Through Cleaner Production and Energy Efficiency).

32 A. G. Putrada, M. Abdurohman, *et al*, "Machine Learning Methods in Smart Lighting Toward Achieving User Comfort: A Survey," in *IEEE Access*, vol. 10, pp. 45137–45178, 2022, doi: https://doi.org/10.1109/ACCESS.2022.3169765.

33 Djenouri, D., Laidi, R. et al. (2019). Machine Learning for Smart Building Applications: Review and Taxonomy. *ACM Comput. Surv.* 52 (2): 24–36. https://doi.org/10.1145/3311950.

34 Dey, M. (2020). A Case Study Based Approach for Remote Fault Detection Using Multi-Level Machine Learning in a Smart Building. *Smart Cities* 3 (2): 401–419. https://doi.org/10.3390/smartcities3020021.

35 B. Qolomany, A. Al-Fuqaha, *et al*, "Leveraging Machine Learning and Big Data for Smart Buildings: A Comprehensive Survey," in *IEEE Access*, vol. 7, pp. 90316–90356, 2019, doi: https://doi.org/10.1109/ACCESS.2019.2926642.

36 G. Revati, J. Hozefa, *et al*, "Smart Building Energy Management: Load Profile Prediction using Machine Learning", in 2021 29th Mediterranean Conference on Control and Automation (MED), Puglia, Italy, 2021, pp. 380–385, https://doi.org/10.1109/MED51440.2021.9480170. Accessed Jun. 22–25, 2021.

37 Alobaidi, M.H., Chebana, F., and Meguid, M.A. (2018). Robust Ensemble Learning Framework for Day-Ahead Forecasting of Household Based Energy Consumption. *Appl. Energy* 212: 997–1012.

38 Kolokotsa, D. (2016). The Role of Smart Grids in the Building Sector. *Energy Build.* 116: 703–708.

39 D. Paul, T. Chakraborty, *et al*, "IoT and Machine Learning Based Prediction of Smart Building Indoor Temperature," In 2018 4th International Conference on Computer and Information Sciences (ICCOINS), 2018, Kuala Lumpur, Malaysia, pp. 1–6, https://doi.org/10.1109/ICCOINS.2018.8510597. Accessed Aug. 13–14, 2018.

40 Pešić, S., Tošić, M. et al. (2019). BLEMAT: Data Analytics and Machine Learning for Smart Building Occupancy Detection and Prediction. *Int. J. Artif. Intell. Tools* 28 (06): 1960005. https://doi.org/10.1142/S0218213019600054.

41 Balakumar, P., Vinopraba, T., and Chandrasekaran, K. (2023). Machine Learning Based Demand Response Scheme for IoT Enabled PV Integrated Smart Building. *Sustain. Cities Soc.* 89: 104260. ISSN: 2210-6707, https://doi.org/10.1016/j.scs.2022.104260.

42 Kaligambe, A., Fujita, G., and Keisuke, T. (2022, 2022). Estimation of Unmeasured Room Temperature, Relative Humidity, and CO_2 Concentrations for a Smart Building Using Machine Learning and Exploratory Data Analysis. *Energies* 15 (12): 4213. https://doi.org/10.3390/en15124213.

43 W. Xia, Y. Jiang, *et al*, "Application of Machine Learning Algorithms in Municipal Solid Waste Management: A Mini Review", *Waste Manag. Res.*, 2022, Vol. 40(6) 609–624. doi: https://doi.org/10.1177/0734242X211033716.

44 Trivedi, A., Bovornkeeratiroj, P. et al. (2021). Phone-based Ambient Temperature Sensing Using Opportunistic Crowdsensing and Machine Learning. *Sustainable Computing: Informatics and Systems* 29 (Part A): 100479. ISSN: 2210-5379, https://doi.org/10.1016/j.suscom.2020.100479.

45 R. Pryss; D. John, *et al*, "Machine Learning Findings on Geospatial Data of Users from the TrackYourStress mHealth Crowdsensing Platform", in 2019 IEEE 20th International Conference on Information Reuse and Integration for Data Science (IRI), 2019, Los Angeles, CA, USA, pp. 350–355, https://doi.org/10.1109/IRI.2019.00061. Accessed Jul. 30, 2019 to Aug. 1, 2019.

46 M. Simsek, B. Kantarci, A. Boukerche, "Knowledge-based Machine Learning Boosting for Adversarial Task Detection in Mobile Crowdsensing", in 2020 IEEE Symposium on Computers and Communications (ISCC), 2020, Rennes, France, pp. 1–7, https://doi.org/10.1109/ISCC50000.2020.9219661.

47 Y. K. Saheed, M. O. Arowolo, "Efficient Cyber Attack Detection on the Internet of Medical Things-Smart Environment Based on Deep Recurrent Neural Network and Machine Learning Algorithms," in *IEEE Access*, vol. 9, pp. 161546–161554, 2021, doi: https://doi.org/10.1109/ACCESS.2021.3128837.

48 Singh, D., Merdivan, E. et al. Convolutional and Recurrent Neural Networks for Activity Recognition in Smart Environment. In: *Towards Integrative Machine Learning and Knowledge Extraction. Lecture Notes in Computer Science*, vol. 10344 (ed. A. Holzinger, R. Goebel, M. Ferri, and V. Palade). Cham: Springer. https://doi.org/10.1007/978-3-319-69775-8_12.

49 Liu, L. and Zhang, Y. (2021). Smart Environment Design Planning for Smart City Based on Deep Learning. *Sustain. Energy Technol. Assessm.* 47: 101425. https://doi.org/10.1016/j.seta.2021.101425.

50 B. Taha and A. Shoufan, "Machine Learning-based Drone Detection and Classification: State-of-the-Art in Research," in *IEEE Access*, vol. 7, pp. 138669–138682, 2019, doi: https://doi.org/10.1109/ACCESS.2019.2942944.

51 A. Yazdinejad, E. Rabieinejad, *et al*, "A Machine Learning-based SDN Controller Framework for Drone Management," In 2021 IEEE Globecom Workshops (GC Wkshps), 2021, Madrid, Spain, pp. 1–6, https://doi.org/10.1109/GCWkshps52748.2021.9682027.

52 M. Z. Anwar, Z. Kaleem, A. Jamalipour, "Machine Learning Inspired Sound-based Amateur Drone Detection for Public Safety Applications", in *IEEE Trans. Vehicul. Technol.*, vol. 68, no. 3, pp. 2526–2534, 2019, doi: https://doi.org/10.1109/TVT.2019.2893615.

53 L. Shan, R. Miura, *et al*, "Machine Learning-based Field Data Analysis and Modeling for Drone Communications," in *IEEE Access*, vol. 7, pp. 79127–79135, 2019, doi: https://doi.org/10.1109/ACCESS.2019.2922544.

54 H. M. Oh, H. Lee, M. Y. Kim, "Comparing Convolutional Neural Network (CNN) Models for Machine Learning-based Drone and Bird Classification of Anti-Drone System," In 2019 19th International Conference on Control, Automation and Systems (ICCAS), Jeju, Korea (South), 2019, pp. 87–90, https://doi.org/10.23919/ICCAS47443.2019.8971699. Accessed Oct. 15–18, 2019.

55 Imani, M., Hasan, M.M. et al. (2021). A Novel Machine Learning Application: Water Quality Resilience Prediction Model. *Sci. Total Environ.* 768: 144459. ISSN: 0048-9697, https://doi.org/10.1016/j.scitotenv.2020.144459.

56 Sophia, S.G.G., Ceronmani Sharmila, V. et al. (2020). Water Management Using Genetic Algorithm-based Machine Learning. *Soft Comput.* 24: 17153–17165. https://doi.org/10.1007/s00500-020-05009-0.

57 Gangrade, S., Lu, D. et al. (2022). Machine Learning Assisted Reservoir Operation Model for Long-Term Water Management Simulation. *JAWRA J. Am. Water Resour. Assoc.* 2: https://doi.org/10.1111/1752-1688.13060.

58 Vallmuur, K. (2015). Machine Learning Approaches to Analysing Textual Injury Surveillance Data: A Systematic Review. *Accident Analysis & Prevention* 79: 41–49. ISSN: 0001-4575, https://doi.org/10.1016/j.aap.2015.03.018.

59 Tejedor, J., Macias-Guarasa, J. et al. (2017). Machine Learning Methods for Pipeline Surveillance Systems Based on Distributed Acoustic Sensing: A Review. *Appl. Sci.* 7 (8): 841. https://doi.org/10.3390/app7080841.

60 Elhoseny, M. (2020). Multi-object Detection and Tracking (MODT) Machine Learning Model for Real-Time Video Surveillance Systems. *Circ. Syst. Signal Process.* 39: 611–630. https://doi.org/10.1007/s00034-019-01234-7.

61 Saeed, F., Paul, A. et al. (2020). Machine Learning Based Approach for Multimedia Surveillance During Fire Emergencies. *Multimed. Tools Appl.* 79: 16201–16217. https://doi.org/10.1007/s11042-019-7548-x.

62 V. Babanne, N. S. Mahajan, *et al*, "Machine Learning Based Smart Surveillance System", in 2019 Third International Conference on I-SMAC (IoT in Social, Mobile, Analytics and Cloud) (I-SMAC), 2019, Palladam, India, pp. 84–86, https://doi.org/10.1109/I-SMAC47947.2019.9032428. Accessed Dec. 2–14, 2019.

63 Poniszewska-Maranda, A., Kaczmarek, D. et al. (2019). Studying Usability of AI in the IoT Systems/Paradigm Through Embedding NN Techniques Into Mobile Smart Service System. *Computing* 101: 1661–1685. https://doi.org/10.1007/s00607-018-0680-z.

64 C. Lim, P. P. Maglio, "Data-Driven Understanding of Smart Service Systems Through Text Mining", https://pubsonline.informs.org. Published Online: May 17, 2018, https://doi.org/10.1287/serv.2018.0208.

65 R. Hirt, N Kuhl, "Service-Oriented Cognitive Analytics for Smart Service Systems: A Research Agenda", in Smart Service Systems: Analytic, Cognition and Innovation, Hilton Waikoloa Village, Hawaii, 2018, https://aisel.aisnet.org/hicss-51/da/smart_service_systems/2/.

66 Minoli, D., Occhiogrosso, B., and Koltun, A. (2022). Situational Awareness for Law Enforcement and Public Safety Agencies Operating in Smart Cities – Part 1: Technology. In: *Springer's Book IoT and WSN Based Smart Cities: A Machine Learning Perspective* (ed. S. Rani, V. Sai, and R. Maheswar). Springer. ISBN: 978-3-030-84181-2.

67 Minoli, D., Occhiogrosso, B., and Koltun, A. (2022). Situational Awareness for Law Enforcement and Public Safety Agencies Operating in Smart Cities – Part 2: Platforms. In: *Springer's Book IoT and WSN Based Smart Cities: A Machine Learning Perspective* (ed. S. Rani, V. Sai, and R. Maheswar). Springer. ISBN: 978-3-030-84181-2.

68 Commercial Buildings Energy Consumption Survey (CBECS), 2018 Commercial Buildings Energy Consumption Survey Final Results, https://www.eia.gov/consumption/commercial/index.php. Accessed Mar. 15, 2023.

69 J. Kennedy R. C. Eberhart, "Particle Swarm Optimization", in Proceedings of the International Conference on Neural Networks, Institute of Electrical and Electronics Engineers, Vol. 4, 1995, pp. 1942–1948. https://doi.org/10.1109/ICNN.1995.488968. Accessed Nov. 27–Dec. 1995.

70 R. C. Eberhart, Y. Shi, "Comparing Inertia Weights and Constriction Factors in Particle Swarm Optimization". In Proceedings of the 2000 Congress on Evolutionary Computation (CEC '00), Vol. 1, 2000, pp. 84–88. https://doi.org/10.1109/CEC.2000.870279. Accessed Jul. 16–19, 2000.

71 Y. Shi, R. C. Eberhart, "A Modified Particle Swarm Optimizer". In Proceedings of the IEEE International Conferences on Evolutionary Computation, 1998, pp. 69–73. https://doi.org/10.1109/ICEC.1998.699146. Accessed May 04–09, 1998.

72 A. Tam, "A Gentle Introduction to Particle Swarm Optimization", Oct. 12, 2021, https://machinelearningmastery.com/a-gentle-introduction-to-particle-swarm-optimization/. Accessed Feb. 28, 2023.

73 Y. Qin, Optimization Control Technology for Building Energy Conservation, U.S. Patent 11,416,739, Aug. 16, 2022. Uncopyrighted material.

74 Staff, https://energyplus.net.

75 Pacific Northwest National Laboratory, "Variable Air Volume (VAV) Systems Operations and Maintenance", 2021, https://www.pnnl.gov/projects/om-best-practices/variable-air-volume-systems. Accessed Mar. 18, 2023.

76 Mtibaa, F., Nguyen, K.K. et al. (2020). LSTM-based Indoor Air Temperature Prediction Framework for HVAC Systems in Smart Buildings. *Neural. Comput. Appl.* 32: 17569–17585. https://doi.org/10.1007/s00521-020-04926-3.

77 Minoli, D. (2011). *Designing Green Networks and Network Operations*. Francis and Taylor.

78 Electric Glossary, Madison Gas and Electric, Madison, Wisconsin, 2009.

79 The Green Grid Organization, Glossary and Other Reference Materials, 2023, https://www.thegreengrid.org/en/resources/glossary.

80 Wikipedia, Extreme Learning Machines, https://en.wikipedia.org/wiki/Extreme_learning_machine. Accessed Mar. 15, 2023.

81 Wikipedia, Spiking Neural Network, https://en.wikipedia.org/wiki/Spiking_neural_network. Accessed Mar. 15, 2023.

6

Current and Evolving Applications to Network Cybersecurity

6.1   Overview

This chapter focuses on cybersecurity challenges in the information and communications technology (ICT, also known as IT) arena and on machine learning (ML) methods to address some of these critical concerns. ML has many applications in cybersecurity including identifying network cyber threats and enhancing host antivirus (AV) software. Beyond a basic introduction, it is not the purpose of this chapter to provide a comprehensive review of cybersecurity, but to highlight some of the many ML techniques that have been leveraged to address these challenges.

More than 4,100 publicly disclosed cybersecurity data breaches occurred worldwide in 2022 resulting in approximately 22 billion records being illegally exposed just in that year [1]. The number of victims in the 2022 breaches was over 422 million (across full breaches and personally identifiable information [PII] exposures), up from 294 million in 2021 [2] (see Figure 6.1). Retail, high technology, and government enterprises are the most frequently attacked institutions; however, small businesses and individuals are not spared from cybercriminal attacks. Quotes such as this were prevalent at press time: "More automation, not just additional tech talent, is what is needed to stay ahead of cybersecurity risks ... Among the chief information security officers (CISOs) surveyed, 100% said they needed additional resources to adequately cope with current IT security challenges" [3]. Resources may include improved ML-based tools.

Cybersecurity concerns deal with at least four environments: (i) ICT corporate networks, intranets, computing resources/databases (DBs) and websites; (ii) (home) personal devices including laptops, smart phones and intelligent appliances;

AI Applications to Communications and Information Technologies: The Role of Ultra Deep Neural Networks, First Edition. Daniel Minoli and Benedict Occhiogrosso.
© 2024 The Institute of Electrical and Electronics Engineers, Inc.
Published 2024 by John Wiley & Sons, Inc.

4,100 publicly disclosed data breaches occurred in 2022. Some of these include:

- Revolut data breach exposes information of more than 50,000 customers
- SHEIN fined US$1.9 million over data breach affecting 39 million customers
- Student loan data breach leaks 2.5 million social security numbers
- Twitter confirms data from 5.4 million accounts was stolen
- Hacker allegedly hits both Uber and Rockstar
- 9.7 million peoples' information stolen in Medibank data leak
- Hacker attempts to sell data of 500 million WhatsApp users on dark web
- Personal and medical data for 11 million people accessed in Optus data breach
- More than 1.2 million credit card numbers leaked on hacking forum
- Twitter accused of covering up data breach that affects millions
- AT&T reveals data breach affecting 9 million wireless accounts (Feb. 15, 2023)
- U.S. Marshals Service suffers 'major' security breach compromising sensitive information (Feb. 17, 2023)
- Dish confirmed that ransomware was to blame for an ongoing outage and warned that hackers exfiltrated data, which "may" include customers' personal information, from its systems (Feb. 28, 2023)
- AT&T data breach at a vendor's system that allowed threat actors to gain access to AT&T's Customer Proprietary Network Information (CPNI). Approximately nine million customers' CPNI was accessed by the threat actors (Mar. 10, 2023)

Figure 6.1 2022 headlines related to security breaches. *Source:* Adapted from Ref. [1].

(iii) distributed Internet of Things (IoT) sensors and devices; and (iv) service provider networks of all types, also including cloud-based services. Below, we refer to any or all of these as "computing systems".[1] A well-understood business goal is to protect the ICT environment – including the constituent computer network – against unauthorized intrusion; cybersecurity deals with establishing *if, how,* and *when* any cyber threats have entered the environment and dealing effectively with the situation.

In particular, in the context of items (i), (ii), and (iii) above, the ubiquity of smarts-enabled embedded systems in practically all types of everyday life devices, and the mission-critical applications in a number of cases (e.g. e-health, grid control), make the concerns for security all-the-more pressing. These challenges are fueled by the following considerations [6]:

- Some computing systems, for example, IoT elements, are relatively new, less broadly understood than traditional ICT systems (on the other hand, the opposite is true of more traditional ICT systems and this, in fact, even fostering security breaches and infractions in that setting);
- Some computing systems, for example, IoT elements, are widely scattered geographically, often in open, uncontrolled environments;

---

1 Portions of this chapter are liberally based on our previous publications in this field [4–15].

- Some computing systems, for example, IoT elements, are widely scattered administratively where multiple, often heterogeneous, environments, processes, technologies, and security mechanisms exist;
- Some computing systems, for example, IoT elements, are widely splintered across vertical applications, many of which tend to be silos of their own (on the other hand, the opposite is true of more traditional ICT systems and this, in fact, even fostering security breaches and infractions in that setting);
- Some computing systems being deployed in the short term, for example, IoT elements, tend to be vendor-specific; wide-ranging, comprehensive standard have not yet been developed, matured, or been implemented (on the other hand, the opposite is true of more traditional ICT systems and this, in fact, even fosters security breaches and infractions in that setting);
- Some computing systems, for example, IoT elements, being deployed at this time tend not to follow an accepted layered architecture, which would enable concept/function simplicity/standardization and the ability to integrate systems (including security) from various vendors, all of this resulting in a fragmented environment (on the other hand, the opposite is true of more traditional ICT systems and this, in fact, even fostering security breaches and infractions in that setting);
- Some computing systems, for example, IoT elements, may use different addressing models and addressing formats, each optimized for a specific application (on the other hand, the opposite is true of more traditional ICT systems and this, in fact, even fostering security breaches and infractions in that setting);
- Some computing systems, for example, IoT elements, tend to employ relatively low complexity (low cost) endpoint platforms with limited memory and computational power; and
- Some computing systems, for example, IoT elements, tend to consume limited electrical power (battery-driven).

Figure 6.2 highlights graphically a typical ecosystem that spans the four categories of computing systems listed above [12].

6.2 General Security Requirements

Devices or intranets connected to a public network, such as the Internet, run the risk of being infected with malicious software or malware such as, to list just a few, viruses, worms, trojans, backdoors, rootkits, exploits, grayware, unsolicited software, privacy-invasive software, spyware, password stealers, adware, and fraudulent dialers. The number of known computer threats is estimated to range into the millions.

Targeted attacks. The attacker's target individuals or organizations to attack, singly or as a group, specifically because of who they are or what they represent; or to access, exfiltrate, or damage specific high-value assets that they possess.

- In contrast, typical malware attacks are more indiscriminate with the typical goal of spreading malware widely to maximize potential profits.

Determined adversaries. The attackers are not deterred by early failures and they are likely to attack the same target repeatedly, using different techniques, until they succeed. These attackers will regroup and try again, even after their attacks are uncovered. In many cases the attacks are consciously directed by well-resourced sponsors. This provides the attackers with the resources to adapt to changing defenses or circumstances, and directly supports the persistence of attacks where necessary.

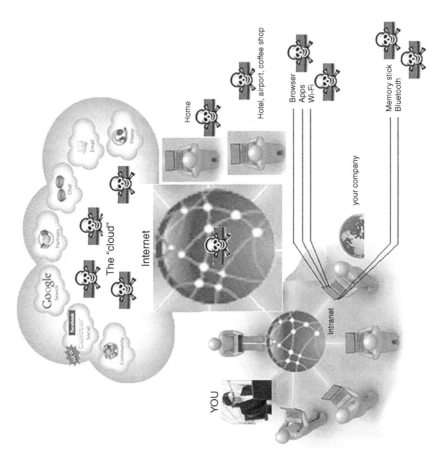

Figure 6.2 Typical computing ecosystem.

One common technique used to compromise a system is "phishing." The term refers to a type of fraud employed to manipulate individuals into activating a link to a malicious website; these malicious websites then may install malware on a user's computing device, or they may impersonate the website of a legitimate merchant or financial institution to deceive the victim into providing sensitive information, such as logins, passwords, or bank account and credit card numbers. The term "phishing" is derived from "fishing" and, like the latter, relies on "bait"; the bait may take the form of an email, text message, or the like purporting to be from a trusted party, such as a bank or other financial institution, or an e-commerce or entertainment platform [16].

The need for security in business applications and corporate intranets is self-evident, especially given that the infection vectors used by hackers tend to evolve over time, becoming difficult to identify and remove. At a broad level, cybersecurity has traditionally addressed the following requirements:

- Confidentiality (C) – ascertaining that the data flow is not intercepted and read; and that the end system is not corrupted such that an intruder can steal data, credentials, or configuration parameters.
- Integrity (I) – ascertaining that the information received (or stored) has not been compromised and/or altered in an unauthorized manner. Infection caused by viruses and worms can be utilized by an inimical agent to alter the original data, thus impacting integrity (among other possible damage). Integrity can also be perceived in the context of authentication: making sure that entities/devices are who they claim to be, and that the identity of the systems or users is not compromised (misappropriated).
- Availability (A) – ascertaining that:
  - Computing systems are not precluded from functioning and/or performing their function in an improper or compromised manner, because they might have been infected with viruses, worms, and other debilitating intrusions and/or have been exploited in terms of various operating systems (OSs), software utilities, packaged microcode, or applications;
  - Computing systems are not hijacked to become rogue devices with camouflaged identity marshalled to compromise other IoT devices and systems either as distributor of viruses or worms or as sources of flooding traffic (thus impacting availability);
  - Computing systems analytics do not become flooded with useless traffic which would cause practical shutdowns and denial of service (DoS); and
  - The data channels or network element (NE) used to support the computing systems are not intentionally swamped (especially when using unlicensed spectrum), or the (normal) traffic overwhelms the system to cause buffer-, packet-, or message-loss.

Security mechanisms covering confidentiality, integrity, and availability are critically needed end to end. As a minimum, encrypted tunnels, encryption of data at rest, and key management are fundamental security imperatives; AV software and firewalls are also part of the needed arsenal. In addition, more specifically, the following mechanisms are typically required:

- Authorization and authentication – this supports part of the Integrity requirement (who is the "user" and what kind of data can this user read/write/modify).
- Encryption and key management – this supports the Confidentiality requirement mentioned earlier.
- Trust and identity management – this supports part of the Integrity requirement (e.g. can the data/user be trusted).

Figure 6.3 provides a high-level overview of the requirements and some of the cited supportive mechanisms. Confidentiality is achieved with encryption, including the use of virtual private networks (VPNs) and the encryption of data at rest. Integrity can be achieved with digital signatures. When the data is transacted among multiple downstream parties (for example, contracts, chain-of-custody, e-health claim processing, and so on), recursive digital signatures as seen in blockchains may be ideal. Availability may, in part, be managed with intrusion detection mechanisms. Firewalls seek to block off penetrations by way of malicious packets or code; the challenge, however, is that with the growing mobility and remote access paradigms, it becomes more difficult to "seal off" or "gate in" the corporate intranet. AV software is used to protect user nodes on an intranet by

Figure 6.3 Security mechanisms.

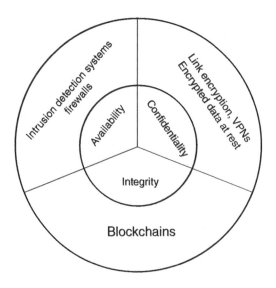

scanning (new) files to determine if they contain hidden code that might match with any malware signature.

Note that the detailed security requirements of distinct computing systems and applications – categories (i), (ii), (iii), and (iv) listed above – may differ to some degree and not all nodes in the ICT ecosystem need to necessarily comply with the tightest security restrictions: security policies will likely be consistent with the application, the risks, and the practical limits of the devices in question. For example, in some applications (e.g. e-health, physical patient monitoring, corporate intranets, and DBs) typically, there is a stringent privacy requirement (many even dictated by regulatory regimens); thus, strong security mechanisms are needed. In some instances, the IoT or other computing systems are used to control physical infrastructure (e.g. grids, process control) – mobility may also be a factor in some of these applications; here, again, strict control is (almost invariably) a requirement. Other applications such as email, videoconferencing, chatting, messaging, or social networking may require less stringent security, however, they do still require reliable security.

6.3 Corporate Resources/Intranet Security Requirements

At the practical level, the following are typical security desiderata:

- Prevention of data breaches;
- Data breach notifications;
- Testing and analysis – protocols of periodic and ongoing tests for the firm in anticipation of potential attacks on system resources;
- Compliance – support of mandatory regulatory compliance to reduce enterprise risk; and
- Adherence to data security standards.

Malicious penetration to data centers and firm's networks are often due to the following:

- The decentralization of user management and access control, including mobile access, remote access, and cloud computing;
- The dynamic changes of systems in the internet environment;
- The existence of multiple systems that require individual management tools and procedures per system; and
- The globalization effects which create a need to connect more users who are not only company personnel, such as service providers, but are clients to the firm's data sources.

Risks faced by firms include (but are not limited to) the following:

- Proprietary business data theft – stealing business information of the firm through an application breach;
- Exposure of sensitive information belonging to customers of the firm;
- Defacement of site pages, editing, and insertion of messages, pictures, and so on;
- Injecting commands for execution at the OS level;
- Escalation of user authentication privileges to perform illegal actions;
- Data disclosure due to file tampering in the DB, leading to forgery of existing information;
- Data disclosure due to file tampering in the OS, leading to forgery of existing information;
- Network breach root compromise, leading to hostile take-over of the app server and then of the main servers;
- Exposure of the application core structure, giving access to and disclosing data; and
- Exposure of the application configuration data, giving access to and disclosing data.

Security Policy and *Defense in Depth* are two important elements to deal with corporate cybersecurity risks. *Security Policy* is a strategy that the ICT department publishes defining how the company will implement information security principles and technologies; it is essentially a business plan that applies to the information security aspects of a business. A security policy (illustrated graphically in Figure 6.4, left side) covers all the company's electronic/computer systems, networks, and data. The security policy is different from security processes and procedures themselves: a policy provides both high level and specific guidelines on how the company plans to protect its data; however, will not specify exactly how that is to be accomplished. A security policy is technology- and vendor-independent – its intent is to set policy only, which the organization can then implement in any manner that accomplishes the specified goals; see Figure 6.4, left side.

Best practices call for multiple, overlapping security solutions. *Defense in Depth* (illustrated graphically in Figure 6.4, right side) is a design philosophy that achieves this layered security approach. The layers of security present in a Defense in Depth deployment should provide redundancy for one another while offering a variety of defense strategies for protecting multiple aspects of an IT environment. Any single points of failure in a security solution should be eliminated, and weak links in the security solution should be strengthened.

Figures 6.5 and 6.6 illustrate some typical processes required to safeguard corporate assets for enterprise security.

Figure 6.7 depicts an example of *end-point* and *centralized* tools that can be used by firms. In particular, intrusion detection systems (IDSs) have been utilized in

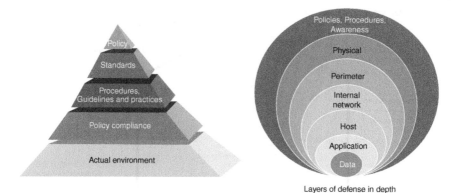

Layers of defense in depth

Figure 6.4 Best practices: policy-based approaches and defense in depth.

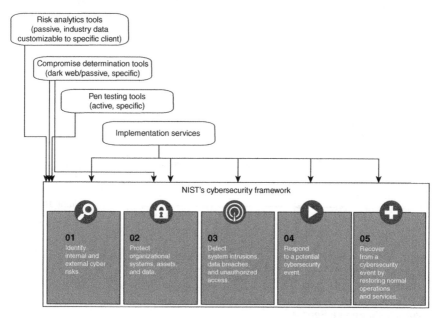

Figure 6.5 Enhanced classical NIST (National Institute of Standards and Technology) methodology.

recent years for security management of firms' intranets. There are a variety of IDSs. Based on their points of placement, they can be categorized into network-based intrusion detection system (NIDS) and host-based intrusion detection system (HIDS) [17]. A NIDS is located at a strategic point in the network such that packets traversing a particular network link can be monitored. NIDSs monitor a given network interface by placing it in promiscuous mode (thus hiding its

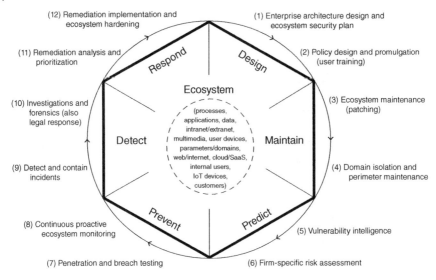

Figure 6.6 Generic process for enterprise security.

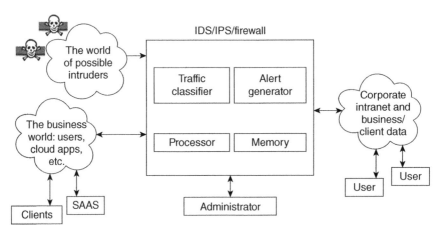

SAAS: Software as a Service

Figure 6.7 Basic IDS environment.

existence from network attackers). HIDSs monitor and reside in individual host machines. HIDSs operate by monitoring and analyzing the host system internals such as OS calls and file systems. HIDSs can also monitor the network interface of the host. Figure 6.8 is a schematic view of a generic intrusion detection ecosystem.

An intrusion protection systems (IPSs) is a signature- and policy-based tool (appliance or software) that checks for well-known vulnerabilities and attack

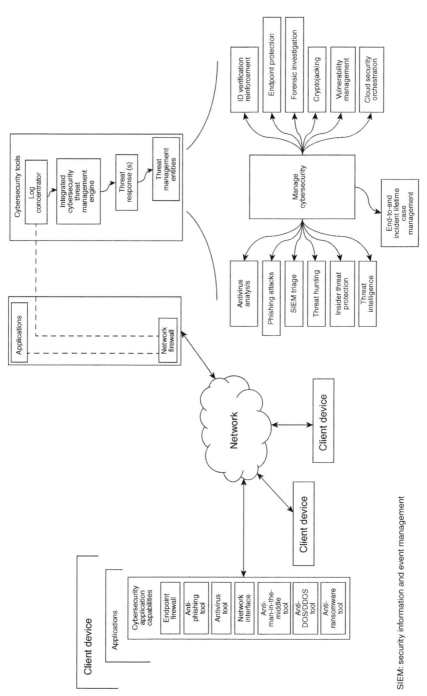

SIEM: security information and event management

Figure 6.8 Example of end-point tools under the protection of an integrated cybersecurity management process. *Source:* Reference [18]/U.S. Patent/public domain.

Table 6.1 Enterprise security mechanism typically utilized by CISOs.

Area	Specific technique and/or approach
Network and client security	Network immunization testing
	External strength testing
	Wireless networks and cellular network testing
	Mobile and cloud
	Overall penetration testing
	Firm-wide disinfection
Application security	Security development lifecycle (SDL)
	Security code review (white box)
	Security design and product review
	Web Application Firewalls (WAF) and database (DB) firewall configuration
	Application threat modeling – based on threat modeling
Compliance	PCI DSS Compliance
	ISO 27799 Health Informatics
Environment security	Testing of sites for physical security
	Testing of sites for availability (redundant power, redundant communications links)

vectors based on a signature DB and established corporate policies. The IPS establishes a standard based on the DB and policies, then generates alerts when traffic deviates from the standard. IPSs typically operate and protect layers 3 and 4 of the protocol model, protecting traffic across several protocol types such as DNS, SMTP, TELNET, RDP, SSH, and FTP. IPSs, however, offer limited protection at the application layer (layer 7).

To maintain a protected environment, after deploying some of these tools, a series of tests can be undertaken on a periodic basis. Testing regimes can be used by firms to strengthen cybersecurity and minimize (exposure) risks. Tests include, but are not limited to, the ones shown in Table 6.1. Some description of the entries of Table 6.1 follows. ML techniques can support many of the constituent elements shown in the figure and in the table.

6.3.1 Network and End System Security Testing

Network immunization testing. Internet and network access channel tests, including company personnel and service provider's access rights. Testing the firm's firewall(s) is a procedure implemented to prevent exposure of the operating environment to the Internet; it enables the organization to obtain insight about the ability of a hacker to gain access to the company servers that contain all the organization's services for day-to-day operations and use these services to access the organization's internal network. The testing simulates an external threat of a hacker that stole the access rights of the organization to reveal the resources of the organization to the outside world.

External strength testing. Testing the strength of threats enables the organization to test the data security protection they have, as viewed from the outside by an internet hacker. External strength examinations include some of the following:

- Communication tests for prevention of access via communication channels to the internet;
- Access tests to the network via a Trojan horse or other spyware;
- Access tests via telephony, known as "War Dialing";
- Eavesdropping tests on wireless networks;
- Examination of hostile take-over via remote workstations – RAS (Remote Access Service), Citrix, and VPN terminals;
- Periodic and permanent monitoring and scanning for alterations in the organization's exposure to the internet; and
- Collection of intelligence on Internet driven attacks.

Wireless networks and cellular network testing. There is a clear need to prevent eavesdropping on wireless networks and identify cellular-enabled and Wi-Fi– enabled elements on the firm's network, particularly considering the ever-growing Bring Your Own Device (BYOD) trend. These tests allow an organization to establish the behavior and characteristics of the wireless portion of the corporate intranet. Typical tests cover the following four aspects:

- Penetration testing of the wireless network and the wireless components;
- Topology tests for connecting existing wireless network components of the organization;
- Identification of wireless cellular components which are unauthorized; and,
- Wireless pulse testing for access from outside the organization's perimeter.

Mobile and cloud. Testing may include the following:

- Assessing present or anticipated exposure to information security risk;
- Penetration testing including misuse and unauthorized use;
- Attacks simulating real-world networks, OSs, applications, and IoT elements (e.g. some described in Chapter 5) to provide vulnerability insight;
- Secured cloud solution design that achieves risk minimization;
- Mobile security controls solutions based on mobile device use cases and available; and
- Application of policies and processes to mobile programs.

Overall penetration testing. Penetration testing typically includes the following (based on the firm's specific environment, concerns, threats):

- Investigation and identification of zero-day attacks in the OS (e.g. Windows, Linux) that are not recognized by the scanning systems;
- Cyber forensics – testing and analysis of workstations and servers, testing of protection and monitoring systems, testing of log files and event analysis of access and data theft;
- Writing spyware and trojan horses that are not identified by known AV or antimalware (AM) systems;
- Reverse engineering of existing threats and research and analysis of threat behavior;
- Identification of trojans in smartphones and in the organization's network;
- Threat surveys to system and manufacturing processes, such as supervisory control and data acquisition (SCADA), human-machine interfaces (HMI), and distributed control system (DCS) or at the terminal level programmable logic controllers (PLCs)[2];
- Modbus and open platform communications (OPC) protocol eavesdropping using Test Anything Protocol (TAP), or via Internet Protocol (IP) networks (e.g. with ARP = Address Resolution Protocol poisoning ethernet bridging).

Firm-wide disinfection. Users' PCs and laptops are often appealing to hackers as they pose a security risk which might enable access to sensitive data via the Internet. At the point of access, the hacker may have a vantage point to enable him to break in to all the organization's data servers and proprietary information. Tests to assess firm's capabilities to prevent this scenario, thus preventing penetration via malware, viruses, spyware, or trojans, are run to prepare for the eventuality of such a break-in and its effect on the firm's network.

6.3.2 Application Security Testing

Attacks to apps are the leading cause of breaches and compromise corporate data. The OWASP = Open Worldwide Application Security Project Top 10 is a list of the most seen application vulnerabilities; currently, these include [20]:

- Injection attacks
- Broken authentication
- Sensitive data exposure
- XML external entities (XXE)
- Broken access control
- Security misconfigurations

---

2 E.g. see Ref. [19].

- Cross site scripting (XSS)
- Insecure deserialization

Security Development Lifecycle (SDL) (typically for larger firms) includes

- Developing a secure software development lifecycle process;
- Building secured code writing standard (best practice).

Security code review (white box) (typically for larger firms). Review the application code to find and fix uncovered security breaches. Code review is focused on testing the implementation of security mechanisms such as:

- Authentication
- Authorization
- DB connection and queries
- Input validation
- Session managements
- Error handling

Security design and product review (typically for larger firms) includes

- Identifying any faulty architectures;
- Implementing security principles;
- Implementing threat mitigation.

Application threat modeling includes

- Performing a security analysis to identify, understand, and mitigate expected threats;
- Finding security bugs early.

Web Application Firewall (WAF) and Database Firewall (DBF) Configuration testing (typically for larger firms).

A WAF protects the application layer and is specifically designed to analyze each HTTPS request at the application layer; it protects web applications from a variety of application layer attacks such as XSS, SQL injection, and cookie poisoning, among others. The WAF protects the web apps by filtering, monitoring, and blocking any malicious Hypertext Transfer Protocol (HTTP)/Hypertext Transfer Protocol Secure (HTTPS) traffic traveling to the web application, and prevents any unauthorized data from leaving the app. The WAF achieves this by adhering to a set of policies that help determine what traffic is malicious and what traffic is safe. Just as a proxy server acts as an intermediary to protect the identity of a client, a WAF operates in similar fashion but in the reverse – called a reverse proxy – acting as an intermediary that protects the web app server from a potentially malicious client [21]. Policies can be customized to meet the specific needs of the web applications. A WAF is typically user-, session-, and application-aware, cognizant of the web apps behind it and what services they offer. Because of this, one can think of a WAF as the intermediary between the user and the app itself, analyzing all communications before they reach the app or the user. Traditional WAFs ensure only allowed actions (based on security policy) can be performed. WAFs are a trusted, first line of defense for applications, especially to protect against the OWASP Top 10 cited above.

Database Firewalls are a kind of application firewall that monitors DB traffic to detect and protect against DB-specific attacks that principally seek to access sensitive data held in the DBs. Database Firewalls also allow for the monitoring and auditing of all access to cloud DBs via logs.

6.3.3 Compliance Testing

Payment Card Industry Data Security Standard (PCI DSS) Compliance (typically for financial firms). Theft of credit card information because of various DB breaches is a phenomenon that has become more prominent in the last few years among financial institutions and gaming companies, insurance organizations, and more. These thefts cause damage of hundreds of millions of dollars in business, loss of capital due to leakages, and the high operational cost of repairing the organizations' destroyed corporate image. To eradicate this trend, credit card companies have joined together to create a data security credit card standard called PCI DSS. Corporations that transfer, store, process, or forward credit card details need to be regulated according to this standard to be licensed by the International Regulatory Committee. Under this definition, all businesses, service providers, third parties and accruers need to adopt this standard. Over time, firms need (i) to perform PCI Compliance gap; and (ii) to identify the best solutions for the gaps.

ISO 27799 Health Informatics Security Standard for the Computational Health of Medical Establishments (typically for healthcare firms). ISO 27799:2008 defines guidelines to support the interpretation and implementation in health informatics of ISO/IEC 27002 and is a companion to that standard. ISO 27799:2008 specifies a set of detailed controls for managing health information security and provides health information security best practice guidelines. By implementing this standard, healthcare organizations and other custodians of health information can ensure a minimum requisite level of security that is appropriate to their organization's circumstances and that will maintain the confidentiality, integrity, and availability of personal health information. ISO 27799:2008 applies to health information in all its aspects; whatever form the information takes (words and numbers, sound recordings, drawings, video, and medical images), whatever means are used to store it (printing or writing on paper or electronic storage) and whatever means are used to transmit it (by hand, via fax, over computer networks or by post), as the information must always be appropriately protected. The Health Informatics Security Standard is applicable for medical establishments, enabling them to keep their patient data private, thereby ensuring patient confidentiality while securing the data and the inter-systems interoperability and eliminating vulnerabilities. These services enable hospitals and other medical facilities to reach and maintain the Health Informatics Security Standard while continuing daily routines which are life-preserving. The key definition of conforming to the Health Informatics Security Standard entails full support for the triad: Confidentiality, Integrity, and Availability; this includes securing (i) data technologies, (ii) data systems, (iii) physical elements (servers, networks), and (iv) safety procedures.

6.4 IoT Security (IoTSec)

Earlier, there was a discussion about security capabilities that can (and should) be implemented in end systems. However, as it was hinted in Section 6.1, some computer systems, particularly IoT devices and sensors are resource-constrained, such that the execution of malware detection and/or removal tasks at normal runtime may cause a noticeable and unacceptable reduction in the performance of the system: the execution of routine malware detection tasks can interfere with the normal services, facilities, or functions of such computer systems. In some cases, such interference can be so considerable that a user of the computer system is effectively denied access to the computer system or the services the computer system supports while the malware detection and/or removal process is undertaken. Examples of such resource-constrained network-connected devices include, among others, mobile computer systems such as mobile telephone handsets; smartphones; tablet computers; smart watches; thin-client laptop computer systems; cloud computing terminal devices; so-called "dumb" terminals; mobile terminals used in various industries such as logistics handsets and utilities handsets; barcode scanners; printers; radio-frequency identification (RFID) scanners; entertainment devices including televisions, set-top boxes, games consoles, handheld games consoles, music or sound players and handheld controllers; alarm systems; industrial controllers and controlling devices; manufacturing robots; vehicular computers such as in-car computer systems, satellite navigation systems, and self-driving vehicle computers [22]. Therefore, in an IoT environment, the security tools have to be tailored to the situation at hand.

In some IoT applications, especially in the case of wireless "fog" connectivity, there is a requirement for "network autonomy," that is, having the ability for *distributed ad hoc self-organization* to support routing, "opportunistic routing," forwarding, and relaying of the content and topology information. In these cases, security considerations are even more critical. In autonomous operations, individual endpoints or even clusters of endpoints are not dependent on a remote central entity (perhaps back in the cloud), to enable them to fulfill their sensing and data collection roles. Thus, a security compromise may not be detected by a centrally located control and audit system (such as an IPS). The expectation is that as more IoT/M2M (machine-to-machine) applications emerge the trend toward autonomous operation of the endpoint devices and toward self-organizing network paradigms will accelerate. Appropriate security mechanisms are required: in an autonomous distributed environment, decisions may be made locally. One distributed application example of many is a system for street light intensity during the night, where the decision is made locally based on the sensed number of people or vehicles in that particular street (some of the data need travel to a central point, although these data may be "accounting" information only). Another example of autonomous operation is a Vehicular Ad Hoc NETwork (VANET).

On the other hand, a *centralized IoT architecture* has a different risk profile than an autonomous distributed environment. The point of integration where all data for decision making, analytics, or storage are aggregated, is a target-rich environment for hackers. Fortunately, more powerful NEs will be typically present (e.g. full-fledged firewalls); thus, it is both critical and more doable to implement strong security measures at those locations.

The security risks of centralized/decentralized design need to be assessed and addressed, preferably through standardized security mechanisms. The security apparatus must support an ecosystem-wide state of secure nodes, secure computing components, secure communications, and secure asset access control. Specifically, security benefits from the availability of an underlying system architecture, on which predictable and reliable functionality can be defined. Architectures serve the purpose of simplifying the discussion of the system's *building blocks* and how they interrelate to each other, systematizing these functional blocks, and fostering standardization, particularly in the context of element-to-element (functional-block to functional-block) interface standardization. Architectures frameworks and standards enable seamless, even plug-and-play connectivity and operation.

A number of existing security measures in the cloud or on premise can be reused in the M2M/IoT environment, but they have to be applied consistently, granularly (at each layer of the architecture model), and in consideration of the IoT nodal constraints listed earlier (e.g. low computing complexity, limited storage, limited power, etc.). In practice, different techniques or mechanisms (possibly of increasing complexity and sophistication) may be utilized at each layer. In fact, this is the way some of the communications links already operate today. Thus, a wireless LAN may include data link layer encryption, however, a mobile user on a wireless laptop working from home with a VPN tunnel to a corporate office will also get Layer 3 encryption. Furthermore, the worker could download a Word file (for example) that had been encrypted on an intranet folder, and so on.

In broad terms, system designs based on decentralized architectures could, in some cases, be preferable from a security perspective; for example, DoS attacks are better weathered by a decentralized model and overall operational functioning is more impervious to breaches and incapacitation; however, many applications require a centralized paradigm.

Some security capabilities are best handled directly at the *hardware level*; a number of such capabilities have emerged in recent years:

- The trusted execution environment (TEE) is of particular importance to (IoT) applications that deal with sensitive user data, including e-health-related data (such as user real estate locations, user real estate contents [e.g. jewelry], medical claims, and so on).
- Intel Trusted Execution Technology (Intel TXT) is a system hardware technology that deals with verifying and/or maintaining a trusted OS.

6.5 Blockchains

The concept of *blockchains* is now receiving considerable research and practical interest. Blockchains provide data integrity across a large number of transactional parties by providing all participants in the ecosystem with a working proof of decentralized trust; classically, this assurance of integrity had to be achieved by utilizing a trusted third party to "escrow" elements of the transaction – a blockchain replaces this trusted third party. A blockchain is a cryptographically-linked list of blocks created by nodes, where each block has a header, the relevant transaction data to be protected, and relevant security metadata (e.g. creator identity, signature, last block number, and so on). It facilitates "decentralized consensus" by being a distributed ledger (which is effectively a distributed DB), that retains a(n) expanding list of records, while simultaneously precluding revision or tampering of such records retrospectively. Because blockchains are intrinsically resistant to modification of the underlying data, they are perceived as embodying a tamper-resistant incorruptible decentralized digital ledger for economic transactions related to virtually anything of value (e.g. see [23–47] among other references). Some characteristics and observations follow:

- A blockchain is a form of distributed ledger (a distributed DB) that retains an expanding list of records, while precluding revision or tampering (an untamperable ledger);
- A blockchain intrinsically provides universal accessibility, incorruptibility, openness, and the ability to store and transfer data in a secure manner;
- Blockchain encompasses a data structure of "child" (aka successor) blocks; each block includes sets of transactions, timestamps, and links to a "parent" (aka predecessor) block; the linked blocks constitute a chain;
- Provides universal accessibility, incorruptibility, openness, and the ability to store and transfer data in a secure manner. The original application was as a ledger for bitcoins;

- Users are able to add transactions, verify transactions, and add new blocks;
- Blockchains offer a mechanism for people (or entities) who do not know or trust each other to create a shared record of asset ownership. A blockchain is an "open platform," a distributed system where the processes are open to examination and elaboration. It is a ledger of data, replicated across a plurality of computers organized in a peer-to-peer (P2P) network;
- A transaction (e.g. a cryptocurrency transaction, a contract, some compiled business form) between two parties is captured by a supporting node (host). The requested transaction is distributed to a P2P network of nodes;
- Proponents see opportunities for the use of blockchains for e-health companies (and banks and many other industries), as it allows a replacement of the common centralized data paradigm, thus fostering additional process disintermediation;
- Possible applications of interest to the e-health industry include claims filing/ processing; claim fraud detection e.g. spot multiple claims from a claimant (medical office) for the same procedure; data decentralization; and cybersecurity management (e.g. data integrity).
- To date the technology has been primarily used to create cryptocurrencies such as Bitcoin, smart contracts, and nonfungible tokens (NFTs) such as digital artworks.

Many applications of blockchains have emerged in the recent past beyond the original applications of cryptocurrency. The data can, in fact, represent a wide variety of elements, documents, facts, packets, transactions, agreements, contracts, monetary transactions, or signatures. A blockchain can support a wide range of tasks, including allowing parties to draw up trustworthy contracts, storing sensitive information, and transferring money safely – all without the intervention of an intermediary. Possible business applications include claims filing/processing; claim fraud detection, for example, to spot multiple claims from a claimant (e.g. medical office) for the same procedure; data decentralization; and cybersecurity management (e.g. data integrity). Additionally, there often are requirements to verify the authenticity of items and systems through multi-stage multinational supply, distribution, and service chains (that might raise concerns about counterfeit items and/or the requirement of tracking legally-controlled items such as medicines, medical devices, controlled pharmaceutical substances, arms, negotiable bonds, and so on.) An important proposed application of blockchains is for cybersecurity, specifically for Integrity.

Thus, a blockchain records the transactions on a multitude of distributed hosts; a replicated, decentralized DB effectively eliminates the possibility of global data corruption (deliberate or accidental). The blockchain is a time-stamped DB that retains the complete logged history of transactions on the system; each transaction processor on the network or system retains their own local copy of this DB and consensus-formation algorithms allow every copy, no matter where such copy is, to remain synchronized. Specifically, a blockchain consists of *blocks* that hold sets of valid transactions; each blockchain block incorporates the hash of the prior block in the blockchain, juxtaposing the two blocks. The linked blocks form a chain.

Members of the network are anonymous entities (processes, individuals, or users) known as *nodes*. Nodes perform a variety of functions depending on the assumed role. A node can create and propose a transaction, validate transactions, and undertake mining to support consensus and establish the integrity of the data. Blockchains make extensive use of hashing [48]. When nodes create transactions, these are signed by nodes using their private key to validate that these nodes are the true owners of the asset that they are transferring to someone else in the blockchain-secured network. In a blockchain, a P2P network is required as well as consensus algorithms to ensure replication across nodes are undertaken. *Peers* support the state of the distributed ledger. The P2P function implies that there is no central control in the blockchain-secured network and all nodes can communicate directly with one another using an appropriate protocol, allowing for transactions (e.g. documents, data, cybercurrency) to be exchanged directly among the peers.

There typically are two types of peers: *endorsing peers* and *committing peers*. Endorsing peers simulate the transaction execution: they execute and endorse the transaction; endorsement policies specify the rules for the transaction endorsement. Committing peers receive transactions endorsed by endorsing peers, verify these transactions, and update the ledger – they may also be *Orderer nodes* that receive transactions from endorsers, sequence them, and forward these transactions to committing peers.

Therefore, nodes can be *miners* or *block signers*. Miners create new blocks containing relevant data (to be protected). Block signers validate and digitally sign the transaction. An important assessment that every blockchain network must make is to determine which nodes are able to append a next-block to the chain. This decision is made utilizing a *consensus mechanism*. Miners are able to add transactions, verify transactions, and add new blocks. Interactions inside the network utilize cryptographic means to securely identify the source and the sink of the data. When a node (miner) wishes to add data to the ledger, a consensus forms in the network to establish where this data should be captured in the ledger. The blockchain encompasses a data structure of "child" (aka successor) blocks; each block includes sets of transactions, timestamps, and links to a "parent" (aka predecessor) block; the linked blocks constitute a chain.

Typically, the consensus mechanism encompasses three steps: (i) the transaction endorsement process where the transaction is simulated by an appropriate process; (ii) the ordering process, which decides the sequence in which endorsed transactions are written into the ledger; and (iii) the validation and commitment process, where committing peers validate the transaction received from the orderers and then commit that transaction to the ledger itself. The P2P messages utilized in the network typically support discovery (initial discovery of other peers in the blockchain-secured network), transaction (querying, invoking, and deploying transactions), synchronization (keeping the blockchain updated on all nodes), and consensus (endorsing the transaction).

Once a miner node connects to the P2P network, the miner must undertake a number of tasks. These include most or all of the following [49].

1) Synchronization with the network: download the pertinent blockchain by requesting historical blocks from other network nodes.
2) Validation of the transaction: transactions that are transmitted over the network are validated by nodes with full functionality by verifying and validating digital signatures and outputs.
3) Validation of the block: validating blocks against established rules; this covers each transaction in the block and the nonce value.
4) Creation of new blocks: as noted, miners can propose a new block by combining validated transactions received over the network.
5) Perform Proof of Work (PoW): miners find a valid block by solving a computational puzzle: miners repeatedly vary the nonce field contained in the header until the hash thus generated is smaller than a predetermined threshold value.
6) Fetch of the reward: once the node has solved the PoW puzzle it broadcasts the results, allowing other nodes to verify and accept the block; if the block is accepted the miner is "somehow" rewarded.

Note that the PoW requires non-trivial computational resources. Some of the steps include retrieving the previous block's header, gathering a set of transactions transmitted over the P2P network into a proposed block; computing the double hash of the previous block's header with a nonce and the proposed block; establishing if the thus-computed hash is smaller than the current threshold difficulty level; and, once the successful PoW problem is solved, broadcasting the block to the network (and fetch the reward); otherwise, repeating the last step.

Combined AI and blockchain applications were just nascent at press time. As noted, a blockchain is a shared, distributed, immutable ledger maintained over a P2P network and used to create an immutable record of transactions; it facilitates transparent exchange of encrypted data simultaneously to multiple parties. Blockchain's ability to decentralize identity and verification may be utilized by organizations as generative AI becomes prevalent in content creation. For example, there is a need to establish how to protect the organization's intellectual property (IP) when it is being used to train the chatbots; another consideration is to determine how to ensure how the organization might infringe on anybody else's IP when using the chatbots to create new content. Blockchain could help ensure that companies get attribution for their IP when it is incorporated into generative AI output and potentially even receive royalties for its reuse. In an authenticity context, blockchains can offer a framework to manage the provenance (with audit trails) of the data, helping improve trust in data integrity and, by extension, in the recommendations that AI provides. Thus, AI can enhance data security. In an augmentation context, AI can expeditiously and comprehensively absorb, interpret and correlate large amounts of endogenous and exogenous data, offering an enhanced level of intelligence to blockchain-based business networks, in turn, providing more actionable insights.

6.6 Zero Trust Environments

Zero trust environment (ZTE), also known as zero trust networking (ZTN), is an evolving security paradigm that operates on a security model establishing trust through continuous authentication and continuous monitoring of every network access attempt. ZTN abrogates the traditional model of assuming everything in an intranet can be trusted: the ZTN policy is "never trust, always verify." In a traditional design, the network perimeters were secured by verifying user identity only the first time a user or device appeared and/or entered an environment. See Figure 6.9.

ZTNs use "microperimeters," where each microenvironment (micro-segment) has its own authentication requirements: microperimeters encapsulate specific assets, such as data, applications, and services by defining the protect surface within a micro-segment (the micro-segment is a smaller, secured area within a larger network that is protected by a micro-perimeter). Utilizing segmentation gateways, authentication is defined not only by user identity but also by parameters such as device, location, time stamp, recent activity, and description of the request. Verification can be supported through multi-factor authentication for users,

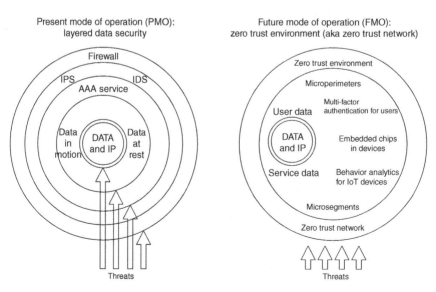

Figure 6.9 Evolution of security mechanism towards zero trust environments. Left: PMO; Right: FMO.

embedded chips in devices, and behavior analytics for connected IoT devices [50–52] (some of the hardware-based methods were also hinted in Section 6.4 above). These multi-instance authentications protect ICT environments from unauthorized users by granting approved users only the specific privileges for which they have a time-defined (e.g. immediate) need. This approach endeavors to ensure that even if attackers gain entry, they cannot operate freely in the network environment. Zero trust can protect devices and data in the ecosystem, expedite anomaly and breach detection, reduce the response time and complexity of the security apparatus, and address the security skills shortage by injecting ML/AI mechanisms in the security process.

6.7 Areas of ML Applicability

Table 6.2 (synthetized from Refs. [53, 54]) identifies various areas where ML techniques have been utilized for cybersecurity application – the table is not exhaustive. Some specific examples follow; many more are documented in the literature. Table 6.3 is a partial *listing* (by title) of US patents and patent applications in the past five years (before this text's publication) that deal with ML usage in cybersecurity settings. The list is included to give a sense of the breath and range of applicability of ML in this setting (these are in addition to the ones cited in the reference section at the end of this chapter).

Table 6.2 Sample of ML applications for cybersecurity.

ML-based cybersecurity tasks	Applicability
AI-based antivirus software	Antivirus software that integrates ML can dynamically identify viruses more effectively than plain old antivirus software.
Analyzing mobile endpoints	Utilizing ML techniques one can identify and analyze threats against mobile endpoints which are now prominent in corporate intranets.
Anomaly detection techniques to identify web robots (bots) that attempt to interoperate with computing services provided over a network	Bots are programs that interact with users or other programs in a manner that can emulate a user, often to nefarious effect. Today, the internet is increasingly crippled by bots. By some estimates, bots consume a full 40% of the capacity of the Internet. In addition, bots perform every sort of mischief from password spraying to attempted subversion of the electoral process. Reliably detecting even previously discovered and documented bots is a difficult problem. Bots deploy all varieties of subterfuge to thwart bot detection strategies [53].

Table 6.2 (Continued)

ML-based cybersecurity tasks	Applicability
Combating AI threats	Hackers themselves are taking advantage of ML technology, but ML can be used to find the deficiencies where cybersecurity issues are detected.
Cybersecurity task automation	ML can be used to automate repetitive cybersecurity-related tasks.
Enhancing human analysis	ML can help administrators to detect malicious attacks, endpoint protection, analyze the network, and vulnerability assessments.
Identification of phishing emails	Phishing emails using feature extraction and ML. Phishing refers to an attempt to fraudulently retrieve sensitive information by masquerading as a trustworthy person or business with a proper need for such information. Although there are several strategies to frustrate phishing, ML-based methods are desirable. Feature extraction is a goal of the ML methods.
Identifying the cyber threats	An ML-based cyber threat identification system can be used to keep an eye on incoming calls and monitoring systems. In particular, IDSs can use ML methodologies.
Intrusion detection systems (IDSs)	IDS may use HMMs that are trained for any incoming network traffic to identify suspicious traffic.
Monitoring emails	Phishing attacks are commonly instituted through emails. Cybersecurity software with ML can be used to avoid these kinds of attacks. Natural language processing (NLP) can also be used to scan emails for any suspicious behaviors.
Network risk scoring	ML can be used to analyze previous cyber-attack datasets and determine what areas of the network are mostly involved in particular attacks. This can help in scoring the attack with respect to a given network area.
User behavior modeling	ML algorithms can be trained to identify the behavior of each user, such as their login and logout patterns, and can alert the cybersecurity team if there are any aberrations.
WebShell monitoring	WebShell is code that is maliciously loaded into a website to provide access to make modifications on the Webroot of the server. This allows attackers to gain access to the database. ML can help in detecting the normal shopping cart behavior and the model can be trained to differentiate between normal and malicious behavior.

Source: Synthetized from Ref. [53, 54].

Table 6.3 Recent US Patents and Patent Applications dealing with ML solutions for cybersecurity.

---

1) Generating risk profile using data of home monitoring and **security** system
2) Forming a **security** network including integrated security system components and network devices
3) Cross-client sensor user interface in an integrated **security** network
4) Internet-of-things (IoT) gateway **tampering detection** and management
5) Systems and methods to **detect abnormal** behavior in networks
6) Training a classifier used to detect network **anomalies** with supervised learning
7) **Malicious** encrypted network traffic identification using Fourier transform
8) Malicious network traffic identification
9) Monitoring user **authenticity** in distributed system
10) Monitoring user **authenticity**
11) Network appliance for dynamic protection from **risky network activities**
12) Verifying **network attack** detector effectiveness
13) **Predicted attack detection** rates along a network path
14) System and method for non-disruptive mitigation of **messaging fraud**
15) Security network integrated with premise **security** system
16) Apparatus and method for detecting **traffic flooding** attack and conducting in-depth analysis using data mining
17) Detecting **spoofing** in device classification systems
18) Intent-based **security** for industrial IoT devices
19) Inception of **suspicious network traffic** for enhanced network security
20) Machine learning based **intrusion detection** system for mission critical systems
21) **Malicious** software identification
22) System and method for tunnel-based **malware** detection
23) Preserving **privacy** in exporting device classification rules from on-premise systems
24) Methods and systems for **encryption** based on cognitive data classification
25) Profiling for malicious encrypted network traffic identification
26) Distributed detection of **security** threats in a remote network management platform
27) **Malicious** encrypted traffic inhibitor
28) Methods for **zero trust** security with high quality of service
29) Prioritizing **policy intent** enforcement on network devices
30) System and method for monitoring and **securing** communications networks and associated devices
31) Forming a **security** network including integrated security system components and network devices
32) **Blockchain** systems and methods for remote monitoring
33) Techniques for auto-remediating **security** issues with artificial intelligence

Table 6.3 (Continued)

34) Determining application **security** and correctness using machine learning based clustering and similarity

35) Security network integrating **security** system and network devices

36) Root-cause analysis and automated remediation for Wi-Fi **authentication** failures

37) Forming a **security** network including integrated security system components and network devices

38) Techniques for **securing** network environments by identifying device types based on naming conventions

39) **Routing** protocol security using a distributed ledger

Source: https://ppubs.uspto.gov/pubwebapp/static/pages/landing.html.

6.7.1   Example of Cyberintrusion Detector

A template-based cyberintrusion detector for preventing cyberattacks on a protected system or processing environment is desirable – especially one that includes neural networks (NNs) for generating (e.g. when in training mode) test inputs for simulated data exchanges between the protected system and the external data sources via interfaces (e.g. ports, transport layers, or other potential channels for cyberintrusion), as described in [55]. In such a system, the NN observes training-mode behaviors of the interfaces based on exchanges of the test inputs, according to data assessment parameters and/or operational modes of the protected system.

Training-mode behaviors may include "gold standard" or optimal behavior sets indicative of optimal performance of the interface (e.g. with respect to a particular test input and/or operational mode) and, therefore, of minimal threat of cyberintrusion over the interface. The NN generates system profiles, or system templates, based on these observed training-mode behaviors and store the system templates to memory which is then accessible to the cyberintrusion detector.

In the real-time operating mode (host-accessible mode), the NN monitors system data exchanged between the protected system and external data sources over the interfaces. For each exchanged set of system data, the NNs apply system templates to replicate, or approximate, the "gold standard" behavior for a particular interface (e.g. based on a particular operational mode and as applied to the exchanged set or sets of system data). A *determined best-fit behavior set* most closely approximating optimal behavior of the interface is achieved by applying a best-fit system template of the stored system templates. For each determined best-fit behavior set, the NN determines the best-fit error, or divergence between the best-fit behavior set and the applied best-fit system template [55]. With this system, if the determined divergence is sufficiently large to meet or exceed policy thresholds for abnormal or anomalous behavior, an event monitor in communication with the NN may take additional preventative or corrective action based on system policies. The preventative or corrective actions available to the event monitor include isolating an interface associated with abnormal or anomalous behavior from the protected system; and/or isolating the protective system or selected components thereof from the interface.

6.7.2 Example of Hidden Markov Model (HMM) for Intrusion Detection

Hidden Markov models (HMMs) have been proposed for intrusion detection [56–59]. A specific method is presented in [17] that (i) employs a dimension reduction technique to extract only important features from network packet data and (ii) applies a decomposition algorithm to lower levels of data, to construct lower level HMMs (representing partial solutions); these lower level HMMs are then combined to form a final, global solution. The multi-layer approach can be expanded to capture multi-phase attacks over longer spans of time. A pyramid of HMMs can resolve disparate digital events and signatures across protocols and platforms to generate actionable information – the lower layers identify discrete events (such as network scan) and higher layers identify new states that are the result of multi-phase events of the lower layers. The discussion that follows is based on the [17] reference.

There are several techniques used by IDSs to gather and analyze data; however, they all have common basic features, namely (i) a detection module that collects data that possibly contain evidence of intrusion and (ii) an analysis engine that processes this data to identify intrusive activity. These engines mainly use two techniques of analysis: (i) anomaly detection and (ii) misuse detection.

- *Anomaly detection* methods can mitigate anomaly predicaments, especially methods based on ML and NNs, statistical modeling, and HMMs. For anomaly detection, a baseline model that is representative of normal system behavior against which anomalous events can be distinguished, is established. When an event indicates anomalous activity, as compared with the baseline, it is considered as malicious. This system characterization can be used to identify anomalous traffic from normal traffic. A useful feature of anomaly-based IDSs is their capability to identify previously unseen attacks. The baseline model in this case is usually automated and it does not require human interference and/or a knowledge base.
- *Misuse detection* utilizes expert systems that are capable of identifying intrusions based on a predefined knowledge base. Consequently, misuse structures are able to reach high levels of accuracy in identifying even very subtle intrusions that might be represented in their knowledge base; when the expert knowledge base is developed carefully, misuse systems produce a minimum number of false positives. There are limitations, however, due to the fact that a misuse detection system is incapable of detecting intrusions that are not represented in

its knowledge base. Therefore, the efficiency of the system is highly dependent on the accurate and thorough creation of the information base. Misuse detection systems require extensive human involvement.

However, at press time, a fair portion of the web-directed traffic is encrypted; this creates a challenge in trying to monitor and identify malicious network traffic. Encryption provides intruders with the ability to hide command-and-control activities. To interdict the intruders, security managers need to deploy additional tools and automation to supplement threat detection, prevention, and remediation by looking for unusual patterns of web traffic that can reveal malicious activity; the messages cannot be decrypted, but the traffic patterns can be monitored. By monitoring the (encrypted) web traffic, specifically by identifying unusual patterns in large volumes of encrypted web traffic, ML can be utilized to automatically detect (i) "previously-known" types threats (i.e. attacks that are generally known and seen before), (ii) "previously-unknown" threats (i.e. previously unknown distinct form of known threats, for example a new malware), and (iii) "completely unforeseen" threats.

Anomaly detection and *misuse detection* can be supported at the IDS. Several ML techniques and models have been explored in the path toward developing IDSs, such as fuzzy logic, artificial neural networks (ANNs), and genetic algorithms (GA). In addition, hybrid intelligent IDSs, such as those based on evolutionary neural networks (ENN) and evolutionary fuzzy neural networks (EFuNN) are also used in practice. As was implied in Chapter 1, NNs are powerful tools for pattern classification and recognition.

The current implementations of HMMs for IDS are mainly based on a single HMM, which are trained for any incoming network traffic to identify anomalous and normal traffic during testing. Other HMM-based IDS implementations rely on multi-HMM profiles, where each of the HMMs is trained for a specific application-type traffic. Posterior probabilities are employed to select network applications using only packet-level information that remain unchanged and observable after encryption, such as packet size and packet arrival time. The limitation with this approach is that it considers only a limited number of features and is unable to detect multistage attacks, which can be crafted to look like normal traffic. Furthermore, HMMs use statistical learning algorithms that suffer in cost exponentially as the volume of data grows: HMMs tend to fail on a large dimension state space (such issue being called the "curse of dimensionality").

Reference [17] describes a system that applies the HMM for intrusion detection in a manner that it becomes capable of providing finer-grained characterization of network traffic using a multi-layer approach; this multi-layer HMM design is capable of detecting multi-stage attacks. The proposed system factors a huge problem of large dimensionality to a discrete set of manageable and reliable elements. A pyramid of HMMs can resolve disparate digital events and signatures across protocols and platforms into actionable information. The method (and realizable system) consists of: (i) receiving network packet data at a processor of a computer-implemented network traffic monitor module; (ii) generating at the processor meaningful HMM observations formatted as data input for one or more first HMMs, the one or more first HMMs forming a first processing layer of HMMs; (iii) generating from the first processing layer of HMMs a first probable sequence of network traffic states; (iv) processing at the processor the first probable sequence of network traffic states to form a feature vector; (v) processing at the processor the feature vector to generate meaningful HMM observations formatted as data input for a second HMM, the second HMM forming a second processing layer; (vi) generating from the second processing layer a second probable sequence of network traffic states; and (vii) on determining that the second probable sequence of network traffic states exhibits a designated probability of a non-normal data traffic state, generating an alert of a likely non-normal data traffic state and transmitting the alert to an administrator.

As noted, IDSs and IPSs are the most important defense tools against sophisticated and ever-growing network attacks. However, due to the lack of reliable test and validation datasets, anomaly-based intrusion detection approaches are suffering from consistent and accurate performance evolutions. Some available datasets suffer from the lack of traffic diversity and volumes, some do not cover the variety of known attacks, while others anonymize packet payload data, which cannot reflect the current trends [60]. It should be noted that the Canadian Institute of Cybersecurity has published a state-of-the-art dataset named CIC-IDS2017, consisting of the recent threats and features. A dataset covers 11 criteria that are necessary in building a reliable benchmark dataset. It contains very common attacks such as XSS, port scan, infiltration, brute force, SQL Injection, Botnet DoS and distributed denial of service (DDoS). This dataset has been used to test the IDS described in Ref. [17].

Figure 6.10 depicts a schematic flow of the IDS. Network packet data is captured, for example through use of a network packet analyzer such as Wireshark. The captured data is then subjected to a series of data processing steps that serve to create meaningful observations formatted as data input for one or more training data-generating HMMs (i.e. a first layer of HMMs in the multi-layer model set forth herein), which data processing steps include (i) feature generation, (ii) feature selection among those generated features or creation of new features by combining the generated features, (iii) using ML algorithms for dimension reduction, and finally (iv) applying vector quantization technique to create meaningful observations for the HMMs. From the network data packet embodied in the captured data, an original set of features is generated that characterize the data. From that original set of features, feature selection and creation take place. Feature selection involves choosing a subset of features from the initial available features, whereas feature creation is a process of constructing new features and is

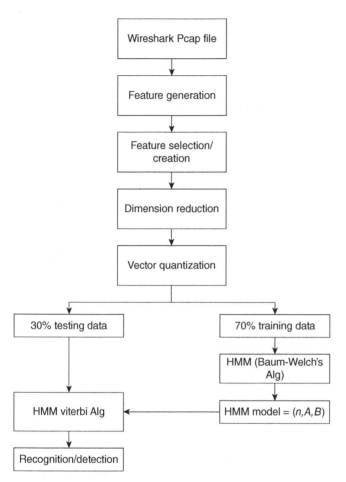

Figure 6.10 Schematic flow chart showing an intrusion detection method configured in accordance. *Source:* Reference [17]/U.S. Patent//public domain.

usually performed after the feature selection process. Feature selection takes a subset of features (*M*) from the original set of features (*N*) where *M* < *N*. To build a robust and high-performance IDS, the features created from the subset of selected features could follow a knowledge-based approach. Other approaches which can be applied to construct new features are data-driven, hypothesis-driven and hybrid.

Features that are discarded may include source port, as source port is part of the categorical encoding that is applied on the following features (Flow_ID, Source_ IP, Destination_IP), resulting in numerical values. In addition, a new feature (label) which identifies the traffic type such as BENIGN (benign traffic), SSH-patator and web-attack-bruteforce is added. The values corresponding to this new feature are also categorically encoded. Dimension reduction is a form of transformation whereby a new set of features is extracted. This feature extraction process extracts a set of new features from the initial features through a functional mapping. However, prior to carrying out such feature extraction, normalization is applied to the original data matrix to standardize the features of the dataset by giving them equal weights. In doing so, noisy or redundant objects will be removed, resulting in a dataset which is more reliable and viable, which in turn improves accuracy. Normalization can be performed using several methods such as, but not limited to min-max, Z-score, and decimal scaling. See [17] for a detailed technical and mathematical discussion of this IDS.

6.7.3 Anomaly Detection Example

As noted above, anomaly detection processes identify samples within a population that possess attributes that depart significantly from the norm. Anomaly detection processes can be useful in a variety of applications, such as fraud, fault, and intrusion detection; specifically, anomaly detection techniques are being leveraged to identify web robots (bots) that attempt to interoperate with computing services provided over a network. As noted in Ref. [53], anomaly detection processes can employ any of a variety of specialized heuristic, statistical, or ML approaches to identify aberrant samples. Each of these approaches has benefits and disadvantages: (i) heuristic approaches are often easy to implement and computationally efficient but can be inaccurate or imprecise; (ii) statistical approaches can be precise, accurate, and computationally efficient but require an understanding of the underlying distribution a priori to be effective; and (iii) ML/ ANN approaches (for bot detection) can also be precise, accurate, and computationally efficient (once trained) but require computationally complex training on large sets of data. While these ANN bot detection engines have met with some success, they also suffer from some disadvantages. The effectiveness of an ANN bot detection engine is limited by its training set which, in turn, is limited to describing bots previously discovered by human experts; thus, even ANN bot detection engines can be overwhelmed by new bot variants. To address these limitations, an *adaptive bot detection system* is described in [53].

This adaptive anomaly detection system is designed to (i) receive a response to a request to verify whether an ostensible client of a service is actually a client or a bot, the response including an indicator of whether the ostensible client is a client or a bot; (ii) receive information descriptive of interoperations between the ostensible client and the service that are indicative of whether the ostensible client is a client or a bot; and (iii) train a plurality of ML classifiers using the information and the indicator to generate a next generation of the plurality of ML classifiers. The information descriptive of the interoperations can include one or more of source IP address, destination IP address, source port, destination port, protocol, total packets exchanged, average inter arrival time of packets, and average time between mouse clicks. ML classifiers can include an ANN, a Bayesian network, or a support vector machine (SVM). The ML classifiers can include a master classifier and one or more community classifiers. The one or more community classifiers can include a peer classifier, an alpha classifier, and an historical classifier. The processor can be further configured to calculate the accuracy of each ML classifier of the next generation.

The processor can be further configured to determine whether any particular ML classifier of the next generation has an accuracy that transgresses a threshold value based on the accuracy of an ancestor classifier of the particular ML classifier; and replace the particular ML classifier with the ancestor classifier where the accuracy of the particular ML classifier transgresses the threshold value. The processor can be further configured to determine whether any particular ML classifier of the next generation has an accuracy that transgresses a threshold value based on the accuracy of the master classifier; and replace the master classifier with the particular ML classifier to generate a new master classifier where the accuracy of the particular ML classifier transgresses the threshold value. The processor is also configured to store the master classifier in a data store. The plurality of ML classifiers can be a first plurality of ML classifiers and the at least one processor can be further configured to add, as a peer classifier, the new master classifier to a second plurality of ML classifiers comprising a second master classifier.

The adaptive bot detection system described in [53] utilizes iterative feedback techniques to converge toward higher bot detection accuracy. Using these techniques, the adaptive bot detection system can mitigate problems introduced by new bot variants by evolving ANN bot detection engines to identify a higher percentage of bot variants and "zero day bots" (e.g. bots of a type not previously discovered) than other bot detection techniques. The adaptive bot detection system includes an ANN engine that operates in-line as an ML classifier, constantly attempting to detect bots (this ANN engine is referred to herein as a master ANN engine). The master ANN engine is associated with and applies an ANN trained to detect bots using the best available training set; this ANN is referred to as a master ANN. Where the master ANN engine detects a bot, the master ANN engine requests that a "Completely Automated Public Turing Test to Tell Computers and Humans Apart" (CAPTCHA) challenge be issued to the detected bot. The response to the CAPTCHA challenge indicates definitively whether the detected bot is an actual bot, or a user, and informs an iterative learning process implemented within the system.

More specifically, the system includes a community of ANN engines. This community includes the master ANN engine and other ANN engines that are each training a respective, associated ANN (these other ANN engines are referred to herein as community engines). In some implementations, the community engines do not detect bots in-line, but rather each community engine trains its associated community ANN, based on the responses to the CAPTCHA challenges and associated information therewith. In these implementations, the community engines train their associated community ANNs as potential replacements for the master ANN. In some implementations, the system seeds initial copies of the associated ANNs from an ANN archive that resides outside the community. This ANN archive can store ANNs associated with other master ANN engines or other ANNs recorded as having a threshold accuracy. In some implementations, to train their associated ANNs, each of the community engines uses a set of parameters relevant to bot detection and associated with the request for the CAPTCHA challenge issued by the master ANN engine. The set of parameters can include, for example, click time, page traversal history, and file type, among others. As each CAPTCHA challenge generates ground truth information (e.g. a bot/no-bot determination), the parameters associated with the request for the CAPTCHA challenge, in conjunction with the bot/no-bot determination, provide sufficient information to execute a learning stroke for ANNs in the community (the set of parameters relevant to bot detection and the ground truth information generated by the CAPTCHA challenge can be collectively referred to a teachable moment).

Figure 6.11 from [53] illustrates a logical architecture of a network computing platform that implements an adaptive bot detection service to identify attempts by bots to interoperate with one or more other services provided by a computer system. The figure depicts a plurality of bots and a plurality of clients. The platform implements a plurality of protected services, including adaptive bot detectors and bot verifiers. Each of the adaptive bot detectors is configured to interoperate with an ANN archive via a trusted network. The ANN archive is configured to store a plurality of ANNs. In certain examples, each protected service includes a service, an adaptive bot detector, and a bot verifier. The service is configured to interoperate with a client via the network and an interface exposed and implemented by the service.

For instance, in some examples, the service includes a web server that exposes and implements an HTTP interface configured to receive HTTP requests and to transmit HTTP responses to the received HTTP requests via the network. In these examples, the clients can include browsers configured to transmit the HTTP requests and to receive the HTTP responses via the network. Alternatively, in some examples, the service includes a web service endpoint that exposes and implements a web service application program interface (API) configured to receive web service requests and to transmit web service responses via the network. In these examples, the clients can include web service clients configured to transmit the web service requests and to receive the web service responses via the network. Regardless of the type of requests received, the service can be configured to process the requests

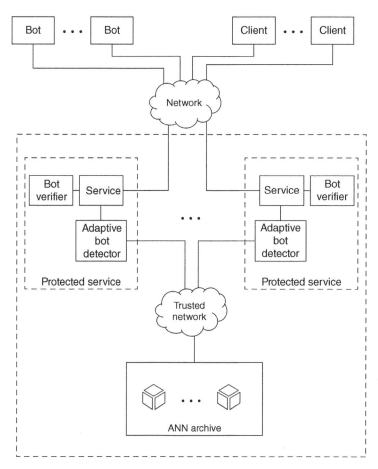

Figure 6.11 Block diagram of a system including a network computing platform with an adaptive bot detector. *Source:* Reference [53]/U.S. Patent/public domain.

to perform a service useful to the client prior to transmitting responses. In some examples, responses include one or more results of the processing performed responsive to the received requests. For instance, where a request is an HTTP request, a response may include, for example, a hypertext markup language (HTML) form that can be rendered by the client so that a user can interact with the client via the form. In some examples, the service is configured to interoperate with the adaptive bot detector while interoperating with an ostensible client (e.g. a bot or a client). In these examples, the protected service is configured to avoid processing requests from an ostensible client that is, in reality, a bot. Figure 6.12 illustrates one instance of an interoperative process that the protected service is configured to execute and/or executes in these examples.

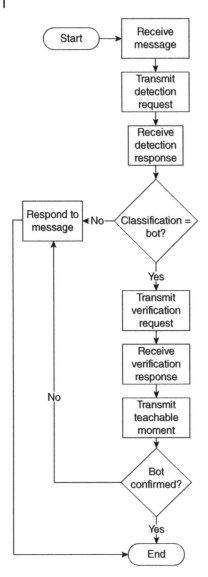

Figure 6.12 Flow diagram of an interoperative process executed by a protected service. *Source:* Reference [53]/U.S. Patent//public domain.

Returning to Figure 6.11, the adaptive bot detector is configured to monitor messages exchanged between the service and an ostensible client during their interoperation with one another to determine whether the ostensible client is a client or a bot. As shown in Figure 6.13, the adaptive bot detector includes and implements a detector interface, ANN engines, an ANN administrator, an accuracy data store, and, over time, multiple generations of ANNs. The ANN engines

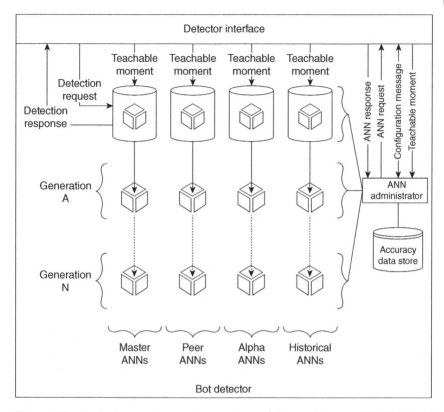

Figure 6.13 Block diagram of an adaptive bot detector. *Source:* Reference [53]/U.S. Patent/ public domain.

include a master ANN engine, a peer ANN engine, an alpha ANN engine, and a historical ANN engine. Refer to [53] for an extensive discussion of the adaptive anomaly detector.

6.7.4 Phishing Detection Emails Using Feature Extraction

One type of fraudulent activity that has become one of the fastest growing online threats is phishing. As noted earlier, phishing refers to an attempt to fraudulently retrieve sensitive information by masquerading as a trustworthy person or business with a proper need for such information. Phishing typically seeks to extract access credentials, bank account information, social security numbers, passwords, credit card information [61–65].

In a phishing attack, an individual may receive a message, commonly in the form of an email directing the individual to perform an action, such as opening an email attachment or following an embedded link (e.g. using a cursor-controlled device or touch screen). If such message were from a trusted source (e.g. a coworker, a known bank or utility company), then such action would carry little risk, but in a phishing attack, such message is from an attacker (e.g. an individual using a computing device to perform a malicious act on another computer device user) disguised as a trusted source, and an unsuspecting individual, for example, opening an attachment to view a "friend's photograph" might in fact install malicious computer software (i.e. spyware, a virus, and/or other malware) on his or her computer. Similarly, an unsuspecting individual directed to a webpage made to look like an authentic login or authentication webpage might be deceived into submitting his or her username, password or other sensitive information to an attacker. Although there are several conventional strategies to frustrate phishing, ML-based methods are desirable: current anti-spam technologies fail to consistently and reliably detect phishing (especially spear phishing) emails, while ML methods can have a high (e.g. 90%) success ratio [66]. Feature extraction is a goal of the ML methods; systems for the continuous monitoring and prevention of phishing emails are an important tool of the organization. A method for detecting a phishing attack on a computer device, can involve scanning email messages; separating email parts from the email message; subjecting the email parts of the email message to a feature extraction operation based on ML algorithms; and analyzing email features extracted from the email parts to determine whether or not any of the email features contain suspected phishing content, confirmed phishing content and benign email content.

Figure 6.14 illustrates a flow chart depicting logical operational steps or instructions of a method for *on-demand detection and scanning of emails* as described in [66] and discussed below. The scanning of emails for malicious attachments, an email body, an email text body, email headers and an analysis of the network for suspicious content that may be harmful for a user's network. The checks can include, but not but are not limited to attachment scanning, email HTML body/text analysis, email headers, and network analysis. This approach can involve the on-demand detection of emails and/or the continuous monitoring and prevention of phishing emails. A suspicious email is reported for detection of a potential phishing attack. Thereafter a step or operation can be implemented in which a tokenizer separates all essential parts of the email for extracting phishing parameters. That is, the tokenizer parses raw email for different objects, namely email headers, email textual part, email html part, and email attachments. These email objects can be further tokenized in a manner, which aims to collect relevant information such as the sender domain, domain-based message authentication, reporting, and conformance (DMARC) signatures from email headers, uniform resource locators (URLs) from HTML part, etc. This data (e.g. items/parameters) can be analyzed to determine whether or not it is a phishing email.

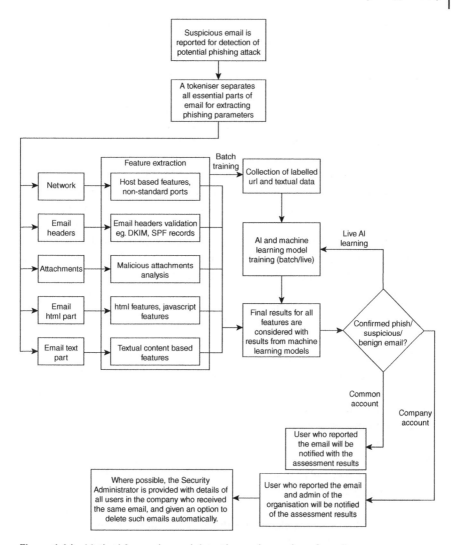

Figure 6.14 Method for on-demand detection and scanning of email.

Feature extraction operations or feature extractions techniques that can be utilized by the aforementioned feature extractor can include, for example, feature extraction techniques/operations such as principal components analysis (PCA), independent component analysis (ICA), linear discriminant analysis (LDA), locally linear embedding (LLE), t-distributed Stochastic Neighbor Embedding (t-SNE), and autoencoders. Feature extraction aims to reduce the number of features in a dataset by creating new features from the existing ones (and then discarding the original features). Feature extraction can involve reducing the number of resources required to describe a large set of data. The first tokenizer can break down the email into different parts, while the second tokenizer can extract the exact information/ parameters that may be required in the feature extraction stage. The number/count of resources can increase after feature extraction. Following processing of the feature extraction operations, batch training data can be subject to an operation involving the collection of labeled URL, images, attachments, textual data and other labeled features. Next, the data can be subjected to ML training (e.g. batch/live).

Examples of ML algorithms that can be utilized may include supervised learning, unsupervised learning, reinforcement learning, self-learning, feature learning, sparse dictionary learning, anomaly detection and association rules (many of these described in Chapter 1). A batch learning algorithm works under the assumption that the entire training data can be available prior to the training task. ML models (MLMs) using batch learning algorithms can be trained on already available data for malicious content present in emails like phishing URLs, attachments containing malwares, spam textual data and so on. Examples of such algorithms include, but are not limited to Bidirectional Encoder Representations from Transformers (BERT) for text classification, SVM algorithms for classifying URLs as malicious or benign followed by multi-label classification on the malicious URLs using Random, a multi-label learning strategy such as the k-Labelset algorithm, a convolutional neural network (CNN) algorithm for logo detection, text extraction for images and CNN-BiLSTM for malware classification in attachments, k-nearest neighbors (k-NN) algorithm for clustering in forensic analysis on log data, and long short-term memory (LSTM) algorithm for anomaly detection for time series log data.

A second type of algorithm that can be utilized to implement ML training is an *online learning* algorithm, which treats the data as a stream of instances, and learns a prediction model by sequentially making predictions and updates. Online learning can increase the scalability of batch learning by updating the weight vectors for classification sequentially by utilizing information with training data. Examples of online learning algorithms that can be adapted for use with varying implementations include, but are not limited to, an online gradient descent algorithm for updating the weight vectors by applying stochastic gradient descent principle only to a single training instance arriving sequentially, and an extended isolated forest algorithm for detecting anomalous user behavior.

6.7.5 Example of Classifier Engine to Identify Phishing Websites

As noted, a (phishing) message may purport to come from a bank or other financial institution, claiming that the person's account has been locked, and providing a link for the person to "unlock" their account. The link will take the person to a website that is designed to mimic the bank's website, with fields for the user to enter their credentials (e.g. username and password, and possibly bank account details). In fact, the website is fraudulent, and once the user has provided their details, these are captured for use by the miscreant operators in conducting illicit transactions with the user's account, which may be drained before the treachery is discovered. Many existing security products are of limited effectiveness in protecting clients from phishing attacks: they take a broad approach, and typically do not prioritize a user's financial accounts, which may create exposure to more sophisticated attacks that are targeted to users of a particular financial institution. Reference [16] describes an ML-based system where salient features are extracted from a training dataset. The training dataset includes, for each of a subset of known legitimate websites and a subset of known phishing websites, URLs and HTML information. The salient features are fed to an ML engine, specifically a classifier engine to identify potential phishing websites is generated by applying the ML engine to the salient features, and parameters of the classifier engine are tuned. This enables identification of potential phishing websites by parsing a target website into URL information and HTML information, and identifying predetermined URL features and predetermined HTML features. The classifier engine is specific to a particular institution and the salient features include at least one institution-specific feature associated with the particular institution. A prediction as to whether the target website is a phishing website or a legitimate website, based on the predetermined URL features and the predetermined HTML features, is received from the classifier engine. Figure 6.15 shows an illustrative distributed architecture. Refer to [16] for detailed information of this type of system and how it can be fine-tuned to specific environments.

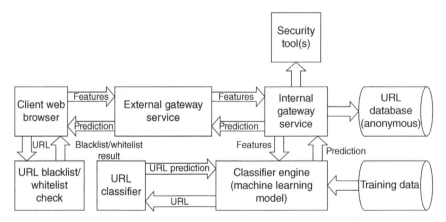

Figure 6.15 Illustrative classifier engine to identify potential phishing websites.
Source: Reference [16]/U.S. Patent/public domain.

6.7.6 Example of System for Data Protection

Protecting data and ensuring data integrity and resiliency in enterprise environments provides numerous technical and nontechnical challenges as noted earlier in this chapter. For example, the level of trust placed in users, system administrators, applications, and underlying IT infrastructure varies widely and may change at any point in time. Providing continuous, cross-platform data protection against emerging and advanced, persistent threats requires a holistic approach that leverages advanced user and infrastructure monitoring, as well as multi-level data protection technologies. There are several data protection and trust platforms on the market from providers; however, these approaches have limitations in that the approaches assume that the underlying IT infrastructure and OSs will maintain not having a security risk. Furthermore, these approaches do not provide multi-layer protection of data, which limits their ability to respond to certain threats. Some data protection products focus on securing specific data flows, but these products address only certain phases of the data lifecycle.

Hence, techniques and systems are needed for an efficient method for providing a secure data protection process. A system for data protection discussed in [67] includes a first computing device comprising a security module and a storage device coupled to the first computing device via a network interface. The security module includes a Software Root of Trust (SRoT) and Hardware Root of Trust (HRoT). The security module is configured to establish a trust channel between the computing device and the storage device or storage service; monitor the computing device and the storage device; create and enforce multi-dimensional data access control by tightly binding data access and permissions to authorized computing devices, users, applications, system services, networks, locations, and access time windows; and take over control of the storage device or storage service in response to a security risk to the system. The platform couples distributed hardware- and/or software-based Root of Trust (RoT) technologies with multi-dimensional data binding and multi-layer, multi-domain (user, apps, systems, network, and storage) real-time monitoring, data fusion, and/or ML-powered anomaly detection.

Figure 6.16 is a block diagram of an illustrative platform for data protection and resiliency, as described in [67]. The platform provides multi-modal, multi-layer advanced threat detection. Concurrent access and mining of multisource, user-, software- and hardware-level events allow a platform ML engine to spot non-obvious relationships and data access patterns that cannot be discerned by monitoring user behavior, application, network, or storage media activity independently. As a result, the probability of detecting advanced threats can be increased, while also reducing dwell and reaction times.

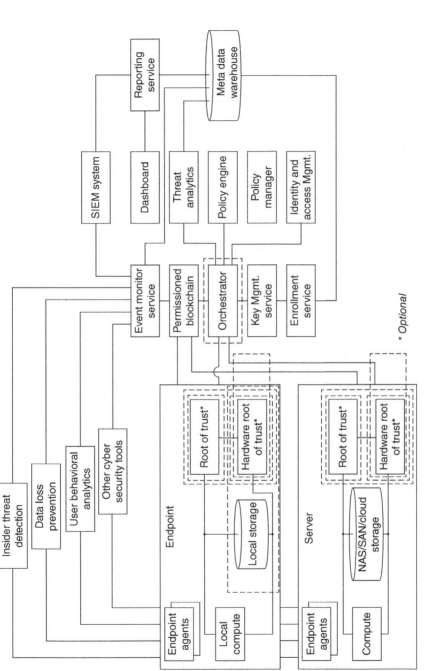

Figure 6.16 Example of platform for data protection and resiliency. *Source:* Reference [67]/U.S. Patent/public domain.

This platform supports the deployment of AI-powered, data access control-related decision making on edge devices (e.g. servers, desktop, laptops, tablets, or mobile devices). Depending on the sensitivity of the data they provide access to or host locally, edge devices can be equipped with an embedded or attached, cyber-hardened HRoT device can be used as cyber sensors and actuators. The highly secure, TEE inside these HRoT devices can be used to run AI-powered threat classifiers outside of their host system, providing an additional layer of protection. Upon detection a local threat (or an external trigger), the HRoT-hosted decision engine could autonomously enter its host system into self-defense mode and execute a preauthorized or dynamically generated playbook. The scripted or dynamically generated response could trigger a number of HRoT-initiated actions, such as locking down attached storage media instantly; changing the drive keys of attached self-encrypting drives (SED) media; securely wiping specific files, folders, or entire drives; or setting specific drives or folders into decoy mode. The concepts described above extend to HRoT-equipped servers as well. Allowing trusted system components to make certain, data protection-related decisions without having to rely on a centralized authority (which could have been disabled or compromised) has the potential to significantly increase an organization's data protection capabilities and resiliency as it reduces a single point of failure.

6.7.7 Example of an Integrated Cybersecurity Threat Management

As discussed earlier, effective and thorough management of cybersecurity threats is critical to the continuous and reliable computing operations of organizations including businesses, financial institutions, hospitals, government agencies, retailers, universities. Organizations are painfully aware of these myriad threats, and actively configure and implement state-of-the-art best practices to secure their ICT infrastructure against the threats. While preventative measures such as application and OS updates, former user deactivation, and other housekeeping activities are critical to ICT operations, these measures alone cannot provide universal protection. Cybersecurity threats are constantly evolving and becoming significantly more sophisticated, therefore constant vigilance and action are required. As soon as a solution is found that identifies, responds to, and eradicates a threat such as a virus, thwarts a trojan horse program, or detects and deletes a phishing attack, the malefactors behind the threats adapt their techniques by using new attack vectors, advanced social engineering ploys, and other deceptions. Reference [18] discusses techniques that enable integrated cybersecurity threat management.

The cybersecurity threat protection applications can include one or more data management schemas. The plurality of threat protection applications can include cybersecurity threat protection application capabilities. The cybersecurity threat protection application capabilities can incorporate endpoint protection, anti-phishing protection, AV protection, firewall protection, man-in-the-middle protection, DoS protection, DDoS protection, and ransomware protection. A plurality of heterogeneous log files is ingested, wherein the log files are generated by at least two of the plurality of cybersecurity threat protection applications. The log files can include text, numbers, codes, and similar data. Each of the plurality of log files is sorted, wherein the sorting enables identification of cybersecurity threat protection elements among the plurality of log files.

The cybersecurity threat protection elements can include an element name, a log source, an event count, a time, a low-level category, a source IP address and port, a destination IP address and port, a username, a magnitude, and so on. The cybersecurity threat protection elements that were identified can be integrated. Since cybersecurity threat protection applications can provide some but not necessarily all information relevant to a particular threat, the information from the applications can be integrated to provide more thorough insight into the threat. Figures 6.17 and 6.18 that follow from [18] depict various aspects of cybersecurity management based on integrated cybersecurity threat management techniques as described in the reference (also see earlier Figure 6.8) – make note of the ML elements.

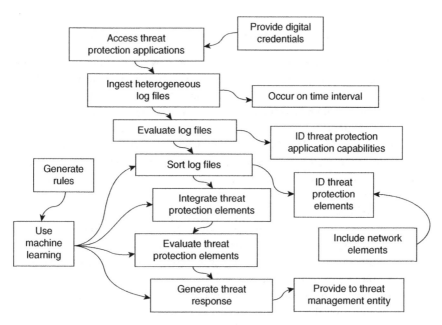

Figure 6.17 Flow diagram for integrated cybersecurity threat management.
Source: Reference [18]/U.S. Patent/public domain.

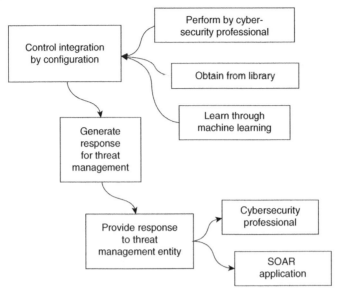

SOAR: security orchestration automation and response

Figure 6.18 A flow diagram for threat response. *Source:* Reference [18]/U.S. Patent/ public domain.

6.7.8 Example of a Vulnerability Lifecycle Management System

As noted earlier in this chapter, thousands of new vulnerabilities impacting applications, OSs, and hardware systems pop up every year. It is fairly difficult for those responsible for their environment's vulnerability lifecycle management to keep track of all existing and new vulnerabilities, maintain an accurate inventory of their endpoints, attribute these vulnerabilities to their endpoints, keep track of vendor's publication of fixes for vulnerabilities and apply patches to the affected endpoints.

Due to the time, cost and complexity of administering software upgrades, in many instances patches necessary to remediate vulnerabilities go unapplied. ML processes have become increasingly available for cyber security management. Reference [68] discusses a system that includes (i) a security and vulnerability analysis processor, (ii) one or more endpoint devices in communication with the security and vulnerability analysis processor through a communication network, and (iii) a vulnerability data ingestion processor configured to obtain, from one or more data sources, security data associated with the one or more endpoint devices. The security and vulnerability analysis processor includes an MLM configured to generate predictions about the risk impact of conducting vulnerability remediations to a particular endpoint device of the one or more endpoint devices, wherein the MLM is trained using a training set comprising the security data associated with the one or more endpoint devices. This system includes the ability to support automated application of one of a security patch or software update to the one or more endpoint devices. The endpoint devices may generate instruction execution telemetry data representing the operating status of processes running on the endpoint devices, and the telemetry data may be fed back to the MLM to generate modified predictions about the risk impact of conducting vulnerability remediations to the particular endpoint device. Figure 6.19 illustrates the environment

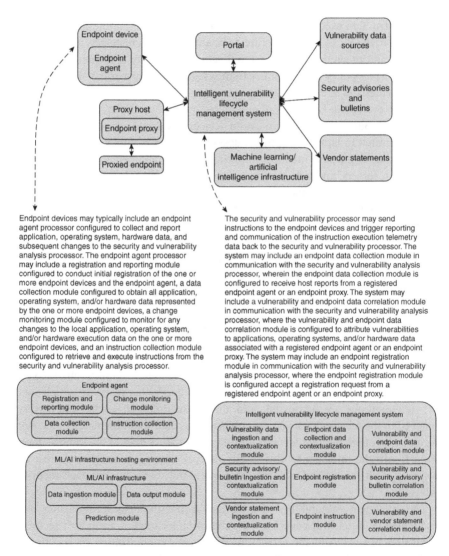

Figure 6.19 Vulnerability lifecycle management system. *Source:* Reference [68]/U.S. Patent/public domain.

where a vulnerability lifecycle management system is applicable. The system is able of (i) ingesting and then contextualizing one or more of security advisories and bulletins which include but are not limited to patches fixing vulnerabilities and/or bugs in vendor software, OSs, and/or hardware; (ii) ingesting and then contextualizing one or more vendor statements regarding vulnerability applicability to the vendor's software, OS, and/or hardware; and (iii) correlating the contextualized vulnerability data with the contextualized application, OS, and/or hardware data of the endpoints to generate a set of vulnerability afflictions.

The instruction execution telemetry data received from an endpoint agent is passed on to the ML/AI infrastructure, to be analyzed and create new and updated learnings and predictions into the intelligent vulnerability management system and the portal's policy engine for use in the creation of new manual or automated patching rules and to inform existing manual or automated patching rules of new data to be evaluated. Examples of endpoints or endpoint devices include a wide range of computing and communication devices connected to a network, such as, but not limited to laptop computers, desktop computers, mobile phones, smart phones, special purpose mobile computing devices, tablet computers, wearable computers, servers, and virtual computing environments.

For endpoints that are unable to receive the installation of an agent, for example but not limited to some network devices such as switches, routers, firewalls, and wireless access points and back-office equipment such as Voice over Internet Protocol (VoIP) phones, or printers, an *endpoint proxy* can be hosted in a proxy host environment either as a virtualized appliance on a hypervisor or a physical appliance. The endpoint proxy will either actively query or receive data from each proxied endpoint for application, OS, and hardware data over well-known information gathering protocols, such as but not limited to Simple Network Management Protocol (SNMP) and Representational State Transfer (REST) queries.

When a fix for a vulnerability is published by the vendor responsible for the patch, availability of the patch will typically be announced via a security advisory or bulletin; security advisory publications usually include, at a minimum, the version or release number of the affected binary, OS, and/or hardware for which a set of vulnerabilities are addressed, and the enumeration of each vulnerability patched in the release. Each vendor publishes this data about the patch in its own format, and often different formats for different product lines of the same vendor. As a result, the intelligent vulnerability lifecycle management system must be aware of each vendor and its format(s) to accurately contextualize each security advisory.

The intelligent vulnerability lifecycle management system accomplishes this awareness by initially discovering the publication location of a vendor's security advisories and analyzing their structure, format, and content either via traditional web scraping tools, ML algorithms, such as but not limited to natural language processing, or a combination of the two, depending on the complexity of the security advisory. After the intelligent vulnerability lifecycle management system ingests and contextualizes new vulnerability data, security advisories and bulletins, and/or vendor statements the system then correlates the new data from the data sources against existing data to update the attribution of new vulnerabilities, patch availability, and vendor statements to each endpoint device and proxied endpoint, which may also be referred to collectively herein interchangeably as an endpoint's overall vulnerability profile.

MLMs used to generate the predictions (about the risk impact of conducting vulnerability remediations to a particular endpoint device) include but are not limited to linear regression, logistic regression, decision tree, AdaBoost, random forest, gradient boosted trees, and feed forward NNs. The MLM is configured to make predictions about the risk and/or impact of conducting vulnerability remediations, for example, applying a software update or a security patch to an endpoint. Refer to Ref. [68] for an extensive discussion of this system.

A. Basic Glossary of Key Security Terms and Concepts

Term	Definition/explanation/concept
Adware	A program that displays advertisements. Adware may display advertisements without user's consent and/or it can infect the organization's computer with malware; for example, if rogue sites are visited.
Availability (A)	Making sure that devices are not prevented from functioning properly and/or performing their function; or, making sure they are not made to operate in an improper or compromised manner. For example, the devices might become infected with viruses or worms or degraded via other debilitating intrusions and/or exploited through weaknesses in the operating system (OS), software utilities, packaged microcode or applications. The term "no repudiation" has also been used in the context of availability.
CNN BiLSTM	A hybrid bidirectional LSTM and CNN architecture. In the original formulation applied to named entity recognition, it learns both character-level and word-level features. The CNN component is used to induce the character-level features [69].
Confidentiality (C)	Making sure that the data packets are not intercepted and examined; also, making sure that the host is not corrupted to the point that a hacker can appropriate data, credentials, information, or configuration parameters (keeping the data safe from being divulged by/to unauthorized agents).
Curse of dimensionality	Issue impacting HMMs where the statistical learning algorithms suffer in cost exponentially as the volume of data grows. The HMMs tend to fail on a large dimension state space: as the incoming network traffic present a large hidden state space, the algorithm falls victim to the curse of dimensionality.
Data management schemas	Cybersecurity threat protection mechanisms based on a security domain which can contain one or more database objects. Access to the database objects can be controlled by granting access privileges to each user or role, where a role can include a user, a manager, or an administrator.
Database firewall (DBF)	A kind of application firewall that monitors database traffic to detect and protect against database-specific attacks that principally seek to access sensitive data held in the databases. DBFs also allow for the monitoring and auditing of all access to cloud databases via logs.
Evolutionary neural networks (ENNs)	NNs that support evolutionary computation (EC). EC is an optimization procedure that can mimic natural selection and biological evolution; it utilizes iterative progress, for example, growth or development in a population – thereafter, the population is selected in a guided random search (typically using parallel processing) to pursue a desired end task.

Term	Definition/explanation/concept
Exploit	Malicious code that takes advantage of operating system or application software vulnerabilities (bugs) that were unwittingly created by developers (perhaps because of insufficient testing) to infect a computer or perform some harmful activity.
Genetic algorithm (GA)	A heuristic algorithm suggested, hinted, or inspired by the process of natural selection. It entails a genetic representation of the solution domain and a fitness function to evaluate the solution domain.
Granular enforcement	Granular enforcement is another term for what zero trust accomplishes: authentications for very specific actions.
Hardware Root of Trust (HRoT)	A device used to describe a hardware element that provides RoT (Root of Trust) functions. The HRoT (i) performs device authentication to ensure that hardware has not been tampered with; (ii) verifies the authenticity of software, particularly boot images, to ensure they have not been tampered with; (iii) provides One-Time Programmable (OTP) memory for secure key storage to facilitate encryption; and (iv) ensures that the system is able to be brought into a known and trusted state [67].
Integrity (I)	Making sure that the packets received (or stored) have not been altered in an unauthorized manner (making sure the data is not modified by unauthorized agents).
Intrusion detection	NIST definition: "The process of monitoring the events occurring in a computer system or network and analyzing them for signs of intrusions, defined as attempts to compromise the confidentiality, integrity, availability, or to bypass the security mechanisms of a computer or network."
Intrusion detection system (IDS)	A system which addresses or automates intrusion detection.
Intrusion prevention system (IPS)	A network security tool that continuously monitors a network for malicious activity and takes action to prevent it.
Keylogger	A program that surreptitiously sends keystrokes or screen shots to an attacker.
Layer 7 firewall	A firewall that can examine packet contents to use more of the data within those contents to define authentication criteria.
Least privilege access	The practice of limiting even trusted users to only the specific applications, services, and data for which they have an immediate need.
Malicious cyberattacks types	Malicious cyberattacks include malware attacks, hacking attacks, distributed denial of service (DDoS) attacks, person in the middle attacks, and so on.

(Continued)

(Continued)

Term	Definition/explanation/concept
Malware	Software that may easily end up on the organization's system designed to cause damage to a user's computer, server, or network. Malware includes viruses, worms, and trojans.
Man-in-the-middle attacks	Attacks where the communications between an unwitting victim and a legitimate website are monitored to harvest personal information, usernames, and passwords.
Micro-segment	A micro-segment is a smaller, secured area within a larger network that is protected by a micro-perimeter. Micro-segments can be used to apply granular access control to specific workflows.
Monitoring tool	Software that monitors the activity of the organization's computer user, typically by capturing keystrokes or screen images.
Multi-factor authentication	The requirement for presentation of more than one credential or attributes to gain access to an ICT resource.
Next-generation firewall (NGFW)	A security entity that monitors the traffic going out to the Internet – across web sites, email accounts, and SaaS. It is protecting the user (versus the web application). An NGFW will enforce user-based policies and will add context to security policies in addition to adding features such as URL filtering, anti virus/anti malware, and potentially its own IPS. While a WAF is typically a reverse proxy (used by servers), NGFWs are often forward proxies (used by clients such as a browser) [21].
Password stealer	Malware that works in conjunction with a *keylogger* and is specifically designed to copy and transmit personal information, such as user names and passwords.
Phishing	A popular method of credential theft (for example user IDs, passwords, PINs, credit card numbers and so on), that tricks users into revealing this information. Phishers utilize phony websites or deceptive email messages that mimic some trusted business by the targeted user to appropriate the credentials and then make nefarious use. This may involve web links or attachments embedded in the deceiving emails.
Protect surface	Protect surface refers to any asset that needs to be protected.
Random *k*-labelsets ensemble (RAkEL)	A multi-label learning strategy that integrates many single-label learning models. Each single-label model is constructed using a label powerset (LP) technique based on a randomly generated size-*k* label subset [70].
Rogue Security Software	Software that is advertised to be a security free-ware but that may attempt to socially engineer the user into participating in a fraudulent transaction, or may, of its own, spawn other undesirable system-level activities.

Term	Definition/explanation/concept
Root of Trust (RoT)	A trusted module that provides a set of functions that are trusted by other modules in a computing environment, such as an operating system (OS). It serves as an independent computing module providing authenticity in a computer or mobile device in which it is embedded.
Scope of vulnerabilities	Operating system vulnerabilities, browser, and application vulnerabilities.
Security tools	Tools include antivirus, anti-phishing, and anti-cryptojacking applications, tools for threat hunting and threat intelligence, identity verification, endpoint protection, forensic investigation, and incident management. Many of these can be ML-based.
Segmentation gateway	Segmentation is a term for reorganizing a larger protect surface. An example is dividing an entire network into smaller protect surfaces defined by value, use, workflow traffic, and other factors. A segmentation gateway is (in effect) a firewall that protects a specific segment within a larger network [50].
Self-encrypting drives (SED)	Data storage device with built-in cryptographic processing that may be utilized to encrypt and decrypt the stored data occurring within the device and without dependence on a connected information system.
SMS authentication	A popular additional (multi-) factor added to user authentication today: users receive SMS codes that they provide to a network or service to prove their identity.
Social engineering	A set of techniques that frustrate security mechanisms by exploiting human nature. For example, this may entail receiving a phone call from someone posing as a representative from one's credit card company, or a vendor, or a government agency; or, it may entail email messages that ask the recipient to click the attachment which, in turn, results in malware being installed on the organization's system. The goal of a social engineering attack is to get the targeted user to perform an action of the attacker's choice.
Spam	Unsolicited email sent to a large distribution, often to upload malware, either by attaching the malware to email messages or by sending a message containing a link to the malware.
Spyware	Software that may end-up on the organization's computer that surreptitiously collects information, such as the websites a user visits or other critical data. Installation typically occurs without the user's knowledge.
Tokenizer	A module, feature, or method that can parse raw emails to extract email objects such as headers, textual part, HTML part, attachments, and so on.

(Continued)

(Continued)

Term	Definition/explanation/concept
Trojan	A self-contained program that takes malicious actions on the computer, possibly *facilitating additional penetrations at a later time.*
Trojan downloader/ dropper	A form of trojan that installs other malicious files to a computer of the organization that it has infected, either by downloading these files from a remote computer or by extracting them directly from a copy contained in its own code.
Trusted platform module (TPM)	A specialized device for hardware authentication. The TPM comprises a dedicated microcontroller that provides secure storage, key generation, and cryptographic operations.
Virus	Malware that replicates itself from one system to another. It may infect other files in the computer to facilitate the execution of the malware code when these, perhaps more common files are accessed.
Web application firewall (WAF)	A security entity that protects the application layer. It is specifically designed to analyze each HTTP/S request at the application layer. It is typically user, session, and application aware, cognizant of the web apps behind it and what services they offer. One can think of a WAF as the intermediary between the user and the app itself, analyzing all communications before they reach the app or the user.
Worm	Malware that spreads by autonomously propagating copies of itself through email, instant messaging (IM), or peer-to-peer (P2P) applications.
Zero trust networking (ZTN)	An environment that operates on a security model that establishes trust through continuous authentication and continuous monitoring of every network access attempt. ZTN abrogates the traditional model of assuming everything in an intranet can be trusted: the ZTN policy is "never trust, always verify".

References

1 O. Powell, "The Biggest Data Breaches And Leaks of 2022", Dec. 9, 2022, https://www.cshub.com/attacks/articles/the-biggest-data-breaches-and-leaks-of-2022. Accessed Mar. 1, 2023

2 E. Griffith, "The Worst Data Breaches in 2022: Were You a Victim?", PC Magazine, Jan. 30, 2023, https://www.pcmag.com/news/cybercrime-in-2022-fewer-data-breaches-but-more-victims, Accessed Mar. 1, 2023.

3 C. Caminiti, "More Automation, Not Just Additional Tech Talent, Is What is Needed to Stay Ahead of Cybersecurity Risks", Technology Executive Council,

Mar. 2 2023, https://www.cnbc.com/2023/03/02/cisos-need-more-automation-to-stay-ahead-of-cybersecurity-risks.html, Accessed Mar. 5, 2023.

4 D. Minoli, B. Occhiogrosso, *et al.*, "Security Considerations for IoT Support of E-Health Applications, Chapter 16", in *Internet of Things: Challenges, Advances and Applications*, Chapman & Hall/CRC Computer and Information Science Series, Q. Hassan, A. R. Khan, S. A. Madani, (Editors), CRC Press, Taylor & Francis, 2018. ISBN: 9781498778510.

5 D. Minoli, B. Occhiogrosso, *et al*, "IoT Security (IoTSec) Considerations, Requirements, and Architectures", in 2017 14th IEEE Annual Consumer Communications & Networking Conference (CCNC), Las Vegas, NV, USA, Jan. 8–11 2017. IEEE catalog number: CFP17CCN-CDR, ISBN: 978-1-5090-6195-2, ISSN: 2331-9860.

6 D. Minoli, B. Occhiogrosso, *et al*, "IoT Security (IoTSec) Mechanisms for e-Health and Ambient Assisted Living Applications", in The Second IEEE/ACM International Workshop on Safe, Energy-Aware, & Reliable Connected Health (SEARCH 2017) (collocated with CHASE 2017, Conference on Connected Health: Applications, Systems, and Engineering Technologies), Philadelphia, July 2017.

7 D. Minoli, "Positioning of Blockchain Mechanisms in IoT-powered Smart Home Systems: A Gateway-Based Approach", Elsevier IoT Journal, Special Issue on IoT Blockchains, Nov. 2019, https://www.sciencedirect.com/science/article/pii/S2542660519302525.

8 D. Minoli, B. Occhiogrosso, "Blockchain-enabled Fog and Edge Computing: Concepts, Architectures and Smart City Applications", in *Blockchain-enabled Fog and Edge Computing: Concepts, Architectures and Applications*, M. M. Rehan, M. H. Rehmani, (Editors), Routledge Taylor & Francis Group, Boca Raton, Florida, 2021.

9 D. Minoli, "Editor-In-Chief, Special Issue on Blockchain Applications in IoT Environments, Overview", Elsevier IoT Journal, Nov. 2019, https://www.sciencedirect.com/science/article/pii/S2542660519302537.

10 D. Minoli, B. Occhiogrosso, "Blockchain Mechanisms for IoT Security", Elsevier IoT Journal, Vol. 1, Issue 1, Summer 2018.

11 D. Minoli, B. Occhiogrosso, E. Coffy, System and Method for a Uniform Measure and Assessment of an Institution's Aggregate Cyber Security Risk and of the Institution's Cybersecurity Confidence Index. U.S. Patent 20170134418, May 11, 2017.

12 D. Minoli, "Addressing Cybersecurity Risks", in NJ SBDC Cybersecurity Conference at TCNJ, Ewing Township, NJ, Apr. 20, 2015.

13 D. Minoli, J. Kouns, *Information Technology Risk Management in Enterprise Environments: A Review of Industry Practices and a Practical Guide to Risk Management Teams*, Wiley, 2010.

14 D. Minoli, J. Kouns, *Security in an IPv6 Environment*, Taylor and Francis, 2009.

15 D. Minoli, J. Carovana, *Minoli-Cordovana Authoritative Computer and Network Security Dictionary*, Wiley, 2006.

16 S. Akhter; S. Pandey, *et al*, Detection of Phishing Websites Using Machine Learning, U.S. Patent 20230065787, Mar. 2, 2023. Uncopyrighted material.

17 W. K. Zegeye, R. A. Dean, Method And System For Intrusion Detection, U.S. Patent 11,595,434, Feb. 28, 2023. Uncopyrighted material

18 J. McCarthy, R. Bermans, D. B. McKinley, Integrated Cybersecurity Threat Management. U.S. Patent 20230068946. Mar. 2, 2023. Uncopyrighted material.

19 Staff of dcs-news.com, "PLC, DCS, SCADA, HMI – What Are the Differences?", Dec. 17, 2018, http://dcs-news.com/plc-dcs-scada-hmi-differences. Accessed Apr. 3, 2023.

20 The OWASP® Foundation, https://owasp.org/www-project-top-ten/.

21 Staff, F5, "Web Application Firewall (WAF)", 2023, https://www.f5.com/glossary/web-application-firewall-waf. Accessed Mar. 5, 2023.

22 F. El-Moussa, B. Azvine, Malware Detection, U.S. Patent 11,586,733, Feb 21, 2023. Uncopyrighted material.

23 A. Dorri, S. S. Kanhere, *et al*, "Blockchain for IoT Security And Privacy: The Case Study Of A Smart Home", in Pervasive Computing and Communications Workshops (PerCom Workshops), 2017 IEEE International Conference on, Kona, HI, USA, Mar. 13–17, 2017, https://doi.org/10.1109/PERCOMW.2017.7917634.

24 A. Ouaddah, A. A. Elkalam, A. A. Ouahman, "FairAccess: A New Blockchain-based Access Control Framework for the Internet of Things", Security and Communication Networks, Dec. 2016, Wiley Online Library, Volume 9, Issue 18, Pages: 5943–5964, 2016 10.1002/sec.1748.

25 ISO, Introduction to ISO JTC1/WG10, June 2015, ISO materials, http://iot-week.eu/wp-content/uploads/2015/06/07-JTC-1-WG-10-Introduction.pdf, Accessed Mar. 6, 2023.

26 M. Pilkington, "Blockchain Technology: Principles and Applications", in *Research Handbook on Digital Transformations*, F. X. Olleros, M. Zhegu, (Editors), Edward Elger Publishing, Northampton, MA, 2016.

27 D. Tapscott, A. Tapscott, *Blockchain Revolution: How the Technology Behind Bitcoin is Changing Money*, Penguin Random House LLC, New York, NY, 2016.

28 A. Kosba, A. Miller, *et al*, "Hawk: The Blockchain Model of Cryptography and Privacy-Preserving Smart Contracts", in 2016 IEEE Symposium on Security and Privacy (SP), San Jose, CA, USA, May 22–26, 2016, https://doi.org/10.1109/SP.2016.55.

29 A. Wright, P. De Filippi, "Decentralized Blockchain Technology and the Rise of Lex Cryptographia", Mar. 10, 2015, Available at SSRN: https://ssrn.com/abstract=2580664, Accessed Mar. 7, 2023.

30 X. Xu, C. Pautasso, *et al*, "The Blockchain as a Software Connector", in 2016 13th Working IEEE/IFIP Conference on Software Architecture (WICSA), Venice, Italy, Apr. 5–8, 2016, https://doi.org/10.1109/WICSA.2016.21.

31 K. Brünnler, D. Flumini, T. Studer, "A Logic of Blockchain Updates", in *Logical Foundations of Computer Science. LFCS, Lecture Notes in Computer Science,* Vol. 10703, S. Artemov, A. Nerode, (Editors), Springer, Cham, 2018.

32 S. N. Artemov, "Explicit Provability and Constructive Semantics", *Bullet. Symbolic Logic.* 2001; 7(1): 1–36.

33 S. I. Matsuo, "How Formal Analysis And Verification Add Security To Blockchain-Based Systems", in Formal Methods in Computer Aided Design (FMCAD), Vienna, Austria, Oct. 2–6, 2017.

34 J. Garay, A. Kiayias, N. Leonardos, "The Bitcoin Backbone Protocol: Analysis and Applications", in Proceedings of Eurocrypt, Sofia, Bulgaria, 26–30 Apr. 2015, 2015.

35 R. Dennis, G. Owenson, B. Aziz, "A Temporal Blockchain: A Formal Analysis", in 2016 International Conference on Collaboration Technologies and Systems (CTS), Orlando, FL, USA, 31 Oct.–4 Nov., 2016.

36 B. Huang, Z. Liu, *et al.*, "Behavior Pattern Clustering in Blockchain Networks", *Multimed. Tools Appl.* 2017; 76: 20099, https://doi.org/10.1007/s11042-017-4396-4.

37 M. K. Awan, A. Cortesi, "Blockchain Transaction Analysis Using Dominant Sets", in *Computer Information Systems and Industrial Management. CISIM. Lecture Notes in Computer Science*, Vol. 10244, K. Saeed, W. Homenda, R. Chaki, (Editors), Springer, Cham, 2017.

38 A. Ouaddah, A. A. Elkalam, A. A. Ouahman, "Towards a Novel Privacy-Preserving Access Control Model Based on Blockchain Technology in IoT", in *Europe and MENA Cooperation Advances in Information and Communication Technologies*, Springer International Publishing, 2017 pp. 523–533.

39 X. Yue, H. Wang, *et al.*, "Healthcare Data Gateways: Found Healthcare Intelligence on Blockchain with Novel Privacy Risk Control", *J. Med. Syst.* 2016; 40: 218, https://doi.org/10.1007/s10916-016-0574-6.

40 G. Magyar, "Blockchain: Solving The Privacy And Research Availability Tradeoff for EHR Data: A New Disruptive Technology In Health Data Management", Neumann Colloquium (NC), 2017 IEEE 30th, Budapest, Hungary, Nov. 24–25, 2017, https://doi.org/10.1109/NC.2017.8263269.

41 M. Samaniego, R. Deters, "Blockchain as a Service for IoT", in 2016 IEEE International Conference on Internet of Things (iThings) and IEEE Green Computing and Communications (GreenCom) and IEEE Cyber, Physical and Social Computing (CPSCom) and IEEE Smart Data (SmartData), Chengdu, China, Dec. 15–18, 2016.

42 Y. Zhang, J. Wen, "The IoT Electric Business Model: Using Blockchain Technology for the Internet of Things", *Peer-to-Peer Netw. Appl.* 2017; 10(4): 983–994.

43 S. Huckle, R. Bhattacharya, *et al.*, "Internet of Things, Blockchain and Shared Economy Applications", *Procedia Comput. Sci.* 2016; 98: 461–466. ISSN: 1877-0509.

44 H. M. Kim, M. Laskowski, "Toward an Ontology-driven Blockchain Design for Supply-chain Provenance", Wiley Online Library, Mar. 28, 2018. https://doi.org/10.1002/isaf.1424.

45 J. Sun, J. Yan, K. Z. K. Zhang, "Blockchain-based Sharing Services: What Blockchain Technology can Contribute to Smart Cities", in Financial Innovation, Springer, Dec. 2016, https://doi.org/10.1186/s40854-016-0040-y.

46 S. Huh, S. Cho, S. Kim, "Managing IoT Devices Using Blockchain Platform", in 2017 19th International Conference on Advanced Communication Technology (ICACT), Bongpyeong, South Korea, Feb. 19–22, 2017.

47 M. Samaniego, R. Deters, "Using Blockchain to Push Software-Defined IoT Components Onto Edge Hosts", in BDAW '16 Proceedings of the International Conference on Big Data and Advanced Wireless Technologies Article No. 58, Blagoevgrad, Bulgaria, Nov. 10–11, 2016.

48 R. C. Merkle, "A Digital Signature Based on a Conventional Encryption Function", in Advances in Cryptology, CRYPTO '87. Included in Lecture Notes in Computer Science, 293, p. 369, https://doi.org/10.1007/3-540-48184-2_32.

49 I. Bashir, *Mastering Blockchain*, Packt Publishing, Birmingham, UK, 2017. ISBN: 978-1-78712-544-5.

50 Cisco Staff, "What Is Zero-Trust Networking?", https://www.cisco.com/c/en/us/solutions/automation/what-is-zero-trust-networking.html. Accessed Mar. 21, 2023.

51 R. Rais, C. Morillo, *et al.*, *Zero Trust Networks*, 2nd Edition, O'Reilly Media, Inc., 2023. ISBN: 9781492096597.

52 S.-S. Zhao, L. Xu, M. Iqba, "Zero-trust Security for Industry 4.0", in *Wireless Communications and Mobile Computing*, Wiley/Hindawi, 2023.

53 J. Leo, J. Suganthi, *et al*, Adaptive Anomaly Detector, U.S. Patent 11,593,714, Feb. 28, 2023. Uncopyrighted material.

54 S. Muthyala, "Top 10 Applications of Machine Learning in Cybersecurity", www.analyticsinsight.net, https://www.analyticsinsight.net/top-10-applications-of-machine-learning-in-cybersecurity/. Accessed Mar. 1, 2023.

55 R. D. Bean, G. W. Rice, System and Method for Neural Network Based Detection of Cyber Intrusion Via Mode-Specific System Templates, U.S. Patent 20230068909, Mar. 2, 2023. Uncopyrighted material.

56 P. Larue, P. Jallon, B. Rivet, "Modified k-mean Clustering Method of HMM States for Initialization of Baum-Welch Training Algorithm", in 2011 19th European Signal Processing Conference, Barcelona, Spain, 2011, pp. 951–955.

57 W.K. Zegeye, R. A. Dean, F. Moazzami, "Multi-Layer Hidden Markov Model Based Intrusion Detection System", in Machine Learning & Knowledge Extraction, Dec. 25, 2018, pp. 265–286, Issue Date Jan. 2019.

58 D. H. Lee, D. Y. Kim, J. I. Jung, "Multi-Stage Intrusion Detection System Using Hidden Markov Model Algorithm", in Proc. of IEEE International Conference on

Information Science and Security (ICISS), Seoul, South Korea, Jan. 10–12, 2008, pp. 72–77.

59 D. Ourston, S. Matzner, W. Stump, "Applications of Hidden Markov Models to Detecting Multi-Stage Network Attacks", in 36th Annual Hawaii International Conference on System Sciences, Jan. 2003, Big Island, Hawaii. Jan. 6, 2003 to Jan. 9, 2003. https://doi.org/10.1109/HICSS.2003.1174909

60 Staff Canadian Institute for Cybersecurity, "Intrusion Detection Evaluation Dataset (CIC-IDS2017)", 2017, https://www.unb.ca/cic/datasets/ids-2017.html. Accessed Apr. 5, 2023.

61 X. MahaLakshmi, N. Swapna Goud, G. Vishnu Murthy, "A Survey on Phishing and It's Detection Techniques Based on Support Vector Method (SVM) and Software Defined Networking (SDN)", *Int. J. Eng. Adv. Technol. (IJEAT)*. 2018; 8(2S): 498–503. ISSN: 2249-8958.

62 S. Marchal, K. Saariet, *et al*, "Know Your Phish: Novel Techniques for Detecting Phishing Sites and Their Targets", Apr. 25, 2016, arxiv org.

63 I. Fette, N. Sadeh, A. Tomasic, "Learning to Detect Phishing Emails", CMU-ISRI-06-112, Jun. 2006, also as Carnegie Mellon Cyber Laboratory, Technical Report CMU-CyLab-06-012.

64 E. S. Aung, C. T. Zan, H. Yamana; "A Survey of URL-based Phishing Detection", DEIM Forum, 2019, G2-3.

65 B. Biswas, A. Mukhopadhyay, "Phishing Detection and Loss Computation Hybrid Model, A Machine-learning Approach", *ISACA J.* 2017; 1: 1–8.

66 M. P. Singh, A. Bhardwaj, Methods and Systems for Detecting Phishing Emails Using Feature Extraction And Machine Learning, U.S. Patent 11,595,435. Feb. 28, 2023. Uncopyrighted material.

67 T. Staab, Adaptive Multi-Layer Enterprise Data Protection and Resiliency Platform, U.S. Patent 11,595,411. Feb. 28, 2023. Uncopyrighted material.

68 M. C. Starr, Intelligent Vulnerability Lifecycle Management, U.S. Patent 20230064373 A1, Mar. 2, 2023. Uncopyrighted material.

69 J. P. C. Chiu, E. Nichols, "Named Entity Recognition with Bidirectional LSTM-CNNs", in Transactions of the Association for Computational Linguistics, Jul. 19, 2016, arXiv:1511.08308, https://doi.org/10.48550/arXiv.1511.08308.

70 R. Wang, S. Kwong, *et al.*, "Active k-labelsets Ensemble For Multi-Label Classification", *Pattern Recognition*. 2021; 109: 107583, https://doi.org/10.1016/j.patcog.2020.107583. ISSN: 0031-3203.

7

Current and Evolving Applications to Network Management

7.1 Overview

After a brief review of machine learning (ML) techniques applicable to service provider networks, this chapter looks at examples of ML methods applied to network management, specifically for fault, configuration, accounting, performance, and security (FCAPS) activities [1, 2]. It is not the purpose of this chapter to provide a tutorial on network management, but it aims at assessing some ML methods applied to network management.

FCAPS methods and policies typically include fault management (FM) methods and policies, configuration management (CM) methods and policies, accounting management (AM) methods and policies, performance management (PM) methods and policies, and security management (SM) methods and policies. FM policies may specify policies used to handle device faults. CM policies may specify policies used to configure devices. AM policies may specify policies used for device accounting purposes, such as reporting, billing, and so on. PM policies may specify metrics and policies used to assess the network's behavior from a performance perspective. SM policies may specify policies used to secure devices. Network management systems make use of a variety of tools, applications, and devices to assist network managers in monitoring and maintaining networks, for example, the one shown in Figure 7.1. The Simple Network Management Protocol (SNMP) is typically used to maintain continuous communication with network elements in the intranet or service provider networks and monitor their status.

Prediction, classification, and/or clustering are useful functions being sought by network managers. Table 7.1 is a partial *listing* (by title) of US Patents and Patent Applications in the past five years (before publication) that deal with ML usage in

AI Applications to Communications and Information Technologies: The Role of Ultra Deep Neural Networks, First Edition. Daniel Minoli and Benedict Occhiogrosso.
© 2024 The Institute of Electrical and Electronics Engineers, Inc.
Published 2024 by John Wiley & Sons, Inc.

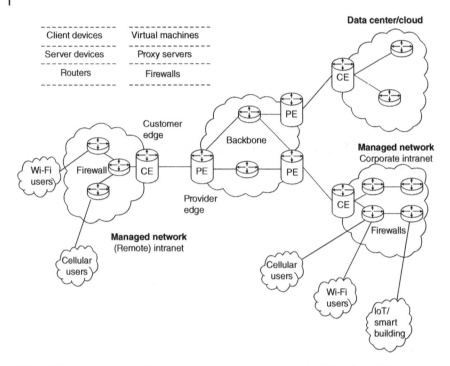

Figure 7.1 General view of a contemporary corporate network: Network and Network Elements.

the network management settings; the list is included to give a sense of the breath and range of applicability of ML in this setting (these are in addition to the ones cited in the reference section at the end of this chapter).

7.2 Examples of Neural Network-Assisted Network Management

This section provides a small sample of some recent examples of ML/neural network (NN)-based systems and tools to assist network managers in their efforts to assure network stability, acceptable performance, and availability. In the network management context, the (labeled) training data used by the NN in the supervised mode include sample network observations that do, or do not, violate a given network health status rule (looking at previously seen patterns that have been labeled as such); an unsupervised model looks instead to whether there are sudden changes in the network behavior. This section is just an exemplary survey and is not intended to be representative or exhaustive of the science.

Table 7.1 Recent US Patents and Patent Applications dealing with ML solutions for network management.

Fault Management (FM)

1) Cross-organizational **network diagnostics** with privacy awareness

2) Adaptive **health status** scoring for network assurance

3) **Monitoring** devices and methods for IP surveillance networks

4) Device **management** for isolation networks

5) Device **profiling** for isolation networks

6) Dynamically adjusting a set of **monitored** network properties using distributed learning machine feedback

7) Intelligent **monitoring** and management of network devices

8) **Automation** system user interface

9) Learning machine–based detection of **abnormal** network performance

10) Integrated cloud system with lightweight gateway for **premises automation**

11) **Alarm** prediction in a telecommunication network

12) Tracking **anomaly** propagation at the network level

13) Method for **data routing** in networks

14) System and method for managing **multimedia communications** across convergent networks

15) Wireless network **monitoring** device

16) Network **node failure** predictive system

17) Network **routing adaptation** based on failure prediction

18) Handling an **event** message in a communications system

19) Automatic **integration** of IoT devices into a network

20) Predicting the impact of network **software upgrades** on machine learning model performance

21) Facilitating temporal data management for **anomalous** state detection in data centers

22) **Telecommunication network** subscriber conversion using cluster-based distance measures

23) Technologies for indicating third party content and resources on **mobile devices**

24) System and method for prescriptive **diagnostics** and optimization of client networks

25) Selective tracking of acknowledgments to **improve network** device buffer utilization and traffic shaping

26) Controller-based **management** of noncompliant power over ethernet devices

27) Multi-application recommendation engine for a remote **network management** platform

---

(Continued)

Table 7.1 (Continued)

28) Dynamic **network management** based on user, device, application, and network characteristics

29) Deployment of deep neural networks (DNN) in embedded devices by means of peer-to-peer **routing** between computational points

30) Sampling traffic telemetry for device classification with distributed probabilistic data structures

31) Intelligent network equipment **failure** prediction system

32) Management of **fault** notifications

33) Intelligent network equipment **failure** prediction system

34) Hard/soft finite state machine (FSM) resetting approach for capturing **network telemetry** to improve device classification

35) Automatic labeling of telecommunication **network data** to train supervised machine learning

36) Deep fusion reasoning engine (DFRE) for prioritizing network **monitoring** alerts

37) **Premises** system automation

38) Machine learning systems and methods to predict **abnormal behavior** in networks and network data labeling

39) Detecting **bug** patterns across evolving network software versions

40) Neural network-assisted computer **network management**

41) Networked computer-system **management** and control

42) Systems and method for replaying and **debugging** live states of network devices

43) Unsupervised learning of local-aware attribute relevance for device classification and clustering

44) Datacenter **power management** through phase balancing

45) Using raw **network telemetry** traces to generate predictive insights using machine learning

46) Determining context and actions for machine learning-detected **network issues**

47) Method and apparatus for intelligent **operation management** of infrastructure

48) Hardware **architecture** determination based on a neural network and a network compilation process

49) Configurable action generation for a **remote network management** platform

50) Intelligent monitoring and testing system for **cable** network

51) Prediction of **network events** via rule set representations of machine learning models

52) Providing customer care based on analysis of **customer care** contact behavior

53) Neural network-assisted computer network **management**

54) Telecommunication network **subscriber conversion** using cluster-based distance measures

55) Automated generation of standard network device **configurations**

Table 7.1 (Continued)

56) Facilitating temporal data management for **anomalous** state detection in data centers

57) Identifying **devices** and device intents in an IoT network

58) **Log classification** using machine learning

59) IoT device application **workload** capture

60) Technologies for configuring and reducing **resource** consumption in time-aware networks and time-sensitive applications

61) Determining a root-cause of a network **access failure** and conducting remediation

62) Detecting and localizing cable plant **impairments** using full band capture spectrum analysis

63) Methods and systems for **managing** facility power and cooling

64) **Control** system user interface

65) Machine learning **command** interaction

Configuration Management (CM)

1) System and method of **admission control** of a communication session

2) **Controlling data routing** among networks

3) Network **provisioning**

4) Cable network **data analytics** system

5) Method and systems for dynamic allocation of **network resources**

6) Premises **management** configuration and control

7) Detection of isolated **changes in network** metrics using smart-peering

8) Access point **registration** in a network

9) **Device type classification** using metric learning in weakly supervised settings

10) Method and system for **routing** user data traffic from an edge device to a network entity

11) **Tracking** of devices across MAC address updates

12) Active labeling of unknown **devices** in a network

13) Network **configuration** change analysis using machine learning

14) Systems, methods and devices for **device fingerprinting** and automatic deployment of software in a computing network using a peer-to-peer approach

15) **Activation** of gateway device

16) Active labeling of unknown **devices** in a network

17) Redrawing **roaming** boundaries in a wireless network

18) **Roaming** and transition patterns coding in wireless networks for cognitive visibility

19) Predicting and forecasting **roaming** issues in a wireless network

(Continued)

Table 7.1 (Continued)

20) **Mobile** management system

21) Edge computing over disaggregated **radio access** network functions

22) Technologies for indicating third party content and resources on **mobile** devices

Accounting Management (AM)

1) Cost optimization of **cloud computing** resources

2) Management of network **configuration** and address provisioning

3) Efficient **allocation** of network resources

4) Generation and management of network **connectivity** information

5) Intelligent data **analytics** collectors

6) Systems and methods for **inventory** discovery in a network

Performance Management (PM)

1) System and method providing **network optimization** for broadband networks

2) Method and a system for controlling traffic **congestion** in a network

3) Latency-based **routing** and load balancing in a network

4) Load **balancing** Internet-of-Things (IoT) gateways

5) Forecasting **network KPIs**

6) Efficient machine learning for network **optimization**

7) Machine learning-based approach to **network planning** using observed patterns

8) **Discovery** and migration planning techniques optimized by environmental analysis and criticality

9) Traffic class-specific congestion signatures for improving **traffic shaping** and other network operations

10) **Degradation** estimation apparatus, computer program, and degradation estimation method

11) Method and apparatus for **load control** of an enterprise network on a campus based on observations of busy times and service type

12) System and method for the collection, generation, and distribution of synthetic **metrics** for computer system management

13) Controlling data **routing** among networks

14) **Optimized** data-over-cable service interface specifications filter processing for batches of data packets using a single access control list lookup

15) Dynamic intent-based **QoS** policies for commands within industrial protocols

16) System and method for prescriptive **diagnostics** and optimization of client networks

Security Management (SM)

(See Chapter 6)

7.2.1 Example of NN-Based Network Management System (Case of FM)

As it should be clear at this juncture, as covered in Chapter 1, and by way of review, NNs comprise a collection of connected artificial "neurons," or cells. Each connection, or "synapse," between cells can transmit a signal from one cell to another. The receiving cell can process the signal(s) and then communicate with other cells connected to it. In typical implementations, the output of each cell, the synapse signal, is calculated by a nonlinear function of its inputs. Cells and synapses may be characterized by weights that vary as learning proceeds, which weights can increase or decrease the strength of the signal that is output. Further, each cell may be characterized by a threshold such that only if the aggregate signal meets the threshold, is the signal output. Typically, cells are organized in layers, with different layers performing different transformations on cell inputs. Signals travel from the first (input) to the last (output) layer within a cell, possibly after traversing the layers multiple times, that is, recurrently. NNs may be "trained" by comparing the network's classification of inputs (which, at the outset, is largely arbitrary) with the known actual classification of the inputs. The errors from each iteration of training may be fed back into the network and used to modify the network's weights. A recurrent neural network (RNN) is a class of NNs where connections between some layers form a directed cycle. This architecture allows an RNN to exhibit dynamic temporal behavior. RNNs can use internal memory to process arbitrary sequences of inputs. When training conventional RNNs, "gradient descent" may be used to minimize the error term by changing each weight in proportion to the derivative of the error with respect to that weight. However, such an approach can encounter the vanishing gradient problem, that is, the gradient can become so small as to effectively prevent the weight from changing its value. Long short-term memory (LSTM) networks are RNNs that avoid the vanishing gradient problem. An LSTM NN can prevent back-propagated errors from vanishing, or conversely, exploding. Instead, errors can flow backward through unlimited layers of the LSTM cell.

Reference [3] presents a network management system that uses LSTMs/RNNs to identify sequences of computer network log entries, indicative of a cause of an event described in a computer network log entry. Sequences of computer network log entries indicative of a cause of an event described in the first type of entry are identified by training LSTM NN to detect computer network log entries of the first type. When it comes to a sequence type of data, for example, computer network logs and computer network telemetry, LSTM networks are capable enough to detect events such as network failures. However, conventional LSTM networks alone leave unanswered which inputs might be related to the detected events and are not helpful in troubleshooting the event from a network management perspective. The system described in [3] provides computer-implemented methods, systems, and computer program products to identify sequences of computer network log entries indicative of a cause of an event described in a computer network log entry. In some implementations, the existence of sequences that indicate an upcoming event can be determined. The system enables one to identify causal chains likely to have led to network failure events and warn of impending network failure events. Such systems may be employed to display such causal chains and warnings to a network operator and also input such causal chains and warnings into downstream automated network management systems that can implement recovery and mitigation strategies, such as (i) changing the configuration of the physical network by disabling certain devices and reassigning the function the disabled devices to other network devices or (ii) creating alternate functionality for the processed performed by the about-to-fail component(s) and then isolating or powering down the about-to-fail components.

This LSTM is characterized by a plurality of ordered cells $F_i = (x_i, c_{i-1}, h_{i-1})$ and a final sigmoid layer characterized by a weight vector w^T. A sequence of log entries x_i is received. A h_i for each entry is determined using the trained F_i. A value of gating function $G_i(h_i, h_{i-1}) = I(w^T(h_i - h_{i-1}) + b)$ is determined for each entry; I is an indicator function, b is a bias parameter – the Indicator function (aka characteristic function) of a subset of a set is a function that maps elements of the subset to 1, and all other elements to 0: **if S is a subset of some set X, one has if and otherwise**. A sub-sequence of x_i corresponding to $G_i(h_i, h_{i-1}) = I(w^T(h_i - h_{i-1}) + b) = 1$ is output as a sequence of entries indicative of a cause of an event described in a log entry of the first type. Figure 7.2 is a block diagram depicting a communications and processing architecture to identify sequences of computer network log entries indicative of a cause of an event described in a computer network log entry.

Managed devices (e.g. network-resident devices), run software that enables them to send alerts, typically in the form of network log entries, when the managed devices detect problems (for example, when one or more user-determined thresholds are exceeded). On receiving these alerts, management entities, executing on the network management system, are programmed to react by executing one, several, or a group of actions, including (i) operator notification, (ii) event logging, (iii) shutdown and replacement of faulty processes and devices, and (iv) other automatic attempts at system repair. Management systems also can poll managed devices over the network to check the values of certain variables, for example using SNMP.

The LSTM RNN is executed as part of the network management function. The call-out cell in Figure 7.2 (bottom) is the "forget gate" cell. It takes as input the outputs h_{t-1} and c_{t-1} of a previous LSTM cell, with the network log entry corresponding to the current cell input x_t, and applies various sigmoid and hyperbolic tangent functions to the inputs and to intermediate products. The first layer of the cell applies a sigmoid gating function below to h_{t-1} and to c_{t-1} and the second layer

Network management typically involves the use of distributed databases, auto-polling of network devices, automatic isolation of problem devices along with replication of the function of troubled devices in other network elements, and high-end workstations generating real-time graphical views of network topology changes, events (including failures), and traffic.

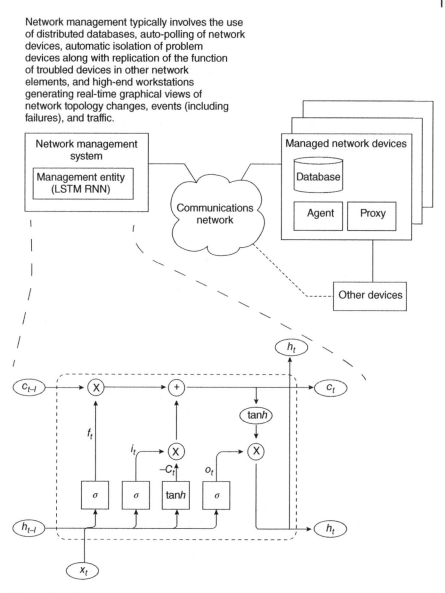

Figure 7.2 Communications and processing architecture to identify sequences of computer network logs. *Source:* Reference [3]/U.S. Patent/public domain.

of the cell applies another sigmoid gating function, as described in detail the reference [3] along with a description of other related computations in the LSTM cell.

The LSTM is trained to detect computer network log entries of the first type in sequences of computer network log entries. Overall, the computer-implemented system comprises the following [3]: (i) training, by one or more computing devices LSTM RNNs to detect computer network log entries of the first type in sequences of computer network log entries, the LSTM network being characterized by a plurality of ordered cells and a final sigmoid layer, the final sigmoid layer receiving the output of the final cell and characterized by a weight vector; (ii) receiving, by one or more computing devices, a sequence of computer network log entries from one or more network devices of the computer network; (iii) determining, by the one or more computing devices, the output value for each log entry in the sequence in accordance with a corresponding trained cell in the plurality of ordered cells; determining, by the one or more computing devices, a value of a gating function based on each cell output, an output of the preceding cell in the plurality of ordered cells, and the weight vector, the gating function yielding "1" for a positive value, and "0" otherwise; and (iv) outputting, by the one or more computing devices, a sub-sequence of the computer network log entries the gating function yielding "1" as a sequence of computer network log entries indicative of a cause of an event described in the computer network log entry of the first type.

7.2.2 Example of a Model for Predictions Related to the Operation of a Telecommunication Network (Case of FM)

Reference [4] defines a processing system that can be used to determine a set of input features via an ML model (MLM) that is deployed in a telecommunication network for a prediction task associated with an operation of the telecommunication network by specifically applying a time series forecast model to a historical dataset to generate a forecast upper bound and a lower bound of the first characteristic of the data source for a time period. The processing system may then detect that the first characteristic exceeds one of the forecasted upper bound or the forecasted lower bound during the time period, and thus generate an alert that an output of the MLM may represent a fault. As is the case in general, the accuracy or performance of the model depends on the data used to train the model. The prediction flow may involve: (i) retrieving data from a database, e.g. a non-Structured Query Language (SQL)-based database, while noting that in many cases, the data obtained may be noisy and need to be preprocessed; (ii) converting data types, e.g. manipulating the data into appropriate form(s) to permit feature engineering to be applied to the data; (iii) feature engineering – a process of manipulating the data retrieved from the database, which may involve removing or adding attributes, normalizing attribute data to a similar scale, and so on; (iv) prediction, e.g. feeding the processed data to the MLM and acquiring results; and (v) constructing a response – the form of the output may depend on the type of the ML task for which the MLM is adapted, as well as the type of response expected by a consuming application.

MLMs may be used for a variety of prediction tasks in telecommunication networks[1]: for example, (i) predicting a level of demand for particular content items or a category of content items (e.g. 4K video, major television broadcast events, such as major national sporting events, a popular television series with a regularly scheduled broadcast times, etc.), which may be used node placement for a content distribution network (CDN), content preplacement at one or more edge nodes, and so on; or (ii) predicting network traffic volumes for backbone links in a nationwide multi-protocol label switching (MPLS) network, which may be used for link load balancing, network traffic rerouting and so forth; or (iii) for intrusion detection, botnet activity detection, denial of service (DoS) attack detection, email or text spam activity detection, fraud detection, subscriber identity module (SIM) hacking and so on, as broadly discussed in Chapter 6; or (iv) to detect fraudulent online purchases, e.g. by providing a propensity (probability) of fraud for each occurrence (broadly, a "fraud score"). See Figure 7.3.

The MLM of [4] entails a machine learning algorithm (MLA) that has been "trained" or configured in accordance with input data (e.g. training data) to perform a particular service (e.g. a prediction task, such as to detect fraud and/or to provide a fraud indicator, or value indicative of a likelihood of fraud). MLAs may include support vector machines (SVMs), e.g. linear or nonlinear binary classifiers, multi-class classifiers, deep learning (DL) algorithms/models, decision tree algorithms/models, e.g. a decision tree classifier, k-nearest neighbor (KNN) clustering algorithms/models, a gradient boosted or gradient descent algorithm/model, such as a gradient boosted machine (GBM), or an XGBoost-based model,

---

1 Telecommunication service provider networks may comprise a core network with components for telephone services, Internet services, and/or video services (e.g., triple-play services, etc.) that are provided to customers and to peer networks. Physical networks may include an IP Multimedia Subsystem (IMS) network; a cellular network such as global system for mobile communications (GSM), long-term evolution (LTE), 5G; an Internet protocol/ multi-protocol label switching (IP/MPLS) backbone network utilizing Session Initiation Protocol (SIP) for circuit-switched and Voice over Internet Protocol (VoIP) telephony services; an Internet Protocol Television (IPTV) network; Internet Service Provider (ISP) network; or a Software Defined Network (SDN) using Network Function Virtualization (NFV); and/or a cloud-services-supporting network, among others. A wireless network that has become available for use by enterprises is a Citizens Broadband Radio Service (CBRS) network which utilizes the radio band of 3550–3700 MHz.

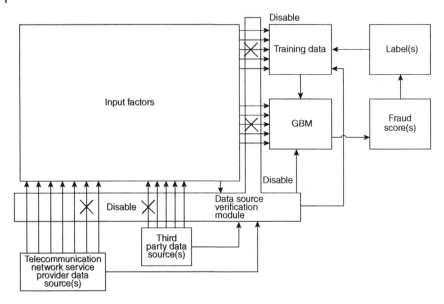

Figure 7.3 Telecommunication network machine learning data source fault detection and mitigation. *Source:* Reference [4]/U.S. Patent/public domain.

and so forth. For each characteristic of interest for a data source, the system may generate a forecasted upper bound and a forecasted lower bound for a time period (e.g. a next and/or upcoming day or other time period) for the characteristic using a time series forecast model. For instance, time series forecast models may be retrained nightly for each characteristic, or for an upper bound and a lower bound for each characteristic, respectively (e.g. these forecast upper and lower bounds may be the "expected characteristics" for the data source). In one example, a time series forecast model may comprise a Seasonal Naïve (S-Naïve) model, a seasonal decomposition model, an autoregressive model, a moving average model, an exponential smoothing model, or a dynamic linear model. A time series forecast model may comprise a Facebook® Prophet model, an Autoregressive Integrated Moving Average Model (ARIMA), a Seasonal ARIMA (SARIMA) model, a Neural Network Auto-Regression (NNETAR) model, an RNN model, and so on. In addition, it should be noted that while a same time series forecast model may be used for forecasting expected upper and/or lower bounds for different characteristics of data of a data source, or different characteristics of different data sources, the parameters (e.g. "tuning" parameters) may be set differently. For instance, for a data source known to not experience a strong seasonality, the seasonality factor of a SARIMA model may be weighted less than for a data source that is known to experience a strong seasonality.

7.2.3   Prioritizing Network Monitoring Alerts (Case of FM and PM)

Typically, a network operator may use a service that monitors a network for detecting anomalies in the network, predicting and/or detecting problems in the network, initiating corrective measures, and/or alerting network administrators and other interested parties as to the health and status of the network. Systems supporting these monitoring services apply analytics to captured network information, to assess the health of the network – for example, a network monitoring service may track and assess metrics such as available bandwidth, packet loss, jitter, and so on, to ensure that the experiences of users of the network are not impaired or deteriorated. As networks continue to evolve, so too will the number of applications and network services present in a given network, as well as the number of metrics available to monitor the network.

Automated systems for monitoring network operations often produce a large number of alerts, many of which are of relatively little value to a network administrator. It has been documented that many network administrators today either ignore important alerts, because of the large volume of alerts presented, or even disable certain alerting functions that could indicate actual problems in the network. Worse, the number of alerts is expected to grow over time, as networks increase in complexity and their corresponding monitoring systems become more capable. There is thus an intrinsic need for prioritizing network monitoring alerts. Reference [5] describes a service that uses data regarding the detected anomalies as input to one or more MLMs. What is needed is a network monitoring system that has the ability to: (i) systematically identify which alerts should actually be reported (i.e. separate the signal-to-noise during alerts generation) and (ii) optimize the available resource capacity that users of the system have to react to alerts by automating its relevancy, potential impact, and/or targets of detected issues. These capabilities can be used to target data to specific users or groups that could be impacted by a network anomaly. For example, problems occurring at the device or component level can be manifested as issues in packet flow, interfaces, call drops, or in any other way (much of the complexity of the network environment is because of the dependencies across the different network layers and the capabilities required to deliver a network-based service). The service described in [5] maps, using a conceptual space, outputs of the one or more MLMs to symbols. The service applies a symbolic reasoning engine to the symbols, to rank the anomalies. The service then sends an alert for a particular one of the detected anomalies to a user interface, based on its corresponding rank. It uses a Deep Fusion Reasoning Engine (DFRE) for prioritizing network monitoring alerts. The service maps, using a conceptual space, outputs of the one or more MLMs to symbols. The service applies a symbolic reasoning engine to the symbols to rank the anomalies. The service sends an alert for a particular one of the detected anomalies to a user interface, based on its corresponding rank.

Figure 7.4 illustrates an example network monitoring system. As shown in the figure, the core of a network monitoring system can be a cloud-based network monitoring service that leverages ML in support of cognitive analytics for the network, predictive analytics (e.g. models used to predict user experience, and so on), troubleshooting with root cause analysis, and/or trending analysis for capacity planning. Generally, architecture may support both wireless and wired networks of all types. Figure 7.5 illustrates an example of the DFRE architecture for prioritizing network monitoring alerts.

Figure 7.6 illustrates an example simplified procedure for prioritizing network alerts using a DFRE.

The ML-based analyzer may include a number of MLMs to perform function of the system such as for cognitive analytics, predictive analysis, and/or trending analytics as depicted in Table 7.2.

DFRE is a Cisco-sponsored project that aims at developing AI solutions combining data fusion, DL, and reasoning; it creates solutions powered with explainability, causal modeling, and tail event handling, taking advantage of state-of-the-art techniques. The system of [5] uses a cognitive automation approach, to build a list of top issues/alerts in a network monitoring service given a semantic measure of the impact of very large numbers of issues generated per hour (alerts), which can be on the order of thousands of alerts raised per hour. A key feature and expected performance measure of the techniques herein is to increase the semantic signal-to-noise ratio (SNR) between true alerts and fake alerts. The DFRE can leverage deep fusion multimodal sensory inputs to produce a sub-symbolic multi-modal data fusion model. The DFRE-based alert prioritization mechanism operates on four basic assumptions:

1) Network devices are sufficiently well instrumented and the network monitoring system (upstream of the DFRE) is able to directly monitor the most essential layers and components of the network. This allows an issue at a certain layer to be detected at the same layer where it was probed. An operator can trade off false positives and/or spurious correlations versus prediction horizon by ignoring the predictions where cause-and-effect have a high degree of separation.
2) Network devices are relatively stable in their design. Hence, they do not require constant re-modeling of their functionalities.
3) The impact of a detected anomaly on affected systems can be expressed by relatively compact and stable functional decompositions.
4) Alerts can be enriched with extra metadata, when available, allowing a more granular and precise ranking and distribution.

Refer to reference [5] for additional information on this model.

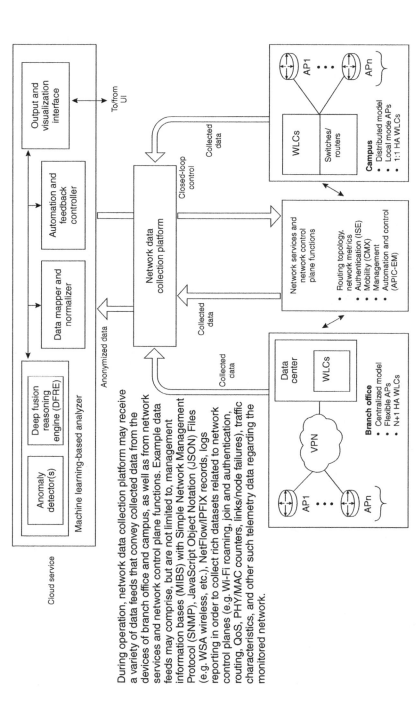

During operation, network data collection platform may receive a variety of data feeds that convey collected data from the devices of branch office and campus, as well as from network services and network control plane functions. Example data feeds may comprise, but are not limited to, management information bases (MIBS) with Simple Network Management Protocol (SNMP), JavaScript Object Notation (JSON) Files (e.g. WSA wireless, etc.), NetFlow/IPFIX records, logs reporting in order to collect rich datasets related to network control planes (e.g. Wi-Fi roaming, join and authentication, routing, QoS, PHY/MAC counters, links/node failures), traffic characteristics, and other such telemetry data regarding the monitored network.

Figure 7.4 Cloud based system for prioritizing network monitoring alerts. *Source:* Reference [5]/U.S. Patent/public domain.

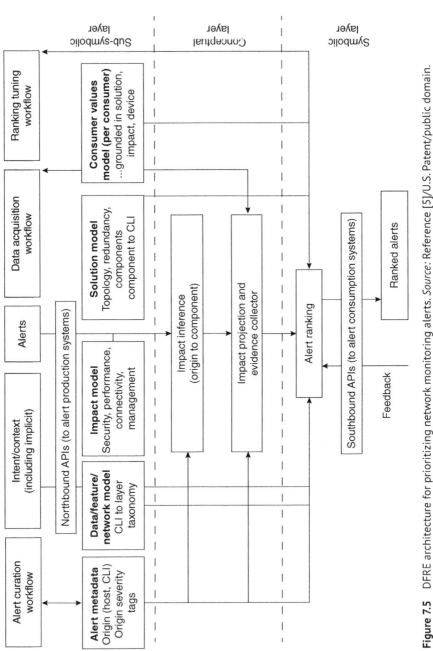

Figure 7.5 DFRE architecture for prioritizing network monitoring alerts. *Source:* Reference [5]/U.S. Patent/public domain.

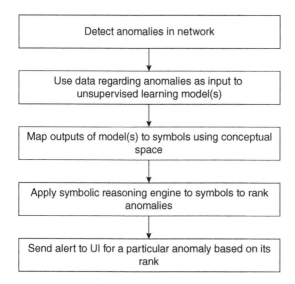

Figure 7.6 Procedure for prioritizing network alerts.

Table 7.2 MLMs to perform function of the system.

Model	Function
Cognitive analytics model(s)	Seek to find behavioral patterns in complex and unstructured datasets. Example: extract patterns of Wi-Fi roaming in the network and roaming behaviors (e.g. the "stickiness" of clients to APs, roaming triggers, etc.). The system may characterize such patterns by the nature of the device (e.g. device type, OS) according to the place in the network, time of day, routing topology, type of Access Point/ Wireless LAN Controller (AP/WLC), etc., and potentially correlated with other network metrics (e.g. application, quality of service [QoS], etc.).
Predictive analytics model(s)	Seek to predict the network status, which is a significant paradigm shift from reactive approaches to network health. For example, in a Wi-Fi network, the analyzer may be configured to build predictive models for the joining/roaming time by taking into account a large plurality of parameters/ observations (e.g. RF variables, time of day, number of clients, traffic load, DHCP/DNS/Radius time, AP/WLC loads, etc.). From this, the analyzer can detect potential network issues before they happen. Furthermore, should abnormal joining time be predicted by the analyzer, the cloud service will be able to identify the major root cause of this predicted condition, thus allowing cloud service to remediate the situation before it occurs.
Trending analytics model(s)	Use multivariate models to predict future states of the network, thus separating noise from actual network trends, and cope with multi-time scale changes. Such predictions can be used, for example, for purposes of capacity planning and other "what-if" scenarios.

7.2.4 System for Recognizing and Addressing Network Alarms (Case of FM)

Network management deals with promptly recognizing and addressing network alarms. The task is complicated when the network elements (e.g. switch, router, hub) are managed by different vendors. A system is needed where the information present in any of the multitude of network alarms is analyzed to determine the operating condition of the various devices. "SNMP Trap" is an SNMP message that is typically used by devices to indicate faults associated with them. Each Trap includes an indicator (such as unique values or error codes) of a specific fault related to a corresponding device. Details related to such indicators are present in management information bases (MIBs) related to the corresponding device. The MIBs related to corresponding devices are different for each device. Additionally, the MIBs may also differ for similar devices provided by different vendors. Thus, the number of MIBs increases to a large number when multiple devices are present in a network, especially when the devices belong to multiple vendors.

Having to refer to a large number of MIBs *manually* to understand the faults associated with the devices becomes a complex, tedious task that is currently performed by a specialized person having relevant technical expertise. When using an *automated system*, enrichment of an alarm deals with presenting the alarm in a desired format having required details. However, the process of providing mapping of network alarms in a user-understandable format is not only a complicated task but also requires significant effort. Initially, a network alarm is received in a raw format; to provide the necessary details to an end user in an understandable format for addressing the network faults, a detailed understanding of different network domains, alarm objects, and parameters is required.

To address such a challenge, the system described in [6] provides a method for recognizing and addressing network alarms in a computer network. Specifically, a trained data model is used to recognize and address the network alarms arising in the network. The trained data model is developed by learning upon multiple network alarms, from multiple devices of multiple vendors, indicating operating conditions of different network devices, and attributes pre-identified to be associated with each of the multiple network alarms. Initially, a network adapter is configured to receive network alarms related to an operating condition of a network device present in the computer network. Thereupon, information present in the network alarm is analyzed to determine elements indicating the operating condition of the network device. The elements indicating the operating condition of the network device may comprise keywords, object identifiers, and values of the object identifiers. Finally, the trained data model is used to analyze and map the network alarm with standard attributes. The network alarm mapped with the standard attribute conveys a network fault that can be easily understood, and thereupon rectified.

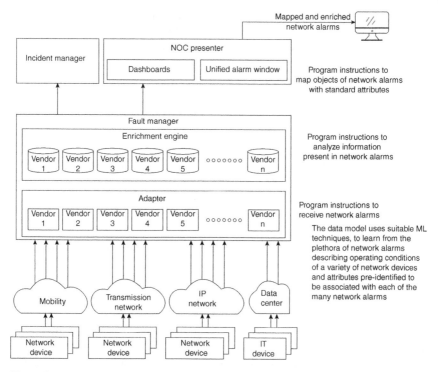

Figure 7.7 Exemplary layered architectural diagram of the system for recognizing and addressing network alarms. *Source:* References [6]/U.S Patent/public domain.

Figure 7.7 (right side) illustrates a block diagram showing different components of the system for recognizing and addressing network alarms in the computer network. The program instructions to receive network alarms may cause the processor to receive SNMP network alarms from different network devices managed by different vendors. The network alarms may indicate the operating conditions of the network devices; the operating condition, for example, may refer to connectivity with other devices, the bandwidth available for communication, or the temperature of the device. The program instructions to analyze information present in network alarms will cause the processor to analyze information present in network alarms to determine elements indicating the operating condition of the network devices. Generally, the elements include keywords, object identifiers, and values of the object identifiers. The program instructions to map objects of network alarms with standard attributes will cause the processor to map objects included in the network alarms with standard attributes using a trained data model. The trained data model may be developed by learning upon several network alarms indicating the operating conditions of several network devices and attributes pre-identified to be associated with each of the several network alarms.

When employed on a network alarm, the trained data model may map objects of the network alarm into at least one standard attribute. In case a relevant attribute is not identified to be present already, a new attribute corresponding to the network alarm may be created.

Figure 7.7 (left side) illustrates an exemplary layered architectural diagram of the system for recognizing and addressing network alarms in the computer network. The layered architectural diagram illustrates a fault manager comprising an enrichment engine and a network adapter. The network adapter may be configured to receive network alarms from different networks such as mobility (e.g. 4G/5G) networks, fiberoptic transmission networks, Internet protocol (IP) networks, and data centers. The network adapter is configured to receive alarms from multiple network device types belonging to multiple vendors.

A data model stored in the network adapter may analyze information present in the network alarms with reference to MIBs related to the network alarms. The data model may be developed using suitable ML techniques, to learn from the plethora of network alarms describing operating conditions of a variety of network devices and attributes pre-identified to be associated with each of the many network alarms. On analysis, the network adapter will map various objects of the network alarms with standard user-understandable attributes, such as a specific problem, severity, probable cause, and unique alarm identifier. Based on information specified in MIBs, the data model present in the network adapter will also (i) create a mapping of enumeration value for specified objects, as per MIBs, (ii) generate an alarm priority mapping, and (iii) generate an alarm clearance mapping. The network adapter then feeds the enrichment engine with the mapping of enumerated values of objects, priority mapping for values received in alarm, and alarm clearance mapping. The enrichment engine leverages the mapping of enumerated values, priority, and clearance received from the network adapter. The enrichment engine translates/replaces the appropriate values received in the network alarm with corresponding mapping values. The network alarms mapped with corresponding attributes and the enriched information are presented to a user at a Network Operation Center (NOC) presenter. The NOC presenter displays such information in real time on dashboards. Further, the NOC presenter provides a unified view of the network alarms mapped with corresponding attributes on a unified alarm window. Such information may also be forwarded to an incident manager for raising tickets for field engineers to address the network alarms.

Figure 7.8 (top) illustrates exemplary instances related to supervised learning techniques used by the trained data model. In one case, a network alarm titled "petTrapTemperaturePredictiveFailureDeasserted" may be received. Keywords present in the name of the network alarm may be identified as Trap, Temperature, Fail, and Deasserted. Such keywords may be searched in a predefined priority matrix Bag-of-Words (BOW), and may be ranked based on the classified keywords.

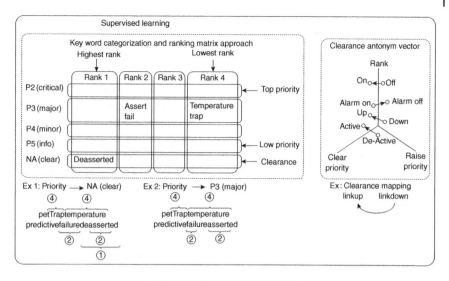

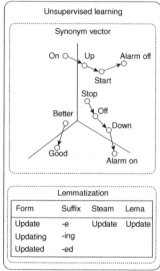

Figure 7.8 Learning technique used by the trained data model. Top: supervised learning. Bottom: unsupervised learning. *Source:* Reference [6]/U.S. Patent/public domain.

A rank of all these keywords may be identified and the highest rank may be selected. Finally, the priority of a keyword having the highest rank may be used for assigning priority to the network alarm. In the current case, priority of the network alarm may be identified as *NA Clear* based on the priority of the keyword "Deasserted" having the highest rank. In another case, a network alarm titled "petTrapTemperaturePredictiveFailureAsserted" may be received. Keywords present in the name of the network alarm may be identified as Trap, Temperature, Fail, and Assert. A rank of all these keywords may be identified and the highest rank may be selected. Finally, the priority of a keyword having the highest rank may be used for assigning priority to the network alarm. In the current case, the priority of the network alarm may be identified as *P3 Major* based on the priority of the keyword "Asserted" having the highest rank. As further illustrated in Figure 7.8 (top), for clearance mapping "Clearance Antonym Vector" may be used. In such case, an alarm clearance BOW may be maintained on one side and raise alarm BOW may be maintained on another side. For clearance mapping, vector relationship may be established between the alarm clearance BOW and the raise alarm BOW. For example, a vector relationship may be established between the keywords "Active" and "DeActive."

Figure 7.8 (bottom) illustrates exemplary instances related to the unsupervised learning technique used by the trained data model. In one implementation, a synonym vector may be utilized by the unsupervised learning technique. Synonyms may be identified for new keywords identified to be associated with the network alarms. For example, as illustrated in Figure 7.8 (bottom), a keyword "good" may be identified as a synonym of the keyword "better." Such newly identified keywords may be added to a BOW referenced by the trained data model. In certain cases, lemmatization technique may be used to establish relationships with new keywords and define rank for the new keywords.

In one exemplary implementation of [6], a trained data model was developed by learning 800 keywords, 120,000 known attributes, and 15,000 known network alarm types, belonging to 60 different vendors. On implementation of the trained data model on a computer network, for identifying network alarms, accuracy ranging from 90.5% (for clearance) to 99.8% (for probable cause) was achieved in mapping of network alarms with the attributes. Refer to [6] for additional details on this system.

7.2.5 Load Control of an Enterprise Network (Case of PM)

The wireless industry has experienced sustained growth in recent years. Wireless technology is rapidly improving, and faster and more numerous broadband communication networks have been installed around the globe. These networks have now become key components of a worldwide communication system that connects people and businesses at speeds and on a scale unimaginable just a couple of decades ago. For example, 4G LTE (fourth generation long-term evolution) has been widely deployed over the past years, and the next generation system, 5G NR (fifth generation new radio) is now being deployed worldwide. In these wireless systems, multiple mobile devices are served voice services, data services, and many other services over wireless connections so they may remain mobile while still connected. See Figure 7.9 for a simple pictorial view. A system

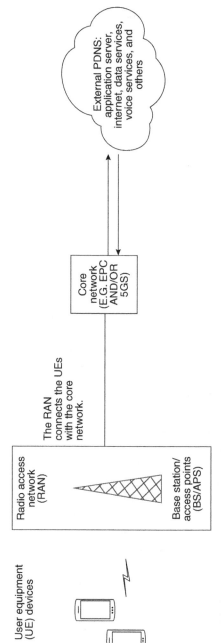

User equipment
(UE) devices

Radio access
network
(RAN)

The RAN
connects the UEs
with the core
network.

Base station/
access points
(BS/APS)

Core
network
(E.G. EPC
AND/OR
5GS)

External PDNS:
application server,
internet, data services,
voice services, and
others

BSs/APs can receive wireless signals from, and send
wireless signals to, the UEs. Transmission between a UE
and the BS/AP occurs on an assigned channel, such as a
specific frequency.

BSs/APs include base stations and access points,
including at least an evolved NodeB (eNB) of an LTE
network or gNodeB of a 5G network, a cellular base
station (BS), a Citizens Broadband Radio Service
Device (CBSD) (which may be an LTE or 5G device), a
Wi-Fi access node, a Local Area Network (LAN) access
point, a Wide Area Network (WAN) access point.
Typically, the BS/APs are used as transceiver hubs,
therefore, the BS/APs transmit at a higher power than
the UEs.

A function of the Core Network is to provide control of
wireless signaling between the UEs and the RAN, and
another function is to provide access to other devices and
services either within its network, or on other networks
such as the External PDNs or the Public Switched
Telephone Network (PSTN) or the Internet. Data
transmission between the BS/AP and the Core Network
utilizes any appropriate communication means, such as
fiberoptic, microwave wireless, or satellite links.

Figure 7.9 Basic configuration for a 4G LTE or 5G NR network. *Source:* Reference [7]/U.S. Patent/public domain.

for managing network loading and adjusting admission, and/or congestion and service types based on real-time analytics in an enterprise wireless communication network that includes base stations/access points (BSs/APs) in communication with a number of user equipment (UEs) devices within the coverage area at a campus location is discussed in [7]. The system dynamically learns the parameterizations for the control system, and adapts the parameters as the network usage evolves, which is useful for an efficient self-managed network. Such a system can provide a greater quality of service (QoS) across users and other efficiencies. The system simplifies deployment at the enterprise's campus locations because load heatmaps and service patterns are learned based on the actual campus locations.

Conventional admission control of individual users and services is performed using static resource reservations, which typically involves allocating the maximum resource needs associated with a given service. Based on the algorithm of choice and the headroom allocated for variabilities in the services, conventional admission control has the drawback that it either overestimates or underestimates in terms of the network capacity planning. A given service can be supported with several configurations; however, because it has no awareness of the potential congestion the network could face, the network's admission control and service configurations are based on fixed configurations. The system of [7] manages network loading, dynamically adjusting admission/congestion control based on real-time analytics. An effective admission/congestion control system is an important element in realizing an effective network. Dynamically learning the parameterizations for the control system, and adapting the parameters as the network usage evolves, is essential for an efficient self-managed network.

This system can provide a greater QoS across users and other efficiencies. Load heatmaps and service patterns are learned based on the actual campus location. Over time, the patterns are re-learned and the deployment preferences eventually become customized to the particular campus location. On initial deployment, an educated estimate of the heatmaps, service patterns, and other behavior can be made, and different types of connectivity can be deployed based on this estimate. The next step is learning from other campuses that certain patterns and behaviors are useful and/or needed. Using learned quantities from previous campus deployments, the network behavior of new campuses can be initialized based on the similarities with the previous campuses, which allows the new network to be deployed in a more optimum manner. Self-learning optimizes connectivity, resource allocation by network, and mobility from macro network to enterprise network (in and out). In addition, there is greater QoS that can be managed across the network(s), better reliability for the end user, and easier and more effective resource management from an enterprise IT perspective.

The system of [7] simplifies deployment at campus locations, on a campus-by-campus basis. By generating heatmaps and determining service patterns, bursts in activity can be predicted with some accuracy. Generally, it is better to anticipate and proactively respond before a burst in activity; otherwise (if the system just waits until it is already overloaded), calls will be dropped, and service can suffer greatly. Particularly, when cell power is reallocated among cells (i.e. some cells shrink in coverage and others grow) the physical characteristics of the wireless signals, such as signal power and multipath, can change dramatically (nonlinearly); this triggers the generation and forwarding of reported session quantities such as reference signals received power (RSRP), which is a measurement of the received power level in an LTE cell network. The average power is a measurement of the power received from a single reference signal. As channel quality indicators (CQIs) change, and connectivity drops. This can cause the radio resource control (RRC) scheduler to behave erratically, misinterpreting what is happening, and responding sub-optimally (scheduling is the process of allocating resources for transmitting data). Thus, in general, if cell power is reallocated after the system is already overloaded and a large number of active sessions are in progress, then many more of the users' sessions will be adversely affected (e.g. more calls dropped) than if the power had been reallocated proactively before the burst of activity that caused the overload. Therefore, it is important for the network to reallocate resources proactively, before any burst in activity. In an enterprise network, in which the locations of the BS/APs are known in the context of the campus geography (the network footprint), cell breathing, in which some cells shrink in coverage and others grow, becomes feasible, whereas in a macrocell network, it is not feasible. Thus, in an enterprise network at a campus location, cell-breathing can be effectively utilized to reallocate resources and disclosed herein. Figure 7.10 is a block diagram of an implementation of an enterprise network that makes use of the system described in [7] with an artificial intelligence (AI) module, ML units, and an AI planning unit. Refer to the reference for an extensive discussion of the matter.

7.2.6   Data Reduction to Accelerate Machine Learning for Networking (Case of FM and PM)

ML is increasingly being used for various tasks in data center networks because of the benefits from the ML-driven workflows. However, the exploration phase (identifying the right model and the set of features to use for the MLM) is rather time-consuming in the networking domain because of the large volume and complexity of data. Reference [8] describes a system (see Figure 7.11) that uses a data reduction engine to reduce the volume of input data for ML exploration for computer networking-related problems. The system receives input data related to a

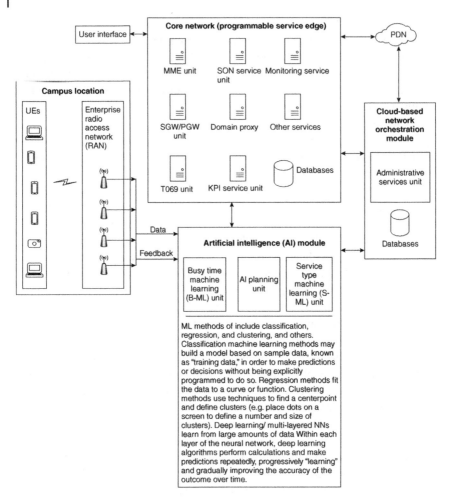

Figure 7.10 Block diagram of an implementation of an enterprise network that makes use of ML/AI mechanisms. *Source:* Reference [7]/U.S. Patent/public domain.

network and obtains a network topology by performing a structured search of a plurality of reduction functions based on a grammar to identify a subset of reduction functions. The system generates transformed data by applying the subset of reduction functions to the input data and may determine whether the transformed data meets or exceeds a threshold (the threshold is a minimum acceptable accuracy for a given computer networking-related problem). The system outputs the transformed data in response to the transformed data meeting or exceeding the threshold.

A particular implementation relates to a data reduction engine. The data reduction engine may include one or more processors; memory in electronic communication with the one or more processors; and instructions stored in the memory. The instructions executable by the one or more processors enable the engine to (i) receive input data related to a network; (ii) obtain a network topology; (iii) perform a structured search of a plurality of reduction functions based on a grammar to identify a subset of reduction functions, where the grammar is based on the network topology and other domain knowledge; (iv) generate transformed data by applying the subset of reduction functions to the input data; (v) determine whether the transformed data achieves a threshold, wherein the threshold is a minimum acceptable accuracy for a given computer networking related problem; (vi) return to a previous transformation of the data if the transformed data does not exceed the threshold; and (vii) output the transformed data in response to the transformed data exceeding the threshold.

Figure 7.11 illustrates a typical ML development workflow. It involves data scientists working with a team of domain experts to receive or access labeled data; identify the set of data sources to use for solving the problem; identify feature extraction that includes merging and creating features from this data (often involving joining multiple tables, and various mathematical operations over the data in those tables); and MLM training that may explore various MLMs accuracy. The typical ML development workflow iterates over these steps until a model converges. After the MLM is trained the MLM may be deployed in production. The cost of ML exploration phase using this workflow is a major factor in the time for the development of network applications. The bandwidth, computation, and human-hour costs result in high turnaround times (often months) for ML development.

The system described in [8] supports a data reduction engine that may speed up the ML exploration for MLMs for network related issues by reducing the volume of data while achieving an acceptable accuracy for the MLM. The data reduction engine finds a smaller dataset that is a subset or transformation of a larger initial dataset. The smaller dataset may provide the same information or roughly the same level of information as the initial dataset and may be easier to work with during ML exploration. Even then, building data reduction engines for networking is challenging because of the large search space, circular dependency, and the tradeoff between cost, fidelity, and granularity of various data sources across the network. For example, the data may be obtained from various vantage points of the network and at various levels of granularity where the various vantage points may have different costs for obtaining the data. In addition, the data reduction engine needs to find a representation of the data such that users may find an MLM and a set of features using that data to achieve sufficient accuracy without knowing the model and the features a priori. Moreover, the search space of

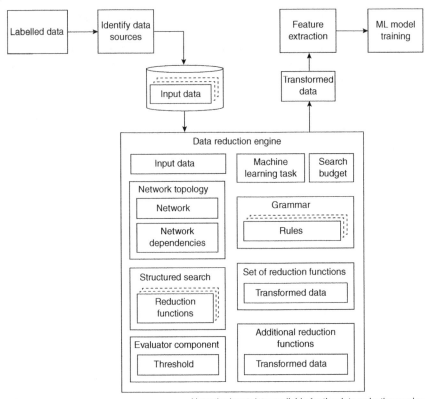

Note: the input data available for the data reduction engine
may be quite large with complex data

Figure 7.11 ML development workflow. *Source:* Reference [8]/U.S. Patent/public domain.

possible strategies for reducing the data is so large that an exhaustive search is too expensive and takes too long. Hence, the data reduction engine of [8] uses network-structural insights from the network topology of the network in combination with a grammar based on the network topology to conduct a structured search of the reduction functions space for reducing the input data (networking related data). The data reduction engine uses the grammar to reduce the search space of the reduction functions. The structured search identifies reduction functions or transformations to apply to the input data. The data reduction engine uses a structured search based on the grammar and the network topology to select a set of reduction functions to apply to the input data to generate transformed data, which is a subset of the input data. Refer to the reference for additional details and applications of this system.

7.2.7   Compressing Network Data (Case of PM)

As is well known and as has been alluded to in earlier sections, there is a need for compressing network data. Radovic et al. [9] describes a method for collecting raw telemetry data from a network environment as time-series datasets. Useful network data is high in volume, making it difficult to store for extended periods. For example, a network with a million data sources may generate 46 GB of one-minute sampled data in one day, which translates to 16 TB of data per year. The number of data sources would even be much higher when considering cloud services, IoT networks with billions of devices, multiple network layers, or high sampling network measurements (e.g. state of polarization, or wireless SNR measurements). Owing to cost, network measurement data is not stored for extended periods, or it is aggregated in larger periods of time (e.g. daily, weekly, monthly, yearly), thus losing fidelity in an important historical record of what has happened in the network. Network measurements of interest could be any number of network observations, such as packet counters, latency, jitter, SNR estimates, state Of polarization (SOP) measurements, wireless CQI reports, alarm states, performance monitoring (PM) data.

The process of aggregation/averaging is a crude way of lossy compression for time-series data. For example, averaging represents a time-series with a single number over a period, so its accuracy is not good. Owing to the large amount of storage typically required for time-series data, a special variation of a database – known as TimeScale Database (TSDB) – has been developed to facilitate the storage, reading, and writing of the time-series data. TSDBs make use of chunks to divide time-series by the time interval and a primary key. Each of these chunks is stored as a table for which the query planner is aware of the chunk's ranges (in time and keyspace), allowing the query planner to immediately identify which chunks an operations data belongs to. In the instance of lossless compression techniques, the compression ratios are typically less than an order of magnitude. Owing to the immense quantity of time-series data that needs to be stored, these compression schemas, even if they are lossless, do not offer a sufficiently large compression ratio to justify their use. This becomes even more apparent when considering the overhead needed to constantly decompress stored time-series.

One attempt at data compression using deep neural networks (DNNs) is known as semantic compression; for example, DeepSqueeze uses autoencoders for data compression of each row in a relational table. The technique, however, requires significant overhead in the form of training multiple models and functions and storing all the weights associated with this multitude of models. The method described in [9] achieves compressing the time-series datasets by deploying the time-series datasets as a DNN in the network environment itself. The time-series datasets are configured to be substantially reconstructed from the DNN using the predictive functionality of the DNN (the network data can be regenerated using the predictive nature of the DNN). The system allows "storing" network data in a DNN by deploying the DNN in the network, which greatly reduces the volume of information that needs to be stored. In this sense, data can be compressed by orders of magnitude, which is very useful for reducing the storage requirements on network data. Figure 7.12 is a diagram of a network data collection infrastructure. Figure 7.13 depicts flowcharts of a process using a DNN to compress (left) and decompress (right) time-series data, such as from network measurements. Refer to reference [9] for an extensive discussion of how this system works and is advantageous.

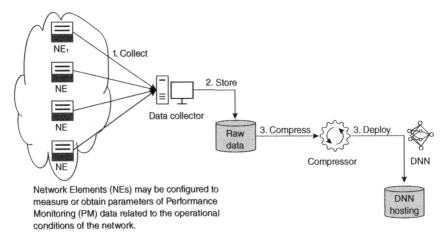

Network Elements (NEs) may be configured to measure or obtain parameters of Performance Monitoring (PM) data related to the operational conditions of the network.

Data from the NEs is collected by a data collector, which is configured to store the raw network data in temporary storage database.

The raw data is taken from the temporary database and is compressed by a compressor

Instead of storing the compressed data in a database according to conventional strategies, the system is configured to deploy the compressed data as a DNN which can be deployed in a DNN hosting system.

Essentially, compressing data includes training a DNN model and then configuring or deploying the compressed data as the DNN, instead of actually storing the compressed data in a database.

The hosting system may be configured as a server which hosts the DNN and allows querying and retrieving of the compressed data.

Figure 7.12 Network data collection infrastructure. *Source:* Reference [9]/U.S. Patent/ public domain.

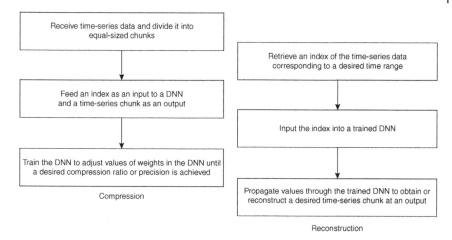

Figure 7.13 Compression of data and reconstruction for the system.
Source: Reference [9]/U.S. Patent/public domain.

7.2.8 ML Predictor for a Remote Network Management Platform (Case of FM, PM, CM, AM)

Conceivably, a network management system can include separate computational instances, using separate sets of computational resources of the remote network management (sub) system, to provide different services. By assigning different sets of computational resources to the end-user instance, the training instance, and the prediction instance, the impact of generating MLMs and/or using such models can be reduced and the level of service provided by the end-user instance can be improved; for example, the latency of serving recorded incident reports and/or knowledgebase articles can be reduced.

Reference [10] describes a remote network management platform that includes: (i) an end-user computational instance comprising the first set of computational resources of the remote network management platform and dedicated to a managed network; (ii) a training computational instance comprising a second set of computational resources of the remote network management platform; and (iii) a prediction computational instance comprising a third set of computational resources of the remote network management platform.

The training computational instance is configured to perform operations including: (i) receiving, from the end-user computational instance, a corpus of textual records; (ii) training, based on the corpus of textual records, an MLM to determine a degree of numerical similarity between input textual records and textual records in the corpus of textual records; and (iii) transmitting, to the end-user computational instance, the MLM. The prediction computational instance is configured to perform operations including: (i) receiving, from the end-user computational instance, an additional textual record; (ii) receiving, from the end-user computational instance, the MLM; determining, by the MLM, respective numerical similarities between the additional textual record and the textual records in the corpus of textual records; and (iii) based on the respective numerical similarities, transmitting, to the end-user computational instance, representations of one or more of the textual records in the corpus of textual records. Figure 7.14 depicts graphically a remote network management architecture that includes three main components: remote network management platform, managed network, and data centers.

Incident reports may be created in various ways. For instance, by way of a web form, an email sent to a designated address, a voicemail box using speech-to-text conversion, and so on. These incident reports may be stored in an incident report database that can be queried. A text query may be entered into the web interface. This web interface may be supplied by way of a computational instance of a remote network management platform. A web interface converts the text query into a database query (e.g. an SQL query), and provides the SQL query to the database. The database contains a number of incident reports with problem description fields e.g. "My email client is not downloading new emails," "Email crashed," and "Can't connect to email." The ML methods discussed in Chapter 2 dealing with natural language processing (NLP) can be used to support queries to the database; techniques include determining word and/or paragraph vectors from samples of text, applying DNNs or other DL algorithms, sentiment analysis, or other techniques to determine a similarity between samples of text. The degree of similarity between two samples of text can be determined in a variety of ways. The two samples of text could be a text field of an incident report and a text field of another incident report, a text field of a resolved incident report, a knowledge-base article, or some other sample of text that may be relevant to the resolution, classification, or other aspects of an incident report. Such techniques may be applied to improve text query matching related to incident reports. These techniques may include a variety of ML algorithms that can be trained based on samples of text. The samples of text used for training can include past examples of incident reports, knowledgebase articles, or other text samples of the same nature as the text samples to which the trained model will be applied. This has the benefit of providing a model that has been uniquely adapted to the vocabulary, topics, and idiomatic word use common in its intended application.

The system of [10] uses a particular method for determining similarity values between samples of text using a DNN model that provides compact semantic representations of words and text strings. A "word vector" may be determined for each word present in a corpus of text records such that words having similar meanings (or "semantic content") are associated with word vectors that are near

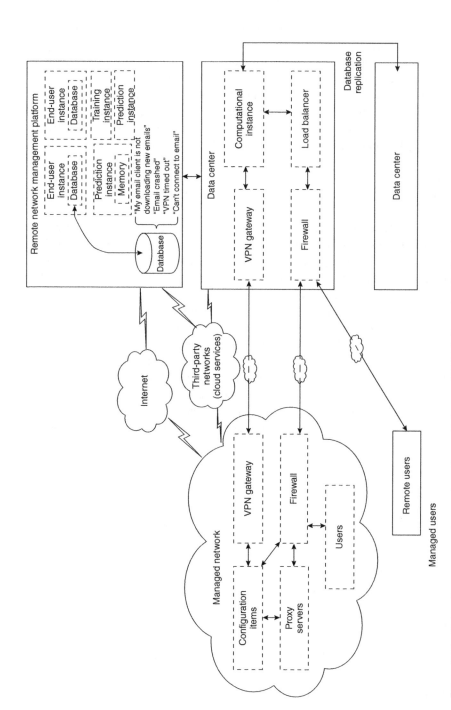

Figure 7.14 Network management ecosystem. *Source:* Reference [10]/U.S. Patent/public domain.

each other within a semantically encoded vector space. Such vectors may have dozens, hundreds, or more elements. These word vectors allow the underlying meaning of words to be compared or otherwise operated on by a computing device. Accordingly, the use of word vectors may allow for a significant improvement over simpler word lists or word matrix methods. Word vectors can be used to quickly and efficiently compare the overall semantic content of samples of text, allowing a similarity value between the samples of text to be determined. This can include determining a distance, a cosine similarity, or some other measure of similarity between the word vectors of the words in each of the text samples. For example, a mean of the word vectors in each of the text samples could be determined and a cosine similarity between the means, then used as a measure of similarity between the text samples.

The DNN (including the matrix of word vectors) can then be trained with a large number of text strings from a database to determine the contextual relationships between words appearing in these text strings. Figure 7.15 (left side) illustrates an example. The DNN includes an input layer, which feeds into a hidden layer, which in turn feeds into an output layer. The number of nodes in the input layer and output layer may be equivalent to the number of words in a pre-defined vocabulary or dictionary (e.g. 20,000, 50,000, or 100,000). The number of nodes in a hidden layer may be much smaller (e.g. 64 as shown in Figure 7.15, or other values such as 16, 32, 128, 512, 1024, etc.). For each text string in the database, the DNN is trained with one or more arrangements of words. For instance, in Figure 7.15 (right side), the DNN is shown being trained with the input word "email" and output (context) words "can't," "connect" and "to." The output words serve as the ground truth output values to which the results produced by the

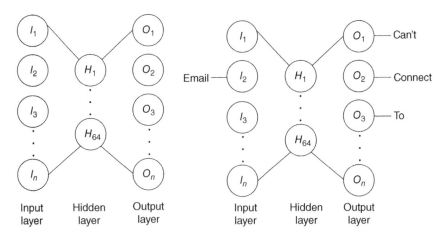

Figure 7.15 DNN configured for learning the contextual meanings of words.

output layer are compared. See Ref. [10] for a detailed description of how the DNN approach is used extensively in this system to parse the log and alert data.

7.2.9 Cable Television (CATV) Performance Management System (Case of PM)

There is interest in having design systems for automatically analyzing spectral power measurements to identify abnormalities in a cable television (CATV) transmission environment. CATV head ends include a separate cable modem termination system (CMTS), used to provide high-speed data services, such as video, cable Internet, Voice over Internet Protocol (VoIP), and so on to cable subscribers. Management of a Random Forest (RF) data network requires periodic measurement of state variables that represent system health or status; such measurements in a hybrid fiber coax (HFC) network include, but are not limited to, full-band spectrum (FBS) capture data, pre-equalization coefficients, impulse noise measurements, histograms, and modulation error ratios (MER). Also, it is desirable to detect and mitigate RF network impairments to prevent the degradation of QoS to customers. One such detection system is described in [11]. This system receives measurements comprising RF power measured over a contiguous range of frequencies, where at least the first portion of the contiguous range is used to transmit signals and at least the second portion of the contiguous range is unused; respective boundaries of the unused portions may be identified and infilled to provide modified measurements. The modified measurements are then automatically analyzed to identify the abnormalities.

Typically, a CMTS will include both Ethernet interfaces as well as RF interfaces so that traffic coming from the Internet can be routed (or bridged) through the Ethernet interface, through the CMTS, and then onto the optical RF interfaces that are connected to the cable company's HFC system. The head end modulates a plurality of QAM channels using one or more EdgeQAM units. Downstream traffic is delivered from the CMTS to a cable modem (CM) in a subscriber's home, while upstream traffic is delivered from a CM in a subscriber's home back to the CMTS. Modern CATV systems have combined the functionality of the CMTS with the video delivery system (EdgeQAM) in a single platform called the Converged Cable Access Platform (CCAP); other CATV systems called Remote PHY (or R-PHY) relocate the physical layer (PHY) of a traditional CCAP by pushing it to the network's fiber nodes – thus, while the core in the CCAP performs the higher layer processing, the R-PHY device in the node converts the downstream data sent by the core from digital-to-analog to be transmitted on radio frequency and converts the upstream RF data sent by CMs from analog-to-digital format to be transmitted optically to the core. CATV systems traditionally bifurcated available bandwidth into upstream and downstream transmissions, i.e. data is only transmitted in one direction across any part of the spectrum. For example, the Data Over Cable Service Interface Specification (DOCSIS) initially assigned upstream transmissions to a frequency spectrum between 5 and 42 MHz and assigned downstream transmissions to a frequency spectrum between 50 and 750 MHz; later the standard expanded the width of the spectrum reserved for each of the upstream and downstream transmission paths, but the spectrum assigned to each respective direction did not overlap; more recently, proposals have emerged by which portions of spectrum may be shared by upstream and downstream transmission, e.g. full duplex and soft duplex architectures [11]. Figure 7.16 depicts an exemplary HFC network from a head end to a node that serves a plurality of home subscribers.

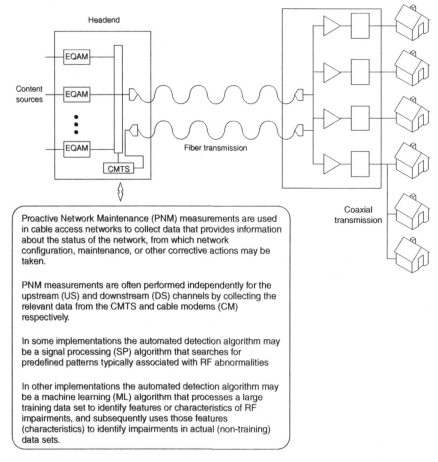

Figure 7.16 Exemplary HFC network from a head end to a node that serves a group of home subscribers. *Source:* Reference [11]/U.S. Patent/public domain.

In recent years, CableLabs DOCSIS standards have introduced a variety of Proactive Network Measurement (PNM) tests for the collection of operational data from various network elements such as the CMs and the CMTSs. PNM measurements are used in cable access networks to collect data that provides information about the status of the network, from which network configuration, maintenance, or other corrective actions may be taken. PNM measurements, for example, include FBS capture data that measures signal quality in both upstream and downstream directions across the full network spectrum. Such measurements may be used, for example, to arrange or rearrange cable modems into interference groups in full duplex architectures, adjust modulation profiles in specific subcarriers, and so on. Other PNM measurements may measure signal quality in only specific subcarriers, and in either case, signal quality may be measured using any of a number of metrics, e.g. SNR MER, impulse noise measurements, and so on. Other PNM measurements may measure distortion products from which pre-equalization coefficients may be derived, which are used to pre-distort transmitted signals to compensate for optical distortion that occurs in the fiber portion of the network. Other PNM measurements may include impulse noise measurements, histograms, and any other metric relevant to a state of the transmission network.

PNM measurements are often performed independently for the Upstream (US) and Downstream (DS) channels by collecting the relevant data from the CMTS and cable modems respectively. The operational data available in the network can be extremely large when taken at time intervals sufficient to allow proactive as opposed to reactive network management. Historically, these sorts of data available in the cable network require one skilled in the domain of radio frequency engineering to interpret the spectral data available from the system to identify abnormalities or defects in the RF spectrum. Typical RF impairments that occur in the cable coaxial networks are suck-out, tilt, roll-off, etc. FBS capture data, which measures received RF power over the full RF spectrum, is typically used to identify the presence of the aforementioned RF impairments.

Regardless of the particular type of abnormality, human identification of the abnormalities by visual review of the RF spectral signal is time consuming and inefficient. Discerning the quality of an RF spectral signal visually is time consuming and fraught with nuance such that human review of this data is done in a reactive manner where issues are already known as opposed to a proactive manner to determine where issues are arising that may not yet be having major impacts on network performance and QoS. There is interest in improved systems and methods to identify defects or abnormalities in RF spectrum in a communications network. Figure 7.17 shows the system described in [11] that implements ML techniques for the automated detection of RF impairments or abnormalities in spectral capture measurement such as PNM data. In some implementations, the

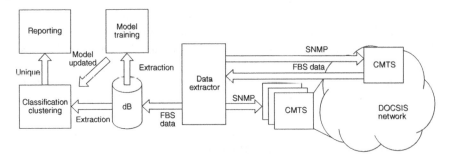

Figure 7.17 System implementing ML techniques for the automated detection of RF impairments. *Source:* Reference [11]/U.S. Patent/public domain.

automated detection algorithm may be a signal processing (SP) algorithm that searches for predefined patterns typically associated with RF abnormalities; in another implementation, the automated detection algorithm is an ML algorithm that processes a large training dataset to identify features or characteristics of RF impairments, and subsequently uses those features (characteristics) to identify impairments in actual (non-training) datasets. Applying ML algorithms to automatically identify RF impairments is not a trivial task. The application of ML techniques on spectral samples has specific challenges when applied for classification. For supervisory-based learning systems, training data must be available to train the algorithm as to what anomalous spectral data may look like. But for spectral-based data, there is unfortunately no systematic method for obtaining labeled data. When using a supervised algorithm, labeling of data can be difficult and costly. For RF spectrum, labeling of data may necessitate subject matter experts manually and meticulously looking through samples and adding labels to those subjectively determined to be abnormal. A lack of precise boundary definitions can confuse ML algorithms and decrease the performance of the ML algorithms.

Much of the prior work done for ML for spectrum analysis is based on the use case of spectrum sensing – i.e. identification of whether a spectral block is available for use. Other related work has looked at Time Series Classification using DL, and some work has been performed in the more specific cases for DOCSIS and Orthogonal Frequency-Division Multiplexing (OFDM). Some prior work discusses the sources of data available in a DOCSIS network useful for ML approaches and the application of DL for classification while still other prior work evaluates the use of convolutional neural networks (CNN) as a multi-label classifier with RxMER data in an OFDM channel. The systems of [11], on the other hand, uses a number of ML algorithms to categorize downstream FBS capture data extracted from cable modems – see Table 7.3 [11] (noting that most of these concepts have also been discussed elsewhere in this text).

Table 7.3 ML algorithms used for CATV RF monitoring.

ML technique	Description
Adaboost	A decision tree-based algorithm that makes use of many simple decision tree classifiers. Adaboost utilizes a base classifier, in this case a decision tree classifier. The Adaboost algorithm builds a final classifier using weighted sample data over a series of training passes. The best decision trees in each pass are added to a final weighted decision tree classifier set.
Logistic regression (LR)	A logit function is used to create a best fit decision boundary between samples labeled as good or bad resulting in a linear decision boundary. The LR algorithm is optimized using the training data to minimize the error of the fitted data. Samples are compared against the constructed decision boundary and labeled accordingly.
Multi-layer perceptron (MLP)	A simple NN that may be used for classification problems. An MLP produces a nonlinear decision boundary using a network of simple nodes.
Residual Network (ResNet)	A CNN architecture developed for DL. The ResNet architecture solves issues associated with vanishing gradients and accuracy degradation with deep learning architectures. ResNet is the only algorithm that uses FBS samples directly without feature extraction and PCA reduction.

Refer to the reference for additional information on this system and its applicability.

A. Short Glossary of Network Management Concepts

Term	Definition/explanation/concept
AdaBoost algorithm	An algorithm that begins by training a decision tree in which each observation is assigned an equal weight. The algorithm identifies the shortcomings by using high-weight data points. After evaluating the first tree (*Tree 1*), one increases the weights of those observations that are difficult to classify and lowers the weights for those that are easy to classify. The second tree (*Tree 2*) is grown on this weighted data. The goal is to improve on the predictions of the first tree using the model *Tree 1 + Tree 2*. One then computes the classification error from this new two-tree ensemble model and grows a third tree to predict the revised residuals. One repeats this process for a specified number of iterations. Subsequent trees help classify observations that are not well classified by the previous trees. Predictions of the final ensemble model is thus the weighted sum of the predictions made by the previous-tree models [12].
Autoencoder (AE)	A type of unsupervised learning NN that is used to learn efficient encodings or representations of unlabeled data by attempting to regenerate inputs from encodings. Also see Appendix A of Chapter 1.
Boosting	A method of converting "weak learners" into "strong learners." When using boosting, each new decision tree is a fit on a modified version of the original dataset.
FCAPS	A model for network management that spans fault, configuration, accounting, performance, and security management. The model is also known as the ISO (or Open Systems Interconnection [OSI]) network management model – ISO is the International Organization for Standardization.
Gradient boosted algorithm (GBA)/Gradient boosted machine (GBM)	An algorithm that trains many models in a gradual, additive, and sequential manner [12]. Similar, yet different from the AdaBoost algorithm; the major difference is how the two algorithms identify the shortcomings of weak learners (e.g. decision trees): GBA uses gradients in the loss function, which is a measure indicating how good the model's coefficients are at fitting the underlying data. A GBM is a model that uses the GBA.
Simple Network Management Protocol (SNMP)	A widely used Internet Engineering Task Force (IETF) network management standardized protocol for collecting and organizing information about managed devices and for modifying that information to affect device behavior. It handles management data in the form of variables on the managed systems. where the variables are defined conceptually in a specified management information base (MIB) that describes the system status and configuration. Variables can be remotely queried and set or reset by managing applications.

Term	Definition/explanation/concept
SNMP Trap	An SNMP message that is typically used by devices to indicate faults associated with them. Each Trap includes an indicator (such as unique values or error codes) of a specific fault related to a corresponding device. Details related to such indicator is present in management information bases (MIBs) related to the corresponding device. To understand the details related to a fault, identifiers carried by the Traps are first extracted, and then details related to such identifiers are looked-up in corresponding MIBs. Different enumeration values may indicate different status of the network alarm enumeration, for example value "1" may indicate critical, "3" may indicate minor, "4" may indicate information, and "6" may indicate clear [6].

References

1 Minoli, D. and Ericson, E.C. (ed.) (1989). *Expert Systems Applications in Integrated Network Management*. Artech House.

2 Minoli, D. (1991). *Telecommunication Technologies Handbook*, 1e. Artech House.

3 S. Verma, G. R. Dash, *et al*, Neural Network-Assisted Computer Network Management, U.S. Patent 20190197397, June 27, 2019. Uncopyrighted material.

4 A. Armenta, L. Savage, Telecommunication Network Machine Learning Data Source Fault Detection and Mitigation, U.S. Patent 20220329328, Oct. 13, 2022. Uncopyrighted material.

5 E. Fenoglio, H. Latapie, *et al*, Deep Fusion Reasoning Engine (DFRE) for Prioritizing Network Monitoring Alerts, U.S. Patent 11,595,268, Feb. 28, 2023. Uncopyrighted Material.

6 P. B. Selokar, P. H. Patil, L. N. Karkala, System and a Method for Recognizing and Addressing Network Alarms in a Computer Network, U.S. Patent 11,588,677, Feb. 21, 2023. Uncopyrighted material.

7 M. Yavuz, P. Natarajan, Method and Apparatus for Load Control of an Enterprise Network on a Campus Based Upon Observations of Busy Times and Service Type, U.S. Patent 11,540,176, Dec. 27, 2022. Uncopyrighted material.

8 B. Arzani, G. Ananthanarayanan, Using Data Reduction to Accelerate Machine Learning for Networking, U.S. Patent 20230062931, March 2, 2023. Uncopyrighted material.

9 I. Radovic, M. Amiri, *et al*, Compressing Network Data Using Deep Neural Network (DNN) Deployment, U.S. Patent 20230057444, February 23, 2023. Uncopyrighted material.

10 B. Jayaraman, A. M. Thakur, *et al*, Centralized Machine Learning Predictor for a Remote Network Management Platform. U.S. Patent 11,595,484, Feb. 28, 2023. Uncopyrighted material.

11 D. E. Virag, S. Chari, S. J. Kraiman, Machine Learning for RF Impairment Detection, U.S. Patent 20230049496, Feb. 16, 2023. Uncopyrighted Material.

12 H. Singh, "Understanding Gradient Boosting Machines", Nov. 3, 2018, https://towardsdatascience.com. https://towardsdatascience.com/understanding-gradient-boosting-machines-9be756fe76ab. Accessed Mar. 8, 2023.

Super Glossary

Term	Chapter citing and/or defining the concept in its glossary
Absolute humidity	5
Absolute pressure	5
Acoustic models	3
Acoustic transfer function (ATF)	3
Acoustic units	3
Activation function	1
Actuator	5
Adaboost algorithm	7
Adaptive beamformer	3
Advanced meter infrastructure (AMI) system	5
Adware	6
Affine transformation	4
Air conditioning airflow efficiency (ACAE)	5
Air handling unit (AHU)	5
Air mixing	5
Air space	5
Air vent valve	5
Air, bypass	5
Air, conditioned	5

(Continued)

AI Applications to Communications and Information Technologies: The Role of Ultra Deep Neural Networks, First Edition. Daniel Minoli and Benedict Occhiogrosso.
© 2024 The Institute of Electrical and Electronics Engineers, Inc.
Published 2024 by John Wiley & Sons, Inc.

(Continued)

Term	Chapter citing and/or defining the concept in its glossary
Air, return	5
Air/liquid cooling	5
Air-cooled system	5
Airside economizer	5
Algorithm	1
American Society of Heating, Refrigerating and Air-Conditioning Engineers (ASHRAE)	5
Artificial intelligence (AI)	1
Artificial neural network (ANN)	1
Atmospheric pressure	5
Atrous convolution	4
Attention models	2
Autoencoder (AE)	1, 7
Availability (A)	6
Backpropagation	3
Backward propagation	1
BACnet (protocol)	5
Base load generation	5
Base Rate	5
Batch normalization (BN)	4
Beamforming	3
Benchmark	5
Bias	1
Bi-directional long-short term memory (BLSTM)	3
Boiling point	5
Boltzmann Machine (BM); also, Deep Boltzmann Machine (DBM); also, Restricted Boltzmann Machine (RBM)	1
Boosting	7
BTU	5
Building automation system (BAS)	5
Building management systems (BMSs)	5
Busy hour drain	5
Bypass	5
Bypass airflow	5

Term	Chapter citing and/or defining the concept in its glossary
Capacity	5
Carbon footprint	5
Carbon neutral	5
Central air conditioner	5
Charge	5
Chatbot	2
Chilled water system	5
Classification	1
Classification process	4
Close-coupled cooling	5
Cloud computing	5
Clustering	1
Clusters	5
CNN BiLSTM	6
CO_2 equivalent	5
Coefficient of Effectiveness (CoE)	5
Coefficient of Performance (COP)	5
Cold spot	5
Cold Supply Infiltration Index (CSI)	5
Commissioning	5
Compressive Transformer	2
Computational fluid dynamics (CFD)	5
Computational linguistics (CL)	2
Computer room air conditioning (CRAC)	5
Computer room air handler (CRAH)	5
Computer vision (CV)	4
Condensation point	5
Condenser	5
Condenser water	5
Conditioned air	5
Confidence scores	3
Confidentiality (C)	6
Consistency models	4
Constant air volume (CAV) HVAC	5

(*Continued*)

(Continued)

Term	Chapter citing and/or defining the concept in its glossary
Constant volume (CV) air conditioner	5
Contactor	5
Control	5
Control device	5
Convolution	4
Convolution dilation rate	4
Convolution value of location (i,j) of the feature map	4
Convolutional (CONV) layers	4
Convolutional neural network (CNN)	1
Convolutional neural networks (CNN)	4
Cooling capacity	5
Cooling tower	5
Cooling, air	5
Cooling, liquid	5
Critical load	5
Cubic feet per minute (CFM)	5
Curse of dimensionality	6
Cutout	5
DALL·E 2	4
Data grid	5
Data-independent filter	3
Data management schemas	6
Database firewalls (DBF)	6
Daylighting	5
dead band	5
Deep belief network (DBN)	1
Deep learning (DL)	1
Deep learning (DL) in natural language processing (NLP)	2
Deep neural network (DNN)	1
Degree-day	5
Delay-and-sum beamformer (DSB)	3
Delta T	5

Term	Chapter citing and/or defining the concept in its glossary
Demand	5
Demand charge	5
Demand interval	5
Demand, average	5
Demand, billing	5
Demand, instantaneous peak	5
Demand, maximum	5
Dew point temperature (DPT)	5
Dialog processing	2
Diffusion models	4
Dimension reduction algorithms	1
Direct expansion (DX) system	5
Discourse	2
Dispatch, dispatching	5
Distant-talking speech communication systems	3
Distributed learning	1
Distribution	5
Dot product	4
Dry-bulb temperature (DBT)	5
Economizer, air	5
Economizer, water	5
Edge detection	4
Efficiency	5
Efficiency, HVAC system	5
Electricity service	5
End-to-end (E2E) learning	3
End-to-end ASR system	3
Energy charge	5
Energy costs	5
Energy efficiency	5
Energy efficiency ratio (EER)	5
Energy, off-peak	5
Energy, on-peak	5

(Continued)

(Continued)

Term	Chapter citing and/or defining the concept in its glossary
Equilibrium	5
Equilibrium chart	5
Equipment room	5
Evacuate	5
Evaporative condenser	5
Evaporator	5
Evolutionary neural networks (ENNs)	6
Exploit	6
Extreme learning machines (ELM)	5
Fan	5
Fast R-CNN model	4
FCAPS	7
Feature map	4
Feedback loop	1
Feed-forward NNs	1
Feed-forward propagation	1
Finite impulse response (FIR) filter	3
Finite-state machines	2
Finite-state transducer (FST)	3
Freezing point	5
Fuel cost adjustments	5
Fully convolution network (FCN)	4
Fully connected NNs (FNN)	4
Gated recurrent models (GRMs)	2
Gated recurrent unit (GRU)	2
Gauge pressure	5
Gaussian mixture model (GMM)	3
Generalization gap	4
Generation, generating plant electric power	5
Generative Adversarial Network (GAN)	1
Generative AI	2
Genetic algorithm (GA)	6
Gigawatt (gW)	5

Term	Chapter citing and/or defining the concept in its glossary
Gigawatt-hours (gWh)	5
Gradient	3
Gradient boosted algorithm (GBA)/Gradient boosted machine (GBM)	7
Gradient descent	3
Granular enforcement	6
Graph neural networks (GNNs)	4
Graphics processing unit (GPU)	4
Green	5
Green buildings	5
Green power, green pricing	5
Greenhouse gases (GHG)	5
Greening	5
Ground-truth	1
Hardware Root of Trust (HRoT)	6
Heat exchanger	5
Heat of condensation	5
Heat pump	5
Heat transfer	5
Heating Seasonal Performance Factor (HSPF)	5
Hidden Markov model (HMM)	3
High temp generator	5
High-flow constraint day	5
High-performance building	5
High-performance data center (HPDC)	5
Horizontal displacement (HDP)	5
Hot spot	5
Humidity	5
Humidity ratio	5
HVAC (heating, ventilation, and air conditioning)	5
Hybrid arbitration	3
Hyperparameter	1, 4
Hyperplane	1
Image classification	4

(*Continued*)

(Continued)

Term	Chapter citing and/or defining the concept in its glossary
Image manipulation	4
Image processing and analysis	4
Image segmentation	4
Image semantic information	4
Imitation algorithms	1
In-context learning	3
Incremental learning	1
Inlet air	5
Inner product	4
Input and output channels of a convolutional layer	4
Input rate	5
In-row cooling	5
Integrated design	5
Integrated energy efficiency ratio (IEER)	5
Integrity (I)	6
Intention analysis	2
Interruptible customer	5
Interruptible service	5
Interspace pruning (IP)	4
Intrusion detection	6
Intrusion detection system (IDS)	6
Intrusion protection system (IPS)	6
Inverted, inverted block rate design	5
i-Vector model	3
Kernel size	4
Keylogger	6
Kilowatt (kW)	5
Kilowatt-hour (kWh)	5
k-Means	3
kVA	5
kWc	5
Language model (LM)	3
Language model(ing) (LM)	2

Term	Chapter citing and/or defining the concept in its glossary
Latent cooling capacity	5
Latent heat	5
Latent heat of condensation	5
Latent heat of vaporization	5
Lattice (ASR context)	3
Layer 7 firewall	6
Least privilege access	6
LEED™ (Leadership in Energy and Environmental Design)	5
Lemmatization	2
Lexis/lexicon	2
Linear regression (LR)	1, 3
Linguistic realization	2
Liquid cooled system	5
Liquid cooling	5
Load	5
Load curve	5
Load factor	5
Load level	5
Load management	5
Load shifting	5
Logistic regression	1, 3
Long short-term memory (LSTM)	1, 2, 3
Loop/looped	5
Loss (losses)	5
M2M (machine-to-machine)	5
M2M communication	5
M2M service provider's domain	5
M2M system	5
Machine (host)	5
Machine learning (ML)	1, 2
Machine learning (ML) in NLP	2
Machine learning (ML) models	1
Machine learning (ML)-based processing	1

(Continued)

(Continued)

Term	Chapter citing and/or defining the concept in its glossary
Machine translation (MT)	2
Machine learning programs (MLPs)	1
Machine-type communications (MTC)	5
Makeup Air Handler (MAH)	5
Makeup Air Unit (MAU)	5
Malicious cyberattacks types	6
Malware	6
Man-in-the-middle attacks	6
Mask	4
Maximum temperature rate of change	5
M-Bus	5
Media access control (MAC)	5
Medical body area network system (MBANS)	5
Megawatt (MW)	5
Megawatt-hour (mWh)	5
Mel frequency cepstral coefficients (MFCCs)	3
MERV	5
Micro-segment	6
Midjourney	4
Minimum Variance Distortionless Response (MVDR) beamformer	3
Model	1
Monitoring tool	6
Morpheme analysis	2
Morphology	2
Multichannel Wiener Filter (MWF)	3
Multi-factor authentication	6
Multi-layer perceptrons (MLP)	1
Multi-lingual TTS	3
Multi-speaker TTS	3
Multi-task learning	1
Natural language generation (NLG)	2
Natural language model	3

Term	Chapter citing and/or defining the concept in its glossary
Natural language processing (NLP)	2
Natural language understanding (NLU)	2
NEBS™	5
Network	5
Neural beamforming	3
Neural network (NN)	1
Neural network layers	1
Neuron	1
Next-generation firewall (NGFW)	6
NNs with fully connected (FC) layers	4
Noise cancellation	3
Nominal cooling capacity	5
Noncondensable gas	5
Non-negative Matrix Factorization (NMF)	3
Object detection	4
Object detection techniques	4
Object instance segmentation	4
Object recognition	4
Ontologies	2
Optimization techniques (DL context)	4
Out of distribution issue	3
Overcooling	5
Parametric machine learning models	1
Particle swarm optimization (PSO)	5
Password stealer	6
Peak day	5
Perceptron	1
Phishing	6
Phone (in phonetics and linguistics context)	3
Phoneme	3
Phonology	2
PID (proportional, integral, and derivative) loop	5
Plenum	5

(Continued)

(Continued)

Term	Chapter citing and/or defining the concept in its glossary
Pole	5
Pooling	1, 3
Pooling (a feature map)	4
Post-occupancy evaluation (POE)	5
Power distribution unit (PDU)	5
Power usage effectiveness (PUE)	5
Power, firm	5
Power, interruptible	5
Power, nonfirm	5
Pressure differential	5
Pre-training	1, 2
Primary distribution, primary distribution feeder	5
Primary voltage	5
Production	5
Protect surface	6
Pruning	4
Psychrometric chart	5
Pump	5
Raised floor	5
Random k-labelsets ensemble (RAkEL)	6
Rate level	5
Rate structure	5
Rates, demand	5
Rates, flat	5
Rates, seasonal	5
Rates, step	5
Rates, time-of-use	5
Receptive fields	1
Recirculation	5
Recirculation air handler (RAH)	5
Recurrent neural network (RNN)	1, 2

Term	Chapter citing and/or defining the concept in its glossary
Referring Expression Generation (REG)	2
Refrigerant	5
Region (on an image)	4
Region-based CNN (R-CNN)	4
Regression	1
Regression/regression analysis	3
Reinforcement learning (RL)	1
Relative humidity (RH)	5
Relief valve	5
Renewable energy	5
Reserve margin	5
Return air	5
Robotic Process Automation (RPA)	1
Rogue security software	6
Room load capacity	5
Root of Trust (RoT)	6
Scene understanding	4
Scheduling	5
Scope of vulnerabilities	6
Seasonal Energy Efficiency Ratio (SEER)	5
Security tools	6
Segmentation gateway	6
Self-attention	2
Self-encrypting drives (SED)	6
Semantic feature map (SFM)	4
Semantic segmentation (process)	4
Semantic segmentation problems	4
Semantic segmentation process (also known as dense prediction)	4
Semantics	2
Semi-supervised learning	1
Sensible cooling capacity	5

(Continued)

(Continued)

Term	Chapter citing and/or defining the concept in its glossary
Sensitivity	5
Sensor network/ wireless sensor network (WSN)	5
Sensors	5
Sentence planning	2
Sequence-to-sequence (S2S) models	2
Service drop	5
Service entrance	5
Service lateral	5
Short cycling	5
Sigmoid layer	1
Simple Network Management Protocol (SNMP)	7
Single-phase service	5
Smart grid (SG)	5
SMS authentication	6
SNMP Trap	7
Social engineering	6
Softmax layer	1
Solution pump	5
Spam	6
Spatial information	4
Spatial transformer network (STN)	4
SPEC	5
Specific heat	5
Specific volume	5
Speech dereverberation	3
Speech distortion weighted multichannel wiener filter (SDW-MWF)	3
Speech enhancement (SE)	3
Speech recognition	2
Speech-act analysis	2
Spiking neural networks (SNNs)	5
Spill point	5
Spyware	6

Term	Chapter citing and/or defining the concept in its glossary
Stable diffusion	4
Standard cubic feet of air per minute (SCFM)	5
Statistical inference	1
Statistical learning methods	1, 2
Step-down	5
Step-up	5
Stochastic gradient descent (SGD) method	3
Strainer	5
Stride	4
Strided convolutions	4
Sub-cooling	5
Sub-floor	5
Submetering	5
Sub-phonetic units	3
Substation	5
Summer peak	5
Superheat	5
Supervised learning	1
Supervisory control and data acquisition (SCADA)	5
Supply air	5
Support vector machines (SVMs)	3
Surface realization	2
Sustainable	5
Sustainable design	5
Syntax	2
Syntax analysis	2
Tariff	5
Temperature	5
Temperature, dew-point	5
Temperature, dry-bulb	5
Temperature, wet-bulb	5
Text-to-speech (TTS) synthesis	2, 5
Therm	5

(Continued)

(Continued)

Term	Chapter citing and/or defining the concept in its glossary
Thermal effectiveness	5
Thin plate spline (TPS)	4
Three-phase service	5
Tokenizer	6
Ton (refrigeration)	5
Training	1
Transformer	2, 5
Transformer architecture	3
Transformers XL (extra long)	2
Translation invariance	4
Transmission	5
Transposed convolution	4
Transposed convolutional layer	4
Trimming the machine	5
Trojan	6
Trojan downloader/dropper	6
Trusted platform module (TPM)	6
Ultra DNN (UDNN)	1
Uninterruptible power supply (UPS)	5
Unpooling (upsampling) (a feature map)	4
Unstructured pruning	4
Unsupervised learning	1
US Green Building Council (USGBC)	5
Vacuum	5
Variable air volume (VAV) AC	5
Variable air volume (VAV) HVAC	5
Variable frequency drive (VFD)	5
Variable-capacity cooling	5
Variational autoencoder (VAE)	4
Video formats	4
Virus	6
Visual computing systems	4
Voiceprint	3

Term	Chapter citing and/or defining the concept in its glossary
Waterside economizer	5
Watts per square foot (WPSF)	5
Web application firewall (WAF)	6
Weigh	4
Weighted finite-state acceptors (WFSAs)	3
Weighted finite-state transducer (WFST)	3
Wet-bulb temperature (WBT)	5
Window of audio data	3
Wireless M-BUS	5
Wireless sensor network (WSN)	5
Word representation	2
Worm	6
Zero-shot TTS	3
Zero-shot voice conversion	3
Zero trust networking (ZTN)	6
ZigBee smart energy	5

Index

a

Accounting management (AM)   407
Acoustic beamforming   129
Acoustic model (AM)   118, 156
Activation function   13, 16, 44
Active learning   32
AdaBoost algorithm   446
Adaptive beamformer   131, 156
Adaptive bot detection system   378, 380
Advanced meter infrastructure
  (AMI)   273, 314
Affine transformation   24, 238
AI application in telecommunications   40
Air handling unit (AHU)   301
Alerts, network monitoring   419, 421
AlexNet   175, 205, 210
Amazon   97
American Society of Heating, Refrigerating
  and Air-Conditioning Engineers
  (ASHRAE)   315
Anomaly detection   370, 374, 378
Antivirus software   370
Apple   98
Application Security Testing   360
Artifacts   216
Artificial general intelligence (AGI)   2

Artificial intelligence control system
  (AICS)   279
Artificial intelligence in medicine   5
Artificial neural networks (ANNs)   7, 8,
  44, 301, 375
ASHRAE standard 55-2017   302
ASR system   98, 133, 142, 143
Attention   77, 153
Attention models   71, 103, 146
Authentication   361
Authorization   361
Autoencoder (AE)   45, 192, 446
Automatic speech recognition
  (ASR)   3, 65, 117
Availability (A)   351, 396
Average pooling   183, 191

b

Backpropagated error   140
Backward propagation (BP),
  Backpropagation   18, 20, 45, 156
BACnet   316
Bard   98
Bayesian methods   35
Bayesian Network (BN)   37
Beamforming, beamformer   129, 156

AI Applications to Communications and Information Technologies: The Role of Ultra Deep Neural Networks, First Edition. Daniel Minoli and Benedict Occhiogrosso.
© 2024 The Institute of Electrical and Electronics Engineers, Inc.
Published 2024 by John Wiley & Sons, Inc.

BERT (Bidirectional Encoder Representation from Transformer) 77, 89, 386

B-frame 198

Bidirectional Encoder Representation from Transformers (BERT) 77, 89, 386

Bi-directional long short-term memory (BLSTM) 147, 157

Binary step function 14

Bing Chat 98

Blockchains 365

BLSTM 147, 157

BMS (Building management systems) 276, 316

BMS with ML/DNN 278

Boiler system 280

Boltzmann Machine (BM) 45

Boosting 446

Bot detection 378

BrainBox AI 301, 305, 312

Building management systems (BMSs) 276, 316

c

Cable modem termination system (CMTS) 441

CAE (convolutional autoencoder) 191, 192

Calibration 286

Carbon neutral 317

CATV Performance Management 441

Chatbot 81, 95–98, 100–103, 235

Chiller pump 280

CIFAR-10 (Canadian Institute For Advanced Research) 220

Classification 27, 46, 193

Classification operation 181

Classification-oriented problems 5

Classification process 182, 238

Classifications of objects in images 202

Classifier Engine 386–387

Classifying Video Data 222

Cloud 359

Clustering 26, 28, 46

CNN (convolutional neural network) 8, 9, 21, 46, 77, 173, 265, 386

CNN architecture 23, 35

CNN BiLSTM 396

CNN Operations 188

Cognitive analytics model(s) 423

Commercial Buildings Energy Consumption Survey (CBECS) 277

Commercially-available off-the-shelf ML-based product for energy management 301

Compliance 362

Compressing Network Data 435

Computational linguistics (CL) 65, 67, 73

Computer room air conditioning (CRAC) 319

Computer vision (CV) 173, 238

Confidentiality (C) 351, 396

Configuration management (CM) 407

Consensus mechanism 367

Constant air volume (CAV) HVAC 301, 318

Content determination 92

Content distribution network (CDN) 417

Continuous Bag-of-Words (CBOW) 90

CONV layers 186, 202

ConvNet (same as CNN) 8, 9, 21, 46, 77, 173, 265, 386

Convolution 238

Convolutional autoencoder (CAE) 190, 191

Convolutional kernel 181

Convolutional (CONV) layer 174, 178, 181, 182, 239

Convolutional neural network (CNN) 8, 9, 21, 46, 77, 173, 265, 386

Convolution kernel   177
Convolution (convolutional) layer
  24, 25, 184
Convolution (convolutional) operation
  181, 189
Convolution process   176, 187
Convolutions, examples   180, 246–249
Cooling tower   280
Curse of dimensionality   396
Cyberintrusion Detector   373
Cybersecurity data breaches   347
Cybersecurity threat management   390, 391
Cyber threats   371

d

Data breaches   353
Data-dependent multichannel
  filters   128
Data-independent multichannel
  filters   128
Data protection   26, 388, 389
Data reduction   431
Decision algorithms or methods   35
Decision tree (DT)   35, 37
Deductive inference   31
Deep belief network (DBN)   46
Deep Fusion Reasoning Engine
  (DFRE)   419, 422
Deep learning (DL)   1, 6, 7, 46, 73
Deep neural network (DNN)   7–9, 11,
  47, 291, 440
Deep Reinforcement Learning   298
Defense in Depth   354
Delay-and-sum beamformer
  (DSB)   129, 157
Demand response (DR)   273
Denoising autoencoder (DA)   129
De-noising front-end   128
DenseNet   209, 212
Density estimation   26, 28
Depthwise convolution   202

Determined best-fit behavior   373
DFRE (Deep Fusion Reasoning
  Engine)   419, 422
Diffusion models   236, 239
Dimension reduction algorithms   47
Direct Sequence-to-sequence (S2S)   302
Distant-talking speech recognition
  environment   127
DL (Deep learning)   1, 6, 7, 46, 73
DNN (Deep neural network)   7–9, 11,
  47, 291, 440
DNN decoder   138

e

Edge detection   239
Efficiency, HVAC system   322
Employment, AI impact   98
Encoder/decoder methods   88
Encoder/decoder structure   201, 226
Encoding process   198
End-to-end (E2E) ASR system   157
End-to-End Learning   148
Energy Audit Stage   286
Energy balance   308
Energy efficiency control measures   288
Energy efficiency ratio (EER)   322
Energy management ML Example–Qin
  Model   279
EnergyPlus Models   281, 289
Enhanced classical NIST (National
  Institute of Standards and
  Technology) methodology   355
Ensemble learning   32
Envelope model   282
Environmental monitoring   274
Evolutionary neural networks (ENNs)   396
Extreme learning machines (ELM)   323

f

Facial expression recognition   228
Faster R-CNN   194

Fast R-CNN 193, 239
Fast R-CNN architecture 194
Fault, configuration, accounting, performance, and security (FCAPS) 407, 446
Fault detection and mitigation 418
Fault management (FM) 407
FCAPS (Fault, configuration, accounting, performance, and security) 407, 446
FC layer 181, 185, 186
Feature detectors 178
Feature extraction 36, 84, 123, 173, 178, 185, 193, 383, 385
Feature extractors 185
Feature map 177–179, 182
Features 178
Feature selection 36
Feature usage in DL 12
Feature vector 123
Feedback loop 47
Feed-forward network 17, 173
Feed-forward NNs 8, 24, 47, 74
Feed-forward propagation 47
5G NR network 429
Filter 133, 179, 188
2×2 filter 179
Filter-and-sum (FS) 130
Filter bank 133
Finite impulse response (FIR) filters 130
Finite-state machines 103
Flattening operation 181
Fully connected (FC) layer(s) 17, 24–25, 178, 184
Fully connected NNs (FNN) 182, 240
Fully Convolutional Networks (FCN) 190

g

GAN (Generative Adversarial Network) 17, 35, 48, 90, 230

Gated recurrent unit (GRU) 104, 138
Gaussian Mixture Model (GMM) Algorithm 38, 119, 158
Generalized eigenvalue (GEV) beamforming 132
Generative Adversarial Network (GAN) 17, 35, 48, 90, 230
Generative AI 101–102
Generative pre-training (GPT) 77, 89, 98
Gesture recognition 19
Google 97
GoogLeNet 175
Google's 98
GPT-4 98
GPT (Generative pre-training) 77, 89, 98
Gradient 158
Gradient boosted algorithm (GBA) 446
Gradient Boosting Machine Method 35
Gradient descent 158
Gradient of loss function 20
Graphic design generators 102
Graphics processing unit (GPU) 176, 241
Graph neural networks (GNNs) 218
Greenhouse gases (GHG) 324
Ground-truth 48

h

Hardware Root of Trust (HRoT) 388
Healthcare 263
Heat exchanger 324
Heating, ventilation, and air conditioning (HVAC) 275, 280, 282, 290, 307, 326
Heat pump 325
HFC network 442
Hidden layers 10, 17–18
Hidden Markov model (HMM) 37, 69, 119, 121, 158, 376

Hidden Markov Model (HMM) for intrusion detection   374
HMM (Hidden Markov model)   69, 119, 121, 158, 376
Host-based intrusion detection system (HIDS)   355
HVAC (heating, ventilation, and air conditioning)   275, 280, 282, 290, 307, 326
HVAC model   282, 284
Hybrid learning processes   30
Hyperparameter   48, 241
Hyperplane   48

i

I-frame   198
Image classification   199–200, 241
Image fusion model   217
Image generators   102
Image management   195
Image patterns   24
Image reconstruction   225
Image segmentation   199
Imaging   173
Imaging applications of AI   195
Inception architecture/ GoogLeNet   207, 210
In-context learning   159
Inductive learning   31
Information and Communications Technology (ICT)   2, 347
Inner product   241
Input layer   17–18
Integrated energy efficiency ratio (IEER)   326
Integrity, security (I)   351, 397
Intelligent transportation systems (ITSs)   218, 261, 268
Internet of Things (IoT)   65, 176, 348
Intrusion detection method   377

Intrusion detection systems (IDSs)   354, 371
intrusion protection systems (IPSs)   356
IoT applications   257
IoT-enabled devices   259
IoT Security (IoTSec)   363–365
ISO 27799 Health Informatics Security Standard for the Computational Health of Medical Establishments   362

k

Keylogger   397
Keyword-spotting   125
k-Means algorithm   38
K-Nearest Neighbor (KNN)   35, 38

l

Language Model for Dialogue Applications (LaMDA)   81
Language model, language model(ing) (LM)   69, 71, 104, 118, 160
Large language model (LLM)   97
Layers in NN   13
Leaky rectified linear unit (LeakyReLU)   215
Learning methods   26, 30, 32
Learning techniques   32
Lemmatization   104
LeNet-5   205
Lexicalization   92
Likelihood ratio (LR)   145
Linear activation function   14
Linear discriminant analysis (LDA)   385
Linear regression (LR)   60, 149
Linguistic realization   104
Load control, management   328, 428
Local region   187
Logistic regression   36, 49, 445

Long short-term memory (LSTM) 50, 71, 75, 105, 140, 161, 265, 413
Long-term memory 140
LSTM memory cell 76
LSTM-MIMO (multi-input multi-output) architecture 304
Luma (luminance) component 178

m

Machine learning (ML) 1, 5, 7, 50
Machine learning methods 33–34
Machine translation (MT) 105
Macroblocks 196
Malicious network traffic 375
Malicious penetration 353
Management information bases (MIBs) 424
Markov chain models 87
Mask 200, 242
Mathematical definition of a DNN 20
Maximizing rewards 29
Max pooling 183, 191
Mean squared error (MSE) 128
Medical wireless body area network 265
Mel frequency cepstral coefficients (MFCCs) 120, 161
Mel frequency cepstrum (MFC) 121
Mel spectrogram 118
Microsoft 97
Midjourney 242
Minimum Variance Distortionless Response (MVDR) Beamformer 130
Misuse detection 374
ML algorithms 2
ML classifier 379
ML model (MLM) 7, 9, 83, 222
MLP (multi-layer perceptron) 8, 17, 51, 175, 445

ML Predictor for a Remote Network Management 437
M2M (machine-to-machine) 257, 261, 329
M2M communication 329
Mobility 272
Model 51
Model-based methods 136
Monitoring encrypted web traffic 375
Multichannel linear prediction (MCLP) 132
Multichannel Wiener filter (MWF) 161
Multi-factor authentication 398
Multi-instance learning 31
Multi-layer machine 13, 15
Multi-layer perceptron (MLP) 8, 17, 51, 175, 445
Multi-task learning 32
Music Generators 102
MVDR beampatterns 131
MVDR filter coefficients 136

n

NAB (Neural Network Adaptive Beamforming) model 137
Narrow AI 2
Natural language generation (NLG) 65, 71, 86, 91, 105
Natural language processing (NLP) 3, 4, 65, 66, 71, 73, 81
Natural language understanding (NLU) 3, 65, 66, 71
Network Alarms 424
Network-based intrusion detection system (NIDS) 355
Network immunization testing 358
Network management 407
Network management ecosystem 439

Network monitoring alerts   419, 421
Network Operation Center (NOC)   426
Network optimization   40
Network risk scoring   371
Neural beamforming   134, 162
Neural network adaptive
   beamforming   135
Neural network layers   52
Neural networks (NNs)   8, 16, 51, 117,
   279, 373
Neural networks in the Qin
   Model   290
Neuron nodes   11, 52, 180
Neurons in a CNN   25
NIST (National Institute of Standards
   and Technology) cybersecurity
   methodology   355
NLG (Natural language generation)   65,
   71, 86, 91, 105
NLP (Natural language processing)   3,
   4, 65, 66, 71, 73, 81
NLU (Natural language
   understanding)   3, 65, 66, 71
NN algorithms   37
NN-based network management
   system   413
NN training   292
Noise cancellation   126, 138–139, 162
Non-adaptive beamformer   131
Nonlinear activation function   14

o

Object detection   182, 183, 242
Object detection applications   218
Object recognition   183, 242
Object validation score   219
Occupants model   282, 284
Online learning   32
OpenAI   97
Optimization Stage   287

Output feature map (OFM)   202
Output layer   17–19

p

Parametric speech   151
Particle swarm optimization
   (PSO)   276, 296, 330
Particle Swarm Optimizer   294
Password stealer   398
Payment Card Industry Data Security
   Standard (PCI DSS) Compliance   362
Penetration testing   359
Performance management (PM)   407
P-frame   198
Phishing   371, 398
Phishing detection   383
Physical security   272
PID (proportional, integral, and
   derivative) loop   306
Plug Load Model   284
Pooling   53, 163
Pooling layer   24, 25, 178, 181, 183
Pooling operation   181
Power management   40
Power-to-thermal load relationship   310
Predicting energy flow in buildings   311
Predictions   416
Predictive analytics model(s)   423
Predictive maintenance   41
Pre-training   53, 82
Previous layer   18
Processing audio features   144
Proof of Work (PoW)   368
Punishment   29

q

Qin Model, Energy management ML
   Example   279
Quality of service (QoS)   41, 430
Quality of transmission estimation   41

r

Random forest (RF)   38
R-CNN (Region-based CNN)
   193, 243
Real-Time Control Parameters   290
Receptive fields   53
Rectified linear unit (ReLU)   14
Recurrent neural network (RNN)   8, 21,
   53, 65, 74, 77, 122, 218
Region-based CNN (R-CNN)   193, 243
Region, input data   25
Region of Interest (ROI)   194
Regression   27, 53
Regression method   35
Regression-oriented problems   5
Regression/regression analysis   163
Reinforcement learning (RL)   29,
   30, 53, 86
ReLU   33
Representations of words   84
Residual Network (ResNet)   24, 175,
   208, 445
ResNeXt   209, 211
Return air   333
Rewards   29
RNN (Recurrent neural network)   8, 21,
   53, 65, 74, 77, 122, 218
RNN layer   138
RNN-LSTM   301
Root of Trust (RoT)   399

s

SCADA (Supervisory Control and Data
   Acquisition)   273
Scanning of emails   384
Seasonal Energy Efficiency Ratio
   (SEER)   334
Security management (SM)   407
Security mechanisms   352
Security Policy   354

Security requirements   349
Self-supervised (hybrid) learning
   29, 31
Semantic edge detection   213
Semantic feature map (SFM)   243
Semantic Segmentation   213
Semi-supervised learning   30
Semi-supervised (hybrid) learning
   methods   28
Semi-supervised learning Sigmoid
   layer   54
Sensor network/wireless sensor network
   (WSN)   334
Sentence aggregation   92
Sentiment extraction   148
Sequence-to-sequence (S2S)
   models   88, 107
Server room   275
Simple Network Management Protocol
   (SNMP)   394, 414, 446
Simulation models   285
Single-shot detector (SSD)   220–221
Situational awareness   231
Smart building   261, 270, 275
Smart building energy
   optimization   279
Smart cities   258
Smart grid (SG)   269, 273, 335
Smart home   260
SNMP (Simple Network Management
   Protocol)   394, 414, 446
SNMP Trap   447
Social engineering   399
Softmax layer   55
Software Root of Trust (SRoT)   388
Spam   399
Spatial resolution   201
Speech dereverberation   163
Speech enhancement (SE)   164
Speech features   133

Speech processing activation   143
Speech recognition   107
speech recognition models   122
Speech Synthesis   70, 150
Spiking neural networks (SNNs)   335
Spyware   399
Statistical inference processes (aka,
   statistical learning)   31
Statistical learning   55
Statistical methods   108
Stochastic gradient descent
   (SGD)   186
Stride   179, 183, 244
Sub-sampling layer   174
Subscriber identity module (SIM)   417
Supervised learning (SL)   30, 55, 427
Supervisory control and data acquisition
   (SCADA)   336
Supply air   280
Support vector machine (SVM)   7, 37,
   119, 164, 379
Surface realization   72
Surveillance/Intelligence   271
Synapse   7, 14
Syntax analysis   108

t

Template approaches   93
Text generators   102
Text structuring   92
Text to Speech (TTS)   33, 65, 67, 108,
   118, 164
Thermal load model   282, 283
Tiles   197
Tokenization   84
Traditional ML approaches   6
Traffic   272
Training   27, 56, 141, 224, 292
Transcription   146
Transductive learning   31

Transfer learning   32
Transformer   72, 78, 79, 89, 108,
   153, 230
Transformer model   80
Transportation   263, 272
Transpose convolutions   201
Trending analytics model(s)   423
Trojan   400
Trusted platform module (TPM)   400
TTS (Text-to-Speech)   3, 65, 67, 108,
   118, 164

u

Ultra deep neural networks
   (UDNNs)   7, 8, 56, 173, 176
Unpooling (upsampling) (a feature
   map)   244
Unsupervised learning
   methods   26, 30, 427

v

VALL-E   155
Variable air volume (VAV)   300
Variable frequency drive (VFD)
   controllers   281, 338
Variational autoencoder (VAE)   245
VAV HVAC system   302
VGG16   175, 210
VGG-19   24
Video Transcoding   228
Virtual Clothing   216
Virtual private networks (VPNs)   352
Visual Geometry Group's VGG 16,
   VGG-19 (aka ConvNet) (aka
   VGGNet)   206
Voice Agnostic Lifelike Language
   (VALL-E)   152
Voice Generators   102, 120
Vulnerability Lifecycle Management
   System   392–393

w

Wakeword detection   122, 124, 143
Water Management   270
Weigh(s)   13, 17, 188, 245
Weighted connection   13
Weighted finite-state transducer
    (WFST)   119, 144, 165
Wireless networks testing   359
Word error rate (WER)   117

y

YOLO detector   221

z

Zero day bots   379
Zero-shot scenario   118
Zero-shot TTS   152, 166
Zero Trust Environments   369
Zero trust networking (ZTN)   400

Printed and bound by CPI Group (UK) Ltd, Croydon, CR0 4YY

16/04/2025

14658363-0002